Handbook of Engineering and Speciality Thermoplastics

Scrivener Publishing
3 Winter Street, Suite 3
Salem, MA 01970

Handbook of Engineering and Speciality Thermoplastics

Volume 2
Water Soluble Polymers

Johannes Karl Fink
Montanuniversität Leoben, Austria

Scrivener

WILEY

Co-published by John Wiley & Sons, Inc. Hoboken, New Jersey, and Scrivener Publishing LCC, Salem, Massachusetts.
Published simultaneously in Canada.

Cover designed by Russell Richardson.

Library of Congress Cataloging-in-Publication Data:

ISBN 978-1-118-06275-3

Printed in the United States of America

10 9 8 7 6 5 4 3 2 1

Preface

This book focuses on water soluble polymers. The text is arranged according to the chemical constitution of polymers and reviews the developments that have taken place in the last decade.

Most chapters follow the same template. A brief introduction to the polymer type is given and previous monographs and reviews dealing with the topic are listed for quick reference. The text continues with monomers, polymerization and fabrication techniques, and discusses aspects of application as well. Following this, suppliers and commercial grades are presented.

How to Use this Book

Utmost care has been taken to present reliable data. Because of the vast variety of material presented here, however, the text cannot be complete in all relevant aspects, and it is recommended to the reader to study the original literature for complete information.

The reader should be aware that in case of patent literature mostly US patents have been cited if available, but not the corresponding equivalent patents in other countries. For this reason, the author cannot assume responsibility for the completeness and validity of, nor for the consequences of, the use of the material presented here. Every attempt has been made to identify trademarks; however, there were some that the author was unable to locate.

Index

There are four indices: an index of trademarks, an index of acronyms, an index of chemicals, and a general index.

In the index of chemicals, compounds that occur extensively, e.g., "acetone", are not included at every occurrence, but rather when they appear in an important context.

Acknowledgements

I am indebted to our university librarians, Dr. Christian Hasenhüttl, Dr. Johann Delanoy, Franz Jurek, Margit Keshmiri, Dolores Knabl, Friedrich Scheer, Christian Slamenik, and Renate Tschabuschnig for support in literature acquisition. I also want to express my gratitude to all the scientists who have carefully published their results concerning the topics dealt with here. This book could not have been otherwise compiled.

Last, but not least, I want to thank the publisher, Martin Scrivener, for his abiding interest and help in the preparation of the text.

Johannes Fink
20th January 2011

Contents

1

Poly(ethylene oxide)

Poly(ethylene oxide) (PEO) is sometimes addressed as poly(ethylene glycol) (PEG). This came about because it can be considered as being derived from the etherification of ethylene glycol (EG) into the polymer. On the other hand, the industrial synthesis, as explained below, starts with ethylene oxide (EO). We will use both names simultaneously, in the same way, as given in the references.

PEG was first studied by Lourenço in 1861 (1). He reported the synthesis of oligomeric PEGs up to hexaethylene glycol. It seems to be the first example of a condensation polymerization reaction at all (2). The first patents appeared around 1930 (3,4). Soon afterwards PEGs were used as components for poly(urethane)s (5).

1.1 Monomers

The basic monomers for PEO are shown in Table 1.1. The structures

Table 1.1 Monomers for Poly(ethylene oxide) Types

Monomer	Remarks
Ethylene oxide	Basic Monomer
Propylene oxide	Less water soluble
Butylene oxide	Less water soluble
Glycidol	Branched structures (6)
Ethoxy ethyl glycidyl ether	Branched structures (6)

are shown in Figure 1.1.

The basic monomer is EO. According to the nomenclature of heterocycles, EO is also addressed as oxiran. EO is synthesized by

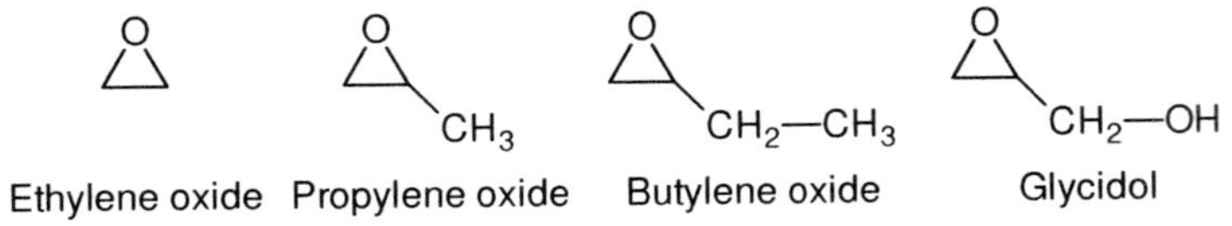

Figure 1.1 Monomers used for Poly(ethylene oxide)

the addition of oxygen to ethene. Propylene oxide or trimethylene oxide may also be used as comonomer together with EO. However, these comonomers should be used only in those small amounts as not to render the resulting copolymer water insoluble. Glycidol is a suitable comonomer for branched structures.

1.2 Polymerization and Fabrication

Water-soluble PEO is prepared by the ring opening polymerization of EO, usually in the presence of a small amount of an initiator such as low molecular weight glycol or triol alcoholate (7). Examples of such initiators include alcoholates of EG, diethylene glycol and other oligomers. Branched types are synthesized with multifunctional alcoholates, such as the potassium salts of glycerol, pentaerythritol, dipentaerythritol, or sorbitol (8). The basic mechanism is shown in Figure 1.2, top.

In Figure 1.2, bottom, the reaction of glycidol is shown. After addition, the negative charge may change its position, which causes the growth from both ends. In the course of the reaction a pendant hydroxyl group may again be activated as it turns into an alcoholate. This leads again to a growth reaction. In this way, branched strictures are formed.

EO or various epoxides, and other cyclic ethers can be polymerized with anionic, cationic, and coordination catalysts. For the commercial production of polymer of such type, the most effective catalysts found are $(CH_3)_3N$ and $SnCl_4$, $CaCO_3$, $FeCl_3$. Other compounds with catalytic activity are $NaNH_2$, ZnO, SrO, and CaO (8).

The living polymerization techniques are preferred in comparison to other methods because molecular weight and polydispersity can be better controlled. The polymerization of EO can be carried

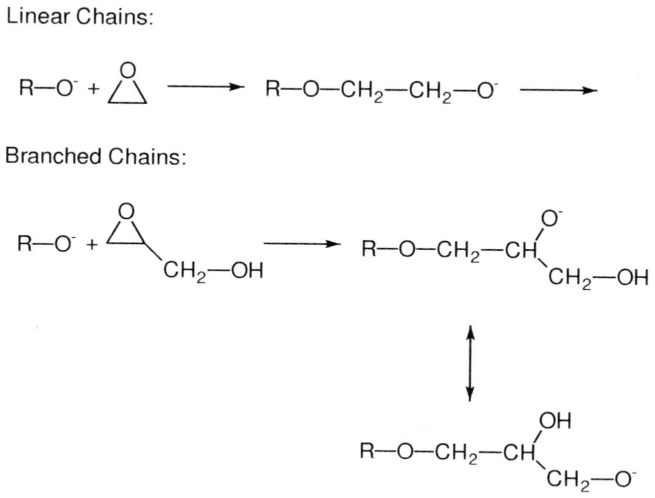

Figure 1.2 Basic Mechanism of Polymerization

out in polar solvents such as tetrahydrofuran (THF), *N,N*-dimethylformamide, dimethyl sulfoxide, etc.

1.2.1 *Functionalization*

The endgroups can be functionalized (9). The functionalization of the endgroups in the course of a living polymerization can be achieved by two different strategies (8):

1. By deactivation of the living species with a suitable electrophile or chain transfer reagent, or
2. By initiation of the living process with an organic anionic species that bears the protected functionalized group.

A disadvantage in the first strategy is that any polymer chain which has been terminated during the propagation for some reason will not react with the electrophile.

In general, functionalized polymeric chains can be obtained by a chemical modification of functional groups, either endgroups or side groups of the polymeric backbone (8).

The effective functionalization can result in end-reactive polymers. This is becoming more important due to the high versatility of the introduced endgroups. One of the most important utilizations of PEG is the construction of polymer brushes, a densely packed layer of tethered polymers anchored on the surface utilizing the end functionality of the polymer chain. Such a PEG brush significantly changes the surface properties. For example, in such a treated surface, the PEG chains are densely packed on a surface and attached by the end of the polymer chain, showing an effective rejection of protein adsorption resulting in a good blood compatibility (8).

Commercially available methoxy-ended PEGs with a methoxy group at one end and a hydroxy group at the other end are used as starting materials for the preparation of monochelic PEG. When PEG is chemically bound to a water-insoluble compound, the resulting conjugate becomes water soluble as well as soluble in many organic solvents.

When PEG is attached to a drug, its activity is commonly retained. Moreover, the bounded drug may display altered pharmacokinetics, which can be favorable. Proteins coupled to PEG exhibit an enhanced blood circulation life time because of reduced kidney clearance and reduced immunogenicity. The lack of toxicity of the polymer and its rapid clearance from the body are advantageous for pharmaceutical applications (8).

In order to couple a PEG chain to a protein or a small drug molecule, it is necessary to activate the hydroxyl end group. For example, the hydroxyl group can be converted into an aldehyde group.

Such a compound can be prepared by reaction of the diethyl acetal of 3-chloropropionaldehyde with PEG alkoxide followed by hydrolysis. A more effective route to the chemically equivalent sulfur analog is shown in Figure 1.3.

PEG aldehydes are inert towards water and react primarily with amines. Thus, eventually, this aldehyde group can then be covalently linked to an amine group of the guest molecule by reductive amination. This approach has been originally proposed for modifying organic or polymer surfaces in water by connecting PEG aldehyde derivatives to exposed amine groups (10).

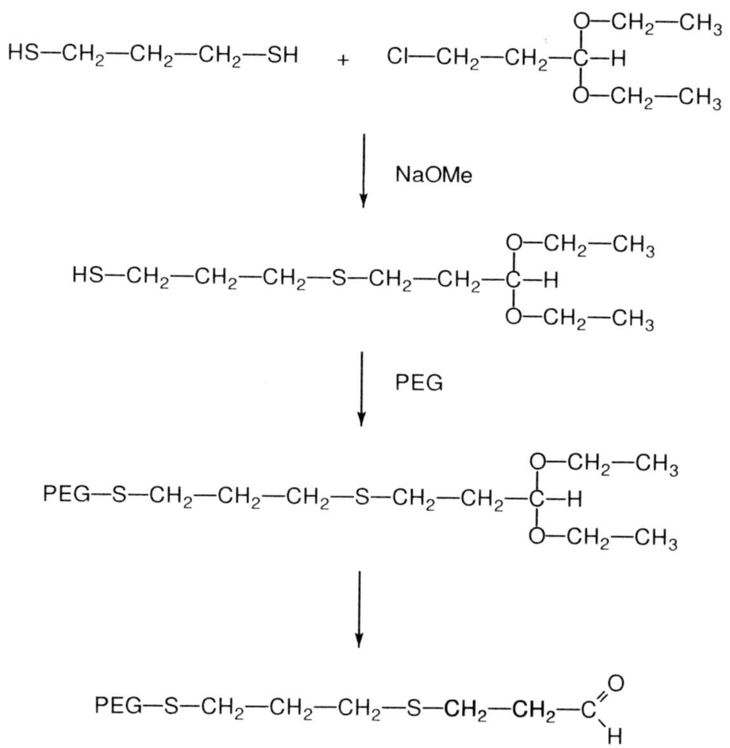

Figure 1.3 Modification of PEG with Aldehyde Functionality

However, the use of acetaldehyde modified PEG is limited by its high reactivity, which leads to condensation side reactions (9). The reactivity of the acetaldehyde can be reduced by protecting this group with acetals. The acetal can be converted back at a pH of 2–3. This procedure allows a synthesis of a PEG type bearing at one end the aldehyde and at the other end a hydroxyl group (11).

Acid functionalized PEG is obtained by the oxidation of the aldehyde end group. However, the selective oxidization of the aldehyde group is problematic, as concomitant degradation reactions of the polymer chain may occur.

The preparation of a heterofunctional PEG in which the polymer has a carboxymethyl group on one end and a hydroxyl group on the other end has been described (12).

Acid functionalized PEG can be obtained by using a hydroxy alcoholate acid salt as initiator, e.g. the sodium salt or the potassium salt of 4-hydroxy butyric acid (8). Both the hydrogens in the carboxyl group and the hydroxyl group are replaced with sodium or potassium. The polymerization is carried out in dry THF.

1.2.2 *Extrusion*

While low molecular weight PEO resins have desirable melt viscosity and melt pressure properties for extrusion processing, they have low melt strength and low melt elasticity which limit their ability to be drawn into films having a thickness of less than about 2 *mil*. Films produced from low molecular weight PEO also have low tensile strength, low ductility, and are too brittle for commercial use (13).

In contrast, high molecular weight PEO resins, would yield films with improved mechanical properties in comparison to those produced from low molecular weight PEO. However, high molecular weight PEO, has a poor processability due to its high melt viscosity. Thus, the melt pressure and melt temperature must be significantly elevated during melt extrusion of high molecular weight PEO. This results in degradation reactions and then severe melt fracture. For this reason, only very thick films of about 7 *mil* or greater can be made from high molecular weight PEO resins.

Commercially available PEO resins can be modified in order to improve their melt-processibility. This is achieved by blending PEO and latex (13). Such a blend has a unique microstructure which can be observed by scanning electron microscopy and atomic force microscopy. The PEO resin of the blend possesses a lamellae structural assembly in which there is an approximately uniform nanoscale dispersion of fine latex particles. In the blend, both individual latex particles, approximately 100–200 *nm* in diameter, and clusters of the particles, approximately a few microns in size, are embedded in the PEO lamellae structural assembly.

Atomic force microscopy illustrates the unique microstructure of a nanoscale dispersion of fine latex particles in the lamellae structure of the PEO resin. Some of the particles form clusters. The particles in the clusters are not tightly packed and do not appear to be coupled. Scanning electron microscopy reveals the approximately uniform dispersion of the latex particles in the PEO resin (13).

The water content is important in making films with improved fracture resistance. PEO and other water-soluble polymers are capable of forming hydrogen bonded complexes with water. The bounded water does not exhibit phase transitions. However, the amount of water which may be bound in such complexes is limited by the number of the ether groups in the polymer chain.

Thus, excess water will remain in the blend as free water. However, free water exhibits phase transitions. These phase transitions can result in microscopic fractures which weaken the blend, resulting in films which are more brittle and possess a lower tensile strength. It has been found that 15% or less of water is allowed (13).

The blends are preferably produced in two steps. In the first step, the latex emulsion is mixed with PEO or is coated on to the PEO. Thereby, the water from the emulsion is absorbed into the PEO, yielding a plasticized PEO with predispersed rubbery particles on the surface of the PEO. In the second step, the plasticized mixture is feeded into a twin screw extruder or high shear mixer to provide melt blending of the PEO and rubber particles (13).

The blending process occurs at 100–200°C. Higher initial temperatures are required to produce the blend from a PEO powder in comparison to pellets due to the greater difficulty in melting the powder. Temperatures in excess of 250°C will result in an excessive thermal degradation of the components.

1.3 Properties

PEG is a clear, colorless, and odorless substance. It is soluble in water, stable to heat, inert to many chemical agents, does not hydrolyze or deteriorate, and is non-toxic.

PEG is considered to be bio-compatible. In other words, PEG is capable to coexist with living tissues or organisms without causing harm. In particular, PEG is not immunogenic, i.e., it has no tendency to produce any immune response in the body. When attached to a moiety having some desirable function in the body, the PEG tends to mask the moiety and can reduce or eliminate any immune response so that an organism can tolerate the presence of the moiety (8).

PEGs are known to be biodegradable aerobically and anaerobically (14). The microbial oxidation of diethylene glycol and poly-(ethylene glycol) with the average molecular weights of 200–2000 *Dalton* have been reported (15).

Triblock copolymers are available that show a reverse thermal gelation (16). A reverse thermal gelation means that a temperature exists below which the block copolymer is soluble in water and above which the block copolymer undergoes phase transition to increase in viscosity to form a semi-solid gel. This temperature is also known as the lower critical solution temperature.

It is interesting to note that acetal resins, e.g. poly(oxymethylene), even when they are related in chemical structure, are not soluble in water. This arises because these materials have high crystallinity.

1.4 Special Additives

PEG is used as an additive for the stabilization of emulsions in emulsion polymerization. Copolymers from vinyl acetate and ethene with an eventually high solids content are prepared by using PEG among other stabilizing ingredients as a stabilizer (17). The stabilizing agent consists essentially of poly(vinyl alcohol) and PEG in an amount of from about 4–6% . If a lesser amount is employed, the system may coagulate because of insufficient stabilization.

For the production of sintered molded parts, either metal based, or ceramic based lubricants are needed. These lubricant compositions contain PEGs and montan waxes (18). It is advantageous to use

PEGs as pressing aids. It is believed that the great advantage of the use of PEGs lies in the fact that they have a relatively low softening point, generally in the range of 40–100°C, which makes it possible to fill the dies used in the metallurgical process with cold material so that lumping or agglomeration can be avoided. When the die is heated in the pressing operation, the PEGs, together with the montan waxes that are used, allow lubrication, so that higher green densities and green strengths of the green compacts are achieved.

Poly(oxyethylene) sorbitan monolaureate is used as an antistatic additive (19). The proposed application is in ionomer copolymers from ethylene and methacrylic acid or acrylic acid. The antistatic agent is uniformly incorporated into the copolymer by conventional melt blending techniques, for example, the antistatic agent is blended with molten copolymer in a manner to form a homogeneous blend, using an extruder. The total amount of the antistatic additive is 1–2%.

The antistatic action of poly(oxyethylene) sorbitan monolaureate is effective almost instantaneously, after less than 1 *h* after production. This almost instantaneous antistatic action is extremely important during the production of films where operators are exposed with static built up on the extrusion equipment (19).

1.5 Applications

PEO types find a wide variety of applications. Some applications of PEO are listed in Table 1.2.

Table 1.2 Applications of Poly(ethylene oxide)

Use	Remarks
Adhesive	For papers (20), denture fixative
Flocculation	Water cleaning
Thickener	Paints, oil field applications, cosmetics (21)
Binder	Ceramic applications
Dispersant	Additive for polymerization processes
Lubricant	Soaps, personal care

PEO can be used as binders for pigments, fillers, metal powders, and ceramics with application in battery electrodes, cathode ray tubes, and fluorescent lamps. The strong hydrogen bonding affinity of PEO accounts for its association with various polar compounds, such as phenolic resins, mineral acids, halogens, ureas, lignin sulfonic acids, and polycarboxylic acids. These complexes have unique properties with their application in batteries, microencapsulated inks, slow-release bacteriostatic agents, or water-soluble adhesives (22). As previously stated, the polymer forms water retentive gels. These gels can be used as absorbent pads and diapers. Moreover, PEO can be used as emollient in cosmetic and hair products.

It can also be formed into flexible films, both by thermoplastic processing and casting techniques. These films can be readily calendered, extruded, molded, or cast. Sheets and films of polyethylene oxide can be oriented in order to get high strength materials. The films are inherently flexible and tough and resistant to most oils and greases. In packaging, PEO can be used to provide heat sealability and hot melt adhesion (22).

Last but not least, end-reactive PEG types are a very important class of material in a variety of fields such as biology, biomedical science, and surface chemistry (8). This arises from their unique properties, such as solubility and flexibility of the chains.

1.5.1 *Textile Impregnation*

A method of preparing a temperature adjustable textile has been reported (23). The textile is treated with PEG and can absorb or release heat at various comfort relevant temperatures depending upon the molecular weight of the PEG used in the formulation added to the textile substrate. The PEG has a molecular weight of 1–1.5 *k Dalton*. Further, the formulation includes a crosslinking agent, an organic acid, and a metal salt.

An excess of PEG may be removed from the wet fabric by vacuum extraction. After treatment, the textile is cured wherein the surface temperature of the textile is raised between 90–115°C for curing. The cured textile is neutralized by washing the cured textile in an alkaline solution.

An example of *thermal cascading* is to create a garment with three individual layers of a PEG-treated substrate, e.g., treated with PEG

1000, 1200 and 1450 respectively. Each of these PEG designations has different temperature ranges at which they melt and solidify thus absorbing and releasing heat (23).

For cold weather wear, the PEG 1450 layer would be worn closest to the body with the PEG 1000 layer positioned in the outermost layer. The PEG 1450 layer, next to the body, would absorb heat and help maintain a desired core temperature of approximately 34°C. If the temperature of this inner layer drops to a temperature of less than 20°C, the substrate would begin to release the stored thermal energy, thus protecting the wearer.

The middle, layer, which consists of PEG 1200, requires less thermal energy to activate the melting phase of the crystalline structure and will absorb thermal energy up to approximately 27°C and begin releasing the thermal energy at temperatures less than about 10°C, thereby slowing the penetration of cold while releasing heat.

Finally, the PEG 1000 layer, i.e., the outer most layer will charge thermally at a still lower temperature and will begin releasing the thermal energy when a cold temperature occurs.

A further advantage of the thermal cascading is that low molecular weight PEG polymers absorb moisture more readily and effectively than high molecular weight PEG polymers. Thus, a more effective moisture management system is present.

The maximum moisture wicking will occur with moisture wicking outward from the body, where is the highest molecular weight PEG treated layer, toward the lowest molecular weight PEG treated layer (23).

1.5.2 *Laundry Detergents*

Besides the indispensable ingredients for the washing process such as surfactants and builders, detergents generally comprise further constituents that can be addressed as *detergent auxiliaries*. These include foam regulators, graying inhibitors, bleaching agents, bleach activators and color transfer inhibitors (24). Further, the auxiliaries also include substances that provide soil-releasing properties. These support the soil-release capability of the formulation.

Due to their chemical similarity to polyester fibers, particularly effective soil-release agents for fabrics made of such materials are copolyesters containing alkylene glycol units and poly(alkylene glycol units). The use of poly(ethylene terephthalate) PEG copolymers has been described (25,26).

Active polyester soil-release polymers include copolyesters of dicarboxylic acids, such as adipic acid, phthalic acid, or terephthalic acid and of diols, such as EG or propylene glycol, and polydiols, e.g., PEG or poly(propylene glycol).

The molar ratio of monomeric diol units to polymeric diol units preferably ranges from 10:1 to 1:10. The degree of polymerization of the polymeric diol units is preferably in the range of 12 to 140. The polyesters can be end blocked with monocarboxylic acids.

The polymer is prepared from styrene, methyl methacrylate (MMA) and PEG by atom transfer radical polymerization (24). Actually, styrene and MMA are vinyl monomers, whereas PEG is not. However, in the course of polymerization, PEG is grafted on to the backbone of the copolymer of styrene and MMA.

Other laundry detergents are copolymers from styrene and maleic anhydride (MA), on to which PEG is added as the MA ring opens (27). The grafting is shown in Figure 1.4.

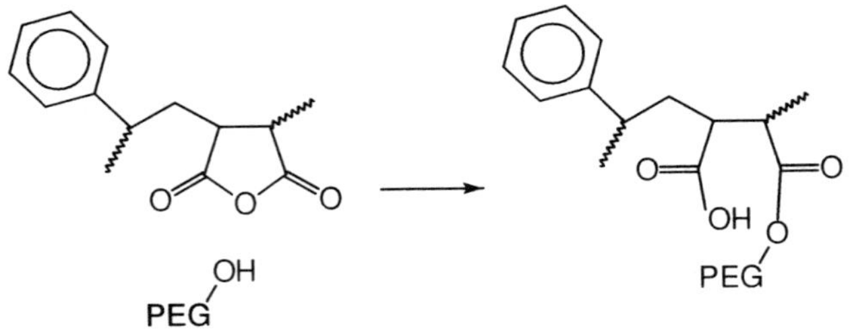

Figure 1.4 Grafting of PEG on to MA units

1.5.3 *Ink Jet Printing Media*

Water-soluble PEOs are used as a binder component for coatings of ink jet printing media (7). The binder in the coating is PEO. This polymer is used because the polymer should not contain ammonium groups. These groups are introduced in the composition from other polymers.

The coatings in the ink receiving layer can be formulated from a blend of binders. These compositions include gelatin, a copolymer from EO and monomer that yields ultimately the vinyl alcohol moiety, and copolymer from styrene, *n*-butyl acrylate), MMA and 2-(*tert*-butylamino) ethyl methacrylate (28).

The use of a poly(vinyl alcohol-ethylene oxide) copolymer in the ink receiving layer provides a number of functional benefits including the control of ink-coalescence, improved humid fastness, and a superior degree of image quality and long-term stability. These benefits are achieved because these copolymers are better compatible with the ink receiving layer and the colorants in the ink (28).

1.5.4 Flocculation and Coagulation

The production of fluoroelastomers occurs via emulsion polymerization methods. The polymer is obtained as a dispersion or latex. These polymers are then separated from the dispersion by addition of a coagulant to form a slurry. Eventually, the slurry is then washed and dried and then shaped into final form for commercial use (29).

Conventionally, the coagulants are salts of inorganic multivalent cations, such as aluminum sulfate, calcium chloride or magnesium chloride. These salts work very well as coagulants. However, residual amounts of these salts remain in the polymer. The presence of these salts renders these polymers unsuitable for their use in sensitive applications, e.g., in the semiconductor industry.

Salts of univalent cations, such as sodium chloride, have been proposed as coagulating agents for the manufacture of fluoroelastomers. Residual amounts of these salts are considered relatively innocuous in some end-use applications. However, excessively large amounts of salts of univalent cations are required to fully coagulate the fluoroelastomer. Moreover, the resulting polymer is difficult to fully dry.

The use of organic coagulants is another method to avoid polymer contamination. Residual amounts of organic coagulants will not contaminate semiconductor processes and in any case may volatilize out of the polymer during the curing process.

Recently, it has been found that PEO homopolymers and copolymers can be used to coagulate fluoroelastomers. To the aqueous dispersion from emulsion polymerization high molecular PEO is added to coagulate the elastomer. When the viscosity-average molecular weight of the PEO is less than 500,000 *Dalton*, either no coagulation occurs or the amount of polymer needed is uneconomically high (29).

The coagulant is preferably added to the dispersion as an aqueous solution. Useful concentrations are 0.005–1.0%. Optionally, preservatives or antioxidants may be added to the solutions in order to extend their shelf life.

1.5.5 *Superabsorbents*

PEG can be used as a crosslinking agent for polysaccharides. These materials find use as superabsorbents. Superabsorbent polysaccharides show a water absorption of 700–5,300 $g\,g^{-1}$ for deionized water and up to 140 $g\,g^{-1}$ for saline solutions.

In detail, the crosslinking agent is a modified PEG, e.g., halogenated mesylated, tosylated, or triflated (30). Preferred polysaccharides are anionic and contain carboxymethyl groups or half esters prepared with maleic anhydride.

Anionic polysaccharides may also include dicarboxylates such as iminodiacetate groups and tricarboxylates such as citrate groups. The preparation of several crosslinking agents and the crosslinking reaction itself has been described in detail (30). The polymers are rather low molecular.

Diglycol dichloride, i.e., 1,5-dichloro-3-oxapentane is prepared from diethylene glycol in benzene solution. To this solution, pyridine is added, followed by the dropwise addition of thionyl chloride. After refluxing for 24 *h* the organic layer is decanted from the pyridinium hydrochloride salt, dried, filtered and evaporated to dryness. In a completely analogous way, triglycol dichloride and tetraglycol dichloride can be prepared.

Carboxymethyl starch is prepared from wheat starch by the reaction with chloroacetic acid. To the reaction product in solution, triglycol dichloride is added and kept at 70°C for 24 *h*. The polymer can be precipitated by means of methanol and isolated to give a white solid (30).

1.5.6 *Food Additive*

The addition of low molecular PEG compounds to the feed has been found to improve the nutritive value of the feed, for instance for poultry, pigs and calves (31). When producing animal feed the PEG is dissolved or suspended in water together with pulverulent or granular nutritious substances and other components.

If the feed contains a liquid hydrophobic component, such as a lipid or a carboxylic acid, this component is suitably added before or after admixing the PEG. A premix is suitably prepared, consisting of, e.g., vitamins, flavorings, minerals, enzymes, antibiotics and probiotics.

Further, it is possible to add to the premix dry components consisting of cereals, animal and vegetable proteins, molasses, and milk products. To this the premix, additional PEG is added and applied to a carrier, which consists of ground cereals, starch or inorganic minerals, such as silicates.

Liquefied lipids that usually consist of slaughter fat and vegetable fat can be also be added. After thorough mixing, a mealy or particulate composition is obtained depending on the degree of grinding of the ingredients. The components of such a tasty meal are summarized in Table 1.3.

210 broilers distributed between 14 cages each containing 15 broilers were feeded with these compositions. The increase in weight, feed intake, feed index, and the relative feed conversion of the broilers has been determined.

The measured results indicate that the test animals, when a given meal of Table 1.3 was served, show a better growth than with comparative feed. At the same time, a lower feed index was obtained for the 40 d value, i.e., a lower feed intake per increase in weight with the feed took place (31). The feed index is the ratio of feed intake and increase in weight, which is a dimensionless number. The lower the number, the better the efficacy of using the offered meals.

Thus, this definition is in contrast to the usual physical definitions as for example in a Carnot cycle. Detailed results about weight increase Δw after 20 d and 40 d, feed intake and feed index are shown in the lower part of Table 1.3.

Table 1.3 Components in Feed and Response of the Animals (31)

Component	Parts by weight				
PEG Type →	RO 200	PEG 6000	PEG 6000	Pluronic	No PEG
Crushed barley	35.0	35.0	35.0	35.0	35.0
Crushed wheat	21.0	21.0	21.0	21.0	21.0
Wheat bran	0.29	0.29	0.27	0.2	3.0
Additive	0.01	0.01	0.03	0.1	–
Tapioca meal	7.0	7.0	7.0	6.0	6.0
Soybean meal	24.0	24.0	24.0	24.0	24.0
Meat meal	5.0	5.0	5.0	5.0	5.0
Feed fat	5.0	5.0	5.0	5.0	5.0
Molasses	2.0	2.0	2.0	2.0	2.0
Premix	2.0	2.0	2.0	2.0	2.0
Response					
Δw 0–20 d /[g]	552	561	574	523	536
Feed intake/[g]	789	797	832	761	766
Feed Index	1.43	1.42	1.45	1.46	1.43
Δw 0–40 d /[g]	1742	1725	1759	1546	1556
Feed intake/[g]	3048	3001	3076	2868	2832
Feed Index	1.75	1.74	1.75	1.86	1.86

1.5.7 *Medical Applications*

1.5.7.1 *Organ Preservation Solutions*

Transplantation of vital organs such as the heart, liver, kidney, pancreas, and lung has become increasingly successful and sophisticated in recent years. Because mammalian organs progressively lose their ability to function during storage, even at low temperatures, transplant operations need to be performed expeditiously after organ procurement so as to minimize the period of time that the organ is without supportive blood flow.

In 1988, the University of Wisconsin (UW) solution has been described (32). This solution contains metabolically inert substances rather than glucose to establish the osmotic pressure, hydroxyethyl starch, and radical scavengers. Initial liver transplant experiments in dogs were performed. This solution subsequently became the standard organ preservation solution for transplant surgery. Improvements in the design of such compositions include (33):

- Modification and simplification of UW solution
- Investigation of organ-specific requirements
- Addition of pharmacologic agents, particularly calcium antagonists for control of acidosis
- The use of a terminal rinse solution
- The use of solutions containing PEG.

The beneficial effect of PEG 8000 on rabbit hearts preserved by oxygenated low pressure perfusion for 24 *h* has been reported (34). The substitution of hydroxyethyl starch with PEG results in excellent cardiac function.

The detailed mechanism of action of PEG activity in organ preservation solution is still unknown. PEG is known to improve tissue viability, reduce ischemic injury by preventing cell swelling, interact with lipids in the cell membrane, and scavenge free radicals (35).

1.5.7.2 *Pharmaceutical Compositions*

Tablets. PEO is known as a component of medicaments in tablet form designed to be administered by the oral route. The general process of making the tablets consists of (36):

- Mixing in the dry state and for a sufficient time, the active ingredient, PEO and additives
- Adding a solvent if desired
- Granulation by passage through a suitable sieve
- Drying the granules thus formed for a sufficient period of time
- Adding further additives, by mixing in the dry state
- Compressing the mixture form from the preceding steps to obtain the desired compressed tablet
- Optionally coating the compressed tablet.

Surface coating can be employed in order to improve the appearance, thus making the drug more readily acceptable to the patient, or for dimensionally stabilizing the compressed tablet. The coating can be a conventional coating suitable for internal use. A surface coating can be obtained using a quick dissolving film.

The PEO in the formulation, forms a hydrogel from contact with water. This hydrogel dissolves more or less rapidly as a function of the molecular weight of the PEO employed. By selecting the molecular weight of the PEO, the kinetics of release of the active ingredient can be controlled.

Surprising results have been reported. In a hydrophilic matrix, when the concentration of the hydrophilic active ingredient increases, it would be expected that the rate of release of the active ingredient would increase. However, the opposite effect was found. This was demonstrated in the case of acyclovir as active ingredient (36). Around 50 examples for formulations have been presented (36). A few formulations are reproduced in Table 1.4.

Films. Conventionally, drugs or pharmaceuticals, may be prepared in a tablet form to allow for accurate and consistent dosing. However, this form of preparing and dispensing medications has some disadvantages: A large proportion of adjuvants that must be added to obtain a size able to be handled. A larger medication form requires additional storage space, and that dispensing includes counting the tablets which has a tendency for inaccuracy.

In addition, many persons, estimated to be as much as 28% of the population, have difficulties in swallowing tablets. While tablets

Table 1.4 Formulations for Tablets (36)

Tablets of Acyclovir Ingredient	*mg*
Acyclovir	200
PEO (MW = 100 000)	700
Magnesium stearate	5
Industrial alcohol	260
Tablets of Nifedipine Ingredient	***mg***
Nifedipine	60
Microcrystalline cellulose	100
PEO (MW = 3 000 000)	336
Colloidal silicon dioxide	2.5
Magnesium stearate	2.5
Industrial alcohol	150
Tablets of Glipizide Ingredient	***mg***
Glipizide	10
PEO	220
Microcrystalline cellulose	55
Hydroxypropyl methyl cellulose	20.0
Lactose	50
Sodium stearyl fumarate	1.7
Coating:	
Methacrylic acid copolymer	10
PEG	2
Talc	2.5
Silicon dioxide	4.5
Tablets of Pentoxiphylline Ingredient	***mg***
Pentoxiphylline	400
PEO	150
Povidone	30
Glycerol behenate	6
Coating:	
Ammonio methacrylate copolymer	20
Lactose	20
Silicon dioxide	8

may be broken into smaller pieces or even crushed as a means of overcoming swallowing difficulties, this is not a suitable solution for many tablet or pill forms. For example, crushing or destroying the tablet or pill form to facilitate ingestion, alone or in admixture with food, may also destroy the controlled release properties.

As an alternative to tablets and pills, films may be used. Historically, films and the process of making drug delivery systems in film form have suffered from a number of unfavorable characteristics that have not allowed them to be used in practice.

Films that incorporate a pharmaceutically active ingredient have been disclosed as far back as 1974 (37,38). The films may be formed into a sheet, dried, and then cut into individual doses. The films are useful for oral, topical, or internal use. However, in the early times of manufacture the films suffered from the aggregation or conglomeration of the active particles, making them inherently nonuniform.

An improvement with this respect appeared when multilayer films instead of monolayer films were proposed (39,40). Namely, a two-sided coating frequently gives advantages because problems due to the warping of the support material and differing hygroscopicity are compensated. Multiple strip coatings and in fact even printing style coatings are possible and offer a considerable variability when processing incompatible active ingredients.

Other approaches to prevent the aggregation of the particles target to additional ingredients, such as gel formers, in order to increase the viscosity of the film prior to drying (41,42). In fact, these methods suffer from requiring additional components.

During film preparation it may be desirable to dry the films at high temperatures. High heat drying produces uniform films and leads to greater efficiencies in film production. However, films containing temperature sensitive components may face degradation problems at high temperatures. Furthermore, highly volatile materials will tend to be quickly released from this film upon exposure to conventional drying methods.

Recently, the problems in the prior art could be overcome by using improved techniques: Film products may be formed by extrusion rather than by casting. Extrusion is particularly useful for film compositions containing PEO components. A single screw extrusion process may be employed (43).

PEO, when used alone or in combination with a hydrophilic cellulosic polymer, achieves flexible, strong films. Additional plasticizers or polyalcohols are not needed for flexibility. To achieve the desired film properties, the content level or the molecular weight of PEO in the polymer component may be varied. Modifying the PEO content affects properties such as tear resistance, dissolution rate, and adhesion tendencies.

One method for controlling film properties is to modify the PEO content. For instance, in some embodiments rapid dissolving films are desirable. By modifying the content of the polymer component, the desired dissolution characteristics can be achieved. High molecular weight PEO of up to 4 *M Dalton* may be desired to increase the mucoadhesivity of the films.

A recipe for forming a film for drug delivery is given in Table 1.5. Simethicone is generally used in the medical field as a treatment for gas or colic in babies. Simethicone is a mixture of fully methylated linear siloxane polymers containing repeating units of poly(dimethylsiloxane) which is stabilized with trimethylsiloxy end blocking units, and silicon dioxide. The mixture is a gray, translucent, viscous fluid which is insoluble in water.

Table 1.5 Film for Drug Delivery (43)

Component	Weight /[*g*]
Poly(ethylene oxide)	227
Sucralose	18.16
Precipitated calcium carbonate	176.38
Orange concentrated flavor	27.24
Tween 80	0.68
Simethicone	4.54
Yellow food coloring: 27 drops	
Red food coloring: 18 drops	

The composition does not only contain the polymer and the active ingredient, but also additional additives for various purposes. In particular, the active ingredients may be taste-masked. An antioxidant may also be added to the film to prevent the degradation of an active ingredient, especially when it is photosensitive.

Also, color additives can be used in preparing the films. Such color additives include food, drug, and cosmetic colors. These colors are dyes, their corresponding lakes, and certain natural and derived colorants. Lakes are dyes absorbed on aluminum hydroxide. Flavors may be chosen from natural and synthetic flavoring liquids. Examples include mint oils, cocoa, citrus oils, and fruit essences (43). Other useful flavorings include aldehydes, such as benzaldehyde, decanal, tolyl aldehyde, and *n*-dodecenal.

Electrostatic Deposition. Coatings from PEG serve to fix active pharmaceutical ingredients deposited by for electrostatic dry deposition on a tablet (44). First a placebo tablet is made by compressing a mixture of 99–99.5% microcrystalline cellulose and 0.5–1% magnesium stearate. The placebo tablets are arranged in a tray of 9 tablets by 9 tablets array for the further deposit of the pharmaceutically active powder.

The charge of the tablet substrate is adjusted so that the desired amount of the pharmaceutically active powder is deposited on the surface of the substrate. Then, micronized PEG 8000 is triboelectrically charged and deposited. Finally, tablets with the surface deposited PEG are placed under infrared lamp until molten. The PEG will form a coating film while cooling. In the similar manner, PEG is deposited at the bottom side of the tablet surface. The PEG dried film coating generally constitutes 2–6% of the total weight of the solid dosage form.

Furthermore, PEG can reverse the negative charge of medicaments so that they can be deposited on a negatively charged substrate. This is achieved by mixing the negatively charged medicament with micronized PEG, and then depositing the mixture on to the negatively charged substrate. Once the charged mixture is deposited on the substrate surface, it is melted and cooled so that a protective film is formed while cooling.

The use of micronized PEG as a protective coating provides several unique advantages. PEG has a low contact angle. Therefore, it can penetrate through the deposited powder and establish a contact with the surface of the substrate. As a result, PEG can form a strong coating even in the presence of loose powder between the coating film and the substrate. A further advantage is that drug substance has a particle size of 5–20 μ.

The dosages are regulated by measuring spectroscopically the amount of medicament that has been deposited. Electrostatic deposition allows the placement of small dosages in the microgram range.

1.5.7.3 *Chemical Coupling*

PEG polymers are neutral polymers which are available in a variety of molecular weights with low polydispersities. These polymers are non-toxic and are useful in biological and pharmaceutical applications.

One such application is the binding of these polymers with sparingly water-soluble small molecules with therapeutic activity. In this way, the resulting conjugates are made water soluble. This process is termed *PEGylation* (45,46). Potential uses and the properties of PEGylated proteins have been reviewed (47,48). Besides PEG, other types of polymers have been used for drug conjugation (49). These are mostly *N*-(2-hydroxypropyl)methacrylamide and poly-(lactide-*co*-glycolide).

The PEGylation of organic molecules enhances the aqueous solubility of the organic molecule and results in other beneficial properties, such as improved plasma half-life, improved biological distribution, and reduced toxicity (50).

The clearance rate of PEGylated proteins is inversely proportional to the molecular weight. Below a molecular weight of approximately 20,000 *Dalton*, the molecule is cleared in the urine. Higher molecular weight PEG proteins are cleared more slowly in the urine and the feaces. The studies were performed using ^{125}I-labeled PEG with different molecular weights (51).

The properties of the drugs that are conjugated to PEG may change significantly. The treatment of diabetes typically requires regular injections of insulin. The use of insulin as a treatment for diabetes dates back to 1922 (52). It was demonstrated that the active extract from the pancreas had therapeutic effects in diabetic dogs. Treatment of a diabetic patient in that same year with pancreatic extracts resulted in a dramatic, life saving clinical improvement.

Due to the inconvenience of insulin injections, insulin has been the focus of massive efforts to improve its administration and bioassimilation. Until recently animal extracts provided all insulin used for treatment of the disease. The advent of recombinant technology allows commercial scale manufacture of human insulin.

Now, a conjugate of insulin, PEG, and oleic acid can be orally administered (53). It was also found that the PEGylation of insulin dramatically increases the activity of insulin. In an insulin PEG lipophile conjugate the PEG lipophile bond is hydrolyzable. In the bloodstream, the hydrolyzable PEG lipophile bond will be hydrolyzed, leaving the highly active insulin PEG compound circulating in the blood.

The clinical development of PEGylated proteins requires the measurement of the pharmacokinetics. Ideally, the concentration of intact PEGylated protein should be measured. Simple methods to measure intact PEGylated conjugates, however, are not available. Sodium dodecyl sulfate poly(acrylamide) gel electrophoresis can be used to measure the relative size of PEGylated proteins, but the mobility of PEG-modified proteins is slower than the expected molecular weight (54, 55).

Conjugates can be indirectly measured by radiolabeling the protein or PEG (56–58). Colorimetric methods based on complex formation between barium-iodide and PEG require that proteins are first removed and have detection limits of around 1–5 μg PEG (59). High performance liquid chromatography can detect PEG with a detection limit around 1–5 $\mu g ml^{-1}$ (60).

A monoclonal antibody that binds to PEG is useful for quantifying the concentration of PEG in a sample in vitro (61). The preparation of PEG conjugates of rapamycin and related compounds has been described (62). Rapamycin and ascomycin are macrocyclic polyketides that are potent immunosuppressants. These compounds have been approved for preventing transplantation rejection.

Rapamycin is attached to PEG by a glycol unit and a thiol acid ester via the gray encircled hydroxyl group. The PEG thio compound is shown in Figure 1.5. The preparation has been described in detail (50).

Several other routes for the introduction of PEG conjugates have been described, as well as the medical indications (63, 64). The medical details are beyond of the scope of this book.

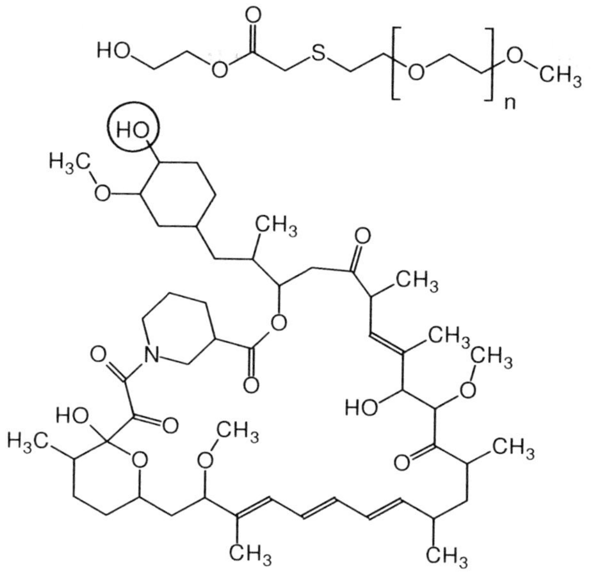

Figure 1.5 Rapamycin and Poly(ethylene glycol) Compound (50)

Therapeutic Proteins. In case of therapeutic proteins, methods to yield highly conjugated proteins have been described. One of the most important features of PEGylation is the reduction of antigenicity and immunogenicity of PEGylated proteins.

A chemical coupling method for adding sites of subsequent PEGylation in a protein is based on a reaction of carbodiimide. This reaction enables the carboxyl groups in proteins to react with additional amino groups of a polyfunctional amine.

In other words, reactive amino groups are added that are suitable for the PEGylation. However, this strategy is interfered by crosslinking reactions that result in polymeric forms of carboxyl amidated proteins (65).

Additional linking sites can be incorporated into a protein for eventual conjugation with activated PEG linkers, without denaturing the protein. This method consists of (66):

1. Coupling a protein with other non-proteinic polymer chains
2. Coupling the thus modified protein with a polyfunctional

amine

3. Coupling the pendent amino groups other non-proteinic polymer chains, to form a multiple of pendent chains.

Subsequently, the method is illustrated with *L*-methionine-α-deamino-γ-mercaptomethane lyase (rMETase). First, rMETase is initially PEGylated with methoxy poly(ethylene glycol) succinimidyl glutarate. Then, the carboxyl groups of the PEGylated protein are reacted with diaminobutane. In this way, the carboxyl groups turn into amide groups.

The amidation is carried out in the presence of a catalyst, such as a water-soluble carbodiimide. A potential crosslinking between the rMETase molecules during the carboxyl amidation is inhibited by the steric hindrance provided by the PEG chains already coupled to the protein.

Finally, the amidated PEGylated rMETase is super-PEGylated by further coupling the pendant amino group resulting from diaminobutane again with methoxy poly(ethylene glycol) succinimidyl glutarate. Biochemical analysis indicated that some 13 PEG chains were coupled to each subunit of rMETase (66). Without amidation, only 6–8 PEG chains are attached. Table 1.6 shows the activity of the protein in the course of the synthesis.

Table 1.6 Protein Activity and Recovery (66)

Step	Specific activity /[Umg^{-1}]	Recovery /[%]
Initial PEGylation	46.4	82.8
Carboxyl amidation	16.7	29.8
Super-PEGylation	7.9	14.1

When unmodified rMETase was directly reacted with dibutylamine without initial PEGylation, rMETase precipitated in the reaction solution apparently due to the crosslinking, leading to a significant loss of its activity. Therefore, a major limit to adding amino groups to proteins using carboxyl amidation is crosslinking of the reacting protein.

Initial PEGylation greatly reduced crosslinking during the carboxyl amidation reaction. With initial PEGylation, there was no difference in molecular weight between PEG-rMETase and super-PEG-rMETase after alkaline hydrolysis to remove all PEG chains (66).

Instead of a linear PEG, hyperbranched variants can be used for coupling (67). The controlled polymerization of glycidol can lead to hyperbranched poly(glycerol)s. The presence of EO in the reaction mixture produces copolymers of ethylene oxide and glycerol (6). The architecture of the copolymer is dictated by the ratio of glycidol to EO, as well as on the order of addition in which monomers are added to the polymerization medium, and their rate of addition to the mixture, which can also include continuous feed of the monomers (67).

The fluorescamine method can used to estimate the degree of PEGylation. This method is based on the reduction in fluorescence intensity due to conjugation of the amino groups by activated PEG (66).

1.6 Suppliers and Commercial Grades

Suppliers and commercial grades are shown in Table 1.7.

Table 1.7 Examples for Commercially Available Poly(ethylene oxide) Polymers

Tradename	Supplier
Breox	BP p.l.c. Corporation United Kingdom
Carbowax®	Union Carbide Corp.
Fomrez®	Chemtura Corp.
Pluracol®	BASF
Pluriol®	BASF
Polyox®	Union Carbide Corp.
Polyox®	Union Carbide Corp.
Ucarfloc®	Union Carbide Corp.

PEO resins are commercially available only in powder form. However, a method has been developed for producing PEO pellets through extrusion, followed by cooling the PEO strands on a fan cooled conveyor belt (13).

1.7 Environmental Impact and Recycling

Disposable personal and medical care products provide the benefit and convenience of one time, sanitary use. Thin films used in such products are typically made from water-insoluble polymers or polymer blends.

However, the disposal of these products is a concern due to limited landfill space. Incineration of such products is not desirable because of increasing concerns about air quality and because of the costs and difficulty associated with separating these products from other disposed articles that cannot be incinerated. Consequently, there is a need for disposable products which may be quickly and conveniently disposed of without dumping or incineration.

Compositions based on PEG have been developed and proposed for such products. These compositions have satisfactory properties during life time, as well exhibit satisfactory properties for disposal (13). A PEO resin can be chemically modified by grafting, reactive extrusion, block polymerization or by branching to improve its processability in the melt (68). Also, blends of modified PEO with water-insoluble polymers have been proposed.

Tradenames appearing in the references are shown in Table 1.8.

Table 1.8 Tradenames in References

Tradename Description	Supplier
Acrawax® Amide wax (18)	Lonza AG, Basel, Switzerland
Aerosil® Fumed Silica (18)	Degussa AG
Airflex® (Series) Vinyl acetate/ethylene copolymer emulsions (17,28)	Air Products and Chemicals, Inc.
Airvol® (Series) Poly(vinyl alcohol)s (17)	Air Products and Chemicals, Inc.
Alcalase® Proteolytic enzyme, detergent (24)	Novo Industries A/S
Alkox™ PEO (22)	Meisei Chemical Works, Ltd.
Atmer® Antistatic agent (19)	Uniquema

Table 1.8 (cont.)

Tradename Description	Supplier
Berset® (Series) Starch and Protein insolubilizer (7)	Bercen Inc.
Bionolle® Poly(butylene succinate) (13)	Showa Highpolymer Co.
Brij® (Series) Ethoxylated fatty alcohols (21)	ICI Surfactants
Brij® 30 Poly(oxyethylene) (4) lauryl ether (21)	ICI Surfactants
Capoten® Angiotensin converting enzyme inhibitor (43)	Bristol-Myers Squibb Co.
Carbowax® (Series) Poly(ethyleneoxide glycol) (PEG) (17)	Union Carbide Corp.
Catapal® Oxo(oxoalumanyloxy)alumane (28)	Sasol Chemical Ind., Ind.
Celluzyme® Detergent enzymes (24)	Novozymes A/S
Cesamet® Nabilone (pharmaceuticum) (43)	Valeant Pharmaceuticals Int.
Ceteareth-25 INCI International Nomenclature of Cosmetic Ingredients, Poly(oxyethylene) cetyl ether (21)	INCI Name
Copaxone® Pharmaceutical preparation (64)	TEVA Pharmaceutical Ind.
Cremophor® EL Ethoxylated castor oil (38, 43)	BASF
Crillet® I Poly(oxyethylene) sorbitan monolaureate (19)	Croda France S.A.
Curesan® Starch and Protein insolubilizer (7)	PPG Industries, Inc.
Disperal® Sol P3 Pseudoboehmite (7)	Condea Chemie GmbH
Esperase® Proteolytic enzyme, detergent (24)	Novozymes A/S Corp.
Fluorad® (Series) Surfactant (7)	3M Comp.
Foamstar® Defoamer (28)	Cognis Corp.
Good Rite® SB 1168 Styrene Butadiene Emulsion (13)	BF Goodrich

Table 1.8 (cont.)

Tradename Description	Supplier
Humulin™ Insulin (53)	Eli Lilly
Hycar® (Series) Amine-terminated butadiene-acrylonitrile (13)	Lubrizol Advanced Materials, Inc., B.F. Goodrich Co.
Irgafos® 168 Tris(2,4-di-*tert*-butylphenyl)phosphite (68)	Ciba Specialty Chemicals Corp.
Irganox® (Series) Hindered phenols, polymerization inhibitor (13)	Ciba Geigy
Irganox® 1010 Pentaerythritol tetrakis(3-(3,5-di-*tert*-butyl-4-hydroxyphenyl)propionate), phenolic antioxidant (68)	Ciba Geigy
Irganox® 1076 Octadecyl-3-(3′,5′-di-*tert*-butyl-4′-hydroxyphenyl) propionate (13,68)	Ciba Geigy
Kenolube® Zinc stearate and amide wax (18)	Hoganas AB
Kraton® Styrenic block copolymer (13)	Shell
Laureth® 4 INCI International Nomenclature of Cosmetic Ingredients, poly(oxyethylene) (4) lauryl ether (21)	INCI Name
Levapren® 600 EVA (68)	Bayer AG
Licowax® Amide wax (18)	Clariant GmbH
Lipase P Amano Lipase enzyme for detergent usage (24,50)	Amano Pharmaceutical Co., Ltd.
Lipolase® Lipase enzyme for detergent usage (24)	Novo Industries A/S
Lodyne® Fluorochemical surfactant (28)	Ciba-Geigy Corp.
Luviskol® VA 64 50% Solution of a copolymer of vinylpyrrolidone and vinylacetate (60:40) in water (28)	BASF AG
Maxatase® Proteolytic enzyme (24)	Gist-Brocades N.V

Table 1.8 (cont.)

Tradename Description	Supplier
Methocel®	Dow
Methylcellulose (24)	
Omyacarb®	Omya AG
Ground limestone (68)	
Pluriol® A 2000	BASF
Poly(ethylene glycol) (24)	
Polyox® WSR	Dow
Poly(ethylene oxide), water soluble resin (22, 68)	
Primal®	Rohm & Haas Comp.
Acrylic polymer (28)	
Rohadon® (Series)	Rohm & Haas
Acrylic polymer (7)	
Rohamere® (Series)	Evonik Rohm GmbH
Acrylic Resins (7)	
Santicizer® (Series)	Solutia, Inc.
Alkyl benzyl phthalates (7)	
Savinase®	Novo Nordisk A/S
Proteolytic enzyme for detergent usage (24)	
Sequarez® (Series)	Omnova Solutions, Inc. Corp.
Insolubilizer for paper coatings (7)	
Slip-Ayd®	Elementis Specialties
Poly(ethylene) wax (28)	
Surlyn®	DuPont
Ionomer resin (19, 68)	
Teflon®	DuPont
Tetrafluoro polymer (28)	
Tinopal®	Ciba-Geigy
Optical brightener (24)	
Triton® X (Series)	Union Carbide Corp. (Rohm & Haas)
Poly(alkylene oxide), nonionic surfactants (7)	
Tween® 20	Uniqema
Sorbitan monolaurate (13)	
Tyril®	Dow
ABS copolymer (68)	
Vazo® (Series)	DuPont
Azonitriles, radical initiators (7)	
Vazo® 67	DuPont
2,2′-Azobis(2-methylbutane-nitrile (7)	

Table 1.8 (cont.)

Tradename Description	Supplier
Zonyl® (Series) Fluorinated nonionic surfactant (29)	DuPont
Zonyl® FS-300 Nonionic fluorosurfactant (7)	DuPont

References

1. A.V. Lourenço, Alcools et Anhydrides polyglycériques, *Comptes Rendus de l'Académie des Sciences, 52*, 52:359–363, 1861. [electronic:] http://gallica.bnf.fr/ark:/12148/bpt6k3009c.image.r=+COMPTES+RENDUS+++DES+S%C3%89ANCES+DE+L.f358.pagination.langFR.
2. B.J. Herold and A. Carneiro, Agostinho Vicente Lourenço 1822–1893, [electronic:] http://www.spq.pt/docs/Biografias/AVLourencoing.pdf, 2007.
3. Process for the production of soaps possessing intensive detergent power, GB Patent 327 393, assigned to IG Farbenindustrie AG, April 01, 1930.
4. Improved cleansing agents, GB Patent 340 232, assigned to IG Farbenindustrie AG, December 17, 1930.
5. Verfahren zur Herstellung von Polyurethanen bzw. Polyharnstoffen, DE Patent 728 981, assigned to IG Farbenindustrie AG, December 07, 1942.
6. P. Dimitrov, E. Hasan, S. Rangelov, B. Trzebicka, A. Dworak, and C.B. Tsvetanov, High molecular weight functionalized poly(ethylene oxide), *Polymer*, 43(25):7171–7178, 2002.
7. R.G. Swisher and H. Li, Inkjet printing media containing substantially water-insoluble plasticizer, US Patent 6 265 049, assigned to Hewlett-Packard Company (Palo Alto, CA), July 24, 2001.
8. S.K. Varshney and J.X. Zhang, Heterofunctional polyethylene glycol and polyethylene oxide, process for their manufacture, US Patent 7 009 033, assigned to Polymer Source Inc. (Dorval, CA), March 7, 2006.
9. S. Zalipsky, Functionalized poly(ethylene glycols) for preparation of biologically relevant conjugates, *Bioconjugate Chem.*, 6(2):150–165, March 1995.

10. J.M. Harris and M.R. Sedaghat-Herati, Preparation and use of polyethylene glycol propionaldehyde, US Patent 5 252 714, assigned to The University of Alabama in Huntsville (Huntsville, AL), October 12, 1993.
11. Y. Nagasaki, T. Okada, C. Scholz, M. Iijima, M. Kato, and K. Kataoka, The reactive polymeric micelle based on an aldehyde-ended poly-(ethylene glycol)/poly(lactide) block copolymer, *Macromolecules*, 31(5): 1473–1479, February 1998.
12. S. Zalipsky and G. Barany, Facile synthesis of α-hydroxy-ω-carboxymethylpolyethylene oxide, *Journal of Bioactive and Compatible Polymers*, 5(2):227–231, 1990.
13. V. Topolkaraev and J.H. Wang, Compositions and process for making water soluble polyethylene oxide films with enhanced toughness and improved melt rheology and tear resistance, US Patent 6 228 920, assigned to Kimberly-Clark Woldwide, Inc. (Neenah, WI), May 8, 2001.
14. F. Kawai and B. Schink, The biochemistry of degradation of polyethers, *Crit. Rev. Biotechnol.*, 6(3):273–307, 1987.
15. S. Matsumura, N. Yoda, and S. Yoshikawa, Microbial transformation of poly(ethylene glycol)s into mono- and dicarboxylic derivatives by specific oxidation of the hydroxymethyl groups, *Makromol. Chem., Rapid Communications*, 10(2):63–67, February 1989.
16. A.-Z. Piao and C. Shih, Mixtures of various triblock polyester polyethylene glycol copolymers having improved gel properties, US Patent 7 135 190, assigned to Macromed, Inc. (Sandy, UT), November 14, 2006.
17. C.D. Smith, Vinyl acetate ethylene emulsions stabilized with poly(ethylene/poly (vinyl alcohol) blend, US Patent 6 673 862, assigned to Air Products Polymers, L.P. (Allentown, PA), January 6, 2004.
18. R. Lindenau, K. Dollmeier, and V. Arnhold, Composition for the production of sintered molded parts, US Patent 7 524 352, assigned to GKNM Sinter Metals GmbH (Radevormwald, DE), April 28, 2009.
19. K. Hausmann, B. Rioux, and J.-M. Francois, Antistatic ionomer blend, US Patent 6 630 528, assigned to E. I. du Pont de Nemours and Company (Wilmington, DE), October 7, 2003.
20. E.D. Mazzarella, L.J. Wood, Jr., and W. Maliczyszyn, Method of sizing paper, US Patent 4 040 900, assigned to National Starch and Chemical Corporation (Bridgewater, NJ), August 9, 1977.
21. P. Hossel, K. Sperling, and V. Schehlmann, Aqueous compositions and their use, US Patent 6 191 188, assigned to BASF Aktiengesellschaft (Ludwishafen, DE), February 20, 2001.
22. T.N. Blanton, D. Majumdar, R.J. Kress, and D.W. Schwark, Reduced crystallinity polyethylene oxide with intercalated clay, US Patent 6 555 610, assigned to Eastman Kodak Company (Rochester, NY), April 29, 2003.

23. J.W. Artley and T.E. Lister, Method of making polyethylene glycol treated fabrics, US Patent 7 585 330, assigned to , and, September 8, 2009.
24. J. Penninger, Boosting the cleaning performance of laundry detergents by polymer of styrene/methyl methacrylate/methyl polyethylene glycol, US Patent 7 431 739, assigned to Henkel Kommanditgesellschaft auf Aktien (Dusseldorf, DE), October 7, 2008.
25. J.A. Moyse, Improvements in the laundering of synthetic polymeric textile materials, GB Patent 1 154 730, assigned to ICI Ltd., June 11, 1969.
26. Detergent compositions, GB Patent 1 377 092, assigned to Unilever Ltd., December 11, 1974.
27. K.A. Rodrigues, Laundry detergents containing styrene-anhydride copolymers grafted with polyethylene glycol, US Patent 6 075 093, assigned to National Starch and Chemical Investment Holding Corporation (Wilmington, DE), June 13, 2000.
28. B.-J. Niu, S. Schuttel, and M. Schaer, Print media products for generating high quality images and methods for making the same, US Patent 7 112 629, assigned to Hewlett-Packard Development Company, L.P. (Houston, TX), September 26, 2006.
29. D.F. Lyons, Coagulation of fluoroelastomer dispersions with polyethylene oxide, US Patent 7 816 468, assigned to DuPont Performance Elastomers LLC (Wilmington, DE), October 19, 2010.
30. C. Couture, D. Bergeron, and F. Picard, Crosslinked polysaccharide, obtained by crosslinking with substituted polyethylene glycol, as superabsorbent, US Patent 7 365 190, assigned to Archer-Daniels-Midland Company (Decatur, IL), April 29, 2008.
31. A.-C. Samuelsson, Animal feed of higher nutritive value, method for production thereof and use of a polyethylene glycol compound, US Patent 6 379 723, assigned to Akzo Nobel, N.V. (Arnhem, NL), April 30, 2002.
32. N.V. Jamieson, R. Sundberg, S. Lindell, K. Claesson, J. Moen, P.K. Vreugdenhil, D.G.D. Wight, J.H. Southard, and F.O. Belzer, Preservation of the canine liver for 24–48 hours using simple cold storage with UW solution, *Transplantation*, 46(4):517, 1988.
33. W. Wicomb, G. Collins, I. Bathurst, and M. Foehr, System for storing polyethylene glycol solutions, US Patent 6 321 909, assigned to Sky High, LLC (Evanston, IL), November 27, 2001.
34. W.N. Wicomb and G.M. Collins, 24-hour rabbit heart storage with UW solution: Effects of low-flow perfusion, colloid, and shelf storage, *Transplantation*, 48(1):6, 1989.
35. G.M. Collins and W.N. Wicomb, New organ preservation solutions, *Kidney Int. Suppl.*, 38:S197, 1992.

36. P. Seth and A. Stamm, Process for manufacturing solid compositions containing polyethylene oxide and an active ingredient, US Patent 6 048 547, April 11, 2000.
37. P. Fuchs and J. Hilmann, Arzneimittelwirkstoffträger in Folienform mit inkorporiertem Wirkstoff, DE Patent 2 432 925, assigned to Schering AG, January 22, 1976.
38. P. Fuchs and J. Hilmann, Medicament carriers in the form of film having active substance incorporated therein, US Patent 4 136 145, assigned to Schering Aktiengesellschaft (DE), January 23, 1979.
39. W. Schmidt, Process for the preparation of an administration and dosage for drugs, reagents or other active substances., EP Patent 0 219 762, assigned to Desitin Arzneimittel GmbH, April 29, 1987.
40. W. Schmidt, Process for producing an administration or dosage form for drugs, reagents or other active ingredients, US Patent 4 849 246, July 18, 1989.
41. M. Horstmann, W. Laux, and S. Hungerbach, Rapidly disintegrating sheet-like presentations of multiple dosage units, US Patent 5 629 003, assigned to LTS Lohmann Therapie-Systeme GmbH & Co. KG (Neuwied, DE), May 13, 1997.
42. H.G. Zerbe, J.-H. Guo, and A. Serino, Water soluble film for oral administration with instant wettability, US Patent 5 948 430, assigned to LTS Lohmann Therapie-Systeme GmbH (Neuwied, DE), September 7, 1999.
43. R.K. Yang, R.C. Fuisz, G.L. Myers, and J.M. Fuisz, Polyethylene oxide-based films and drug delivery systems made therefrom, US Patent 7 666 337, assigned to MonoSol Rx, LLC (Portage, IN), February 23, 2010.
44. S.B. Wei and H. Uang, Polyethylene glycol coating for electrostatic dry deposition of pharmaceuticals, US Patent 6 372 246, assigned to Ortho-McNeil Pharmaceutical, Inc. (Raritan, NJ), April 16, 2002.
45. R.B. Greenwald, Y.H. Choe, J. McGuire, and C.D. Conover, Effective drug delivery by PEGylated drug conjugates, *Adv. Drug Deliv. Rev.*, 55 (2):217–250, February 2003.
46. G. Pasut, A. Guiotto, and F.M. Veronese, Protein, peptide and non-peptide drug PEGylation for therapeutic application, *Expert Opin. Ther. Pat.*, 14(6):859–894, June 2004.
47. C. Delgado, G.E. Francis, and D. Fisher, The uses and properties of PEG-linked proteins., *Crit. Rev. Ther. Drug Carrier Syst.*, 9(3-4):249, 1992.
48. F.M. Veronese, Peptide and protein PEGylation: A review of problems and solutions, *Biomaterials*, 22(5):405–417, March 2001.
49. J. Khandare and T. Minko, Polymer-drug conjugates: Progress in polymeric prodrugs, *Prog. Polym. Sci.*, 31(4):359–397, April 2006.

50. J. Gu, M. Ruppen, T. Zhu, and M. Fawzi, Processes for preparing water-soluble polyethylene glycol conjugates of macrolide immunosuppressants, US Patent 7 605 257, assigned to Wyeth (Madison, NJ), October 20, 2009.
51. T. Yamaoka, Y. Tabata, and Y. Ikada, Distribution and tissue uptake of poly(ethylene glycol) with different molecular weights after intravenous administration to mice, *J. Pharm. Sci.*, 83(4):601–606, April 1994.
52. F.G. Banting, C.H. Best, J.B. Collip, W.R. Campbell, and A.A. Fletcher, Pancreatic extracts in the treatment of diabetes mellitus, *Can. Med. Assoc. J.*, 12(3):141, March 1922.
53. N.N. Ekwuribe, M. Ramaswamy, and J. Rajagopalan, Drug-oligomer conjugates with polyethylene glycol components, US Patent 7 030 084, assigned to Nobex Corporation (Durham, NC), April 18, 2006.
54. T. Suzuki, N. Kanbara, T. Tomono, N. Hayashi, and I. Shinohara, Physicochemical and biological properties of poly(ethylene glycol)-coupled immunoglobuling G, *Biochimica et Biophysica Acta (BBA)-Protein Structure and Molecular Enzymology*, 788(2):248–255, 1984.
55. N.V. Katre, M.J. Knauf, and W.J. Laird, Chemical modification of recombinant interleukin 2 by polyethylene glycol increases its potency in the murine Meth A sarcoma model, *Proc. Natl. Acad. Sci. U.S.A.*, 84 (6):1487, 1987.
56. Y. Kaneda, S. Yamamoto, T. Kihira, Y. Tsutsumi, S. Nakagawa, M. Miyake, K. Kawasaki, and T. Mayumi, Synthetic cell-adhesive laminin peptide YIGSR conjugated with polyethylene glycol has improved antimetastatic activity due to a longer half-life in blood, *Invasion & metastasis*, 15(3-4):156, 1995.
57. T. Cheng, B. Chen, L. Chan, P. Wu, J. Chern, and S. Roffler, Poly(ethylene glycol) modification of β-glucuronidase-antibody conjugates for solid-tumor therapy by targeted activation of glucuronide prodrugs, *Canc. Immunol. Immunother.*, 44(6):305–315, 1997.
58. J.M. Mullin, C.W. Marano, K.V. Laughlin, M. Nuciglio, B.R. Stevenson, and A.P. Soler, Different size limitations for increased transepithelial paracellular solute flux across phorbol ester and tumor necrosis factor-treated epithelial cell sheets, *J. Cell. Physiol.*, 171(2):226–233, 1997.
59. C.E. Childs, The determination of polyethylene glycol in gamma globulin solutions, *Microchem. J.*, 20(2):190–192, June 1975.
60. P.J. Miles, K.V. Langley, C.J. Stacey, and T.L. Talarico, Detection of residual polyethylene glycol derivatives in pyridoxylated-hemoglobin-polyoxyethylene conjugate, *Artif. Cell. Blood Substit. Biotechnol.*, 25 (3):315–326, 1997.
61. S. Roffler, T.-L. Cheng, and P.-Y. Wu, Monoclonal antibody for analysis and clearance of polyethylene glycol and polyethylene glycol-

modified molecules, US Patent 7 320 791, assigned to Academia Sinica (Taipei, TW), January 22, 2008.

62. T. Zhu, S.M. Shah, and R.W. Saunders, Water soluble SDZ RAD esters, US Patent 6 331 547, assigned to American Home Products Corporation (Madison, NJ), December 18, 2001.
63. S.M. Walsh, A.G. Shah, J.J. Mond, A. Lees, and J.J. Drabick, Antimicrobial polymer conjugate containing lysostaphin and polyethylene glycol, US Patent 7 452 533, assigned to Biosynexus Incorporated (Gaithersburg, MD), November 18, 2008.
64. A. Konradi, M.A. Pleiss, J.L. Smith, C.M. Semko, and C. Vandevert, Polyethylene glycol conjugates of heterocycloalkyl carboxamido propanoic acids, US Patent 7 595 318, assigned to Elan Pharmaceuticals, Inc. (San Francisco, CA), September 29, 2009.
65. F.F. Davis, T. Van Es, and N.C. Palczuk, Non-immunogenic polypeptides, US Patent 4 179 337, December 18, 1979.
66. S. Li, Z. Yang, X. Sun, Y. Tan, and S. Yagi, Methods for increasing protein polyethylene glycol (peg) conjugation, US Patent 7 799 549, assigned to Anticancer, Inc. (San Diego, CA), September 21, 2010.
67. F. Ignatious, Heterofunctional copolymers of glycerol and polyethylene glycol, their conjugates and compositions, US Patent 7 196 145, assigned to SmithKline Beecham Corporation (Philadephia, PA), March 27, 2007.
68. B.A. Balogh and V.A. Topolkaraev, Films, fibers and articles of chemically modified polyethylene oxide compositions with improved environmental stability and method of making same, US Patent 6 515 075, assigned to Kimberly-Clark Worldwide, Inc. (Neenah, WI), February 4, 2003.

2

Poly(vinyl alcohol)

The synthesis of poly(vinyl alcohol) (PVA) was described in 1924 by Haehnel and Herrmann (1–3). These researchers showed that the hydrolysis of poly(vinyl acetate) (PVAc) and poly(vinyl propionate) yields PVA. The inventors stated even then that PVA is water soluble. There followed an impressive series of additional inventions belonging to this topic.

Actually, there are only a few monographs on the general topic (4, 5). However, the literature in articles and patents is numerous.

2.1 Monomers

The basic monomer for PVA is vinyl acetate (VA). Copolymers with ethylene are common, although these copolymers are not water soluble in general, with the water solubility dependent on the composition. On the other hand, the hydroxyl group in PVA can be modified by grafting. Monomers and compounds used for grafting are shown in Table 2.1.

2.2 Polymerization and Fabrication

PVA is synthesized from PVAc by hydrolysis. Vinyl alcohol cannot be polymerized, because this monomer is not a stable compound. Instead, it isomerizes, or more correctly, tautomerizes immediately to acetaldehyde by the corresponding rearrangement of the hydrogen atoms. The sequence of reactions is shown in Figure 2.1. The hydrolysis is achieved in acetic acid by the aid of either acidic or basic catalysts. PVA is insoluble in acetic acid and precipitates.

Table 2.1 Monomers and Modifiers

Compound	References
Vinyl Compound	
Vinyl acetate	(6)
N-Vinyl isocyanate	(7)
2-Acrylamido-2-methyl-1-propane sulfonic acid	(8)
Modifier	
Diacetyl-*p*-aminosalicylic ester	(9)
Diethylaminopropylamine	(10)
Diethylaminoethylamine	(10)
Dimethylaminopropylamine	(10)
1,3-Propanesulfone	(11)
1,2-Epoxy-5-hexene	(12)

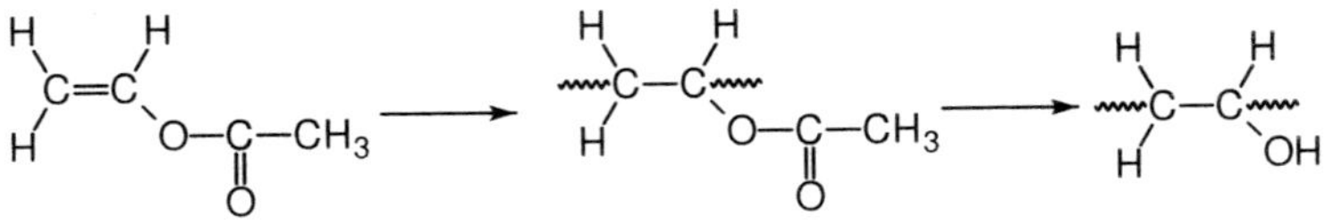

Figure 2.1 Synthesis of Poly(vinyl alcohol)

The pendent hydroxyl group can be further modified. For example, grafting with amines such as diethylaminopropylamine, diethylaminoethylamine, or dimethylaminopropylamine, results in polymers with enhanced cytotoxicity (10).

For the preparation of copolymers from vinyl acetate and vinyl isocyanate, isopropyl percarbonate as the catalyst is dissolved in benzene and to this mixture the monomers are added. After polymerization at 40°C, the solution with the vinyl acetate/vinyl isocyanate copolymer is poured into methanol to form a clear solution of the copolymer. The solution of the copolymer in methanol is evaporated to remove all the unreacted monomers and then redissolved in methanol.

To this solution 6N HCl is added, and the resulting mixture is refluxed in order to hydrolyze the copolymer. Eventually, the copolymer is precipitated by pouring and stirring the refluxed mixture into tetrahydrofuran (7). These types of copolymers find use as flocculants. Other radical initiators suitable for polymerization are summarized in Table 2.2.

Table 2.2 Radical Initiators

Compound	Reference
Di-*n*-propyl peroxydicarbonate	(8)
Isopropyl percarbonate	(7)
Di-*n*-butyl peroxydicarbonate	(8)
Bis(4-*tert*-butylcyclohexyl) peroxydicarbonate	(8)
tert-Butyl peroxyneodecanoate	(8)
2-Ethylhexyl peroxydicarbonate	(8)
Dicetyl peroxydicarbonate	(8)

A poly(vinyl alcohol vinylamine) copolymer can be obtained from a copolymer of VA and *N*-vinyl acetamide by hydrolysis (13). The hydrolysis is achieved by using 0.1 molar sodium methoxide ($Na^{+}\ {}^{-}O{-}CH_3$) in methanol solution at 90°C.

A continuous process for the production of vinyl alcohol copolymers has been described, in particular 2-acrylamido-2-methyl-1-propane sulfonic acid (AMPS) containing copolymers (8). Aqueous dispersions from these polymers can also be obtained (14).

The raw PVA can be purified with an anion exchange resin. A

process to produce a PVA film consists of (15, 16):

- Providing a crude aqueous PVA solution with a pH of 5–6.9
- Directing the crude solution through a column with a cation exchange resin
- Collecting the purified aqueous polymer solution from the column.

The use of a macroreticular polymeric cation exchange resin allows for improved contact of the crude PVA solution with the cation exchange resin as the solution transverses the column, compared with gel form cation exchange resins. Preferred macroreticular polymeric cation exchange resins include so-called strong cation exchange resins, comprising styrene divinylbenzene copolymers functionalized with sulfonic acid groups (16).

2.2.1 *Hydrogels*

PVA hydrogels are important in biomedical and engineering applications (17). Subsequently we discuss some methods for the production of hydrogels.

2.2.1.1 *Physical Crosslinking*

Freeze-thaw cycling of solutions of PVA results in the formation of physical crosslinks. The sites of crosslinking are microcrystals that are formed by hydrogen bonding. Such hydrogels are also named cryogels. In this way, PVA is gelling by physical crosslinking. In other words, cryogels do not require the introduction of chemical crosslinking agents or radiation. Therefore, cryogels are easily produced with low impact on incorporated bioactive molecules.

The incorporated molecules are limited, however, to those that can tolerate the freeze-thaw cycles required to make the gel. Thus, the resulting material can contain bioactive components that will function separately following implantation. In addition, PVA cryogels are highly biocompatible. They exhibit a very low toxicity, which is caused at least partially due to their low surface energy. They contain few impurities and their water content can be made similar to natural tissue of 80–90%.

There is still some discussion on the exact mechanism of gelation of PVA by a freeze-thaw cycle. Three models have been proposed to explain the physical crosslinking in the course of a freeze-thaw cycle (18):

1. Direct hydrogen bonding
2. Direct crystallite formation
3. Liquid-liquid phase separation followed by a gelation mechanism.

The first two models suggest that the gel is formed through a nucleation and growth phase separation. The third model suggests a spinodal decomposition followed by phase separation.

Hydrogen bonding will form nodes and crystallite formation will form larger polymer crystals. However, both of these models demand closely connected crosslinks, with relatively small crosslinking nodes. This observation is supported by studies on the gelation mechanism of PVA.

In contrast, spinodal decomposition causes the redistribution of the polymer into polymer rich and polymer poor regions followed by a gelation process which results in more distantly spaced crosslinks. It is thought that phase separation through spinodal decomposition is likely to be responsible for the improved mechanical properties of PVA after crosslinking. It occurs due to a quenching of the polymer solution.

During the freezing process, the system undergoes a spinodal decomposition, whereby polymer rich and poor phases appear spontaneously in the homogeneous solution. This process occurs because the phase diagram of the quenched PVA at certain temperatures exhibits two coexisting phases with different polymer concentration.

The physical properties of cryogels depend on the molecular weight of the polymer as such, the concentration of the aqueous solution, temperature, the time of freezing, and the number of freeze-thaw cycles. In this way, the properties of a cryogel can be tailored. However, since the properties of the material change dramatically at every freeze-thaw step, the control over the properties of the finished gel is somewhat limited (18).

In general, the modulus of a PVA cryogel increases with the number of freeze-thaw cycles. For example, thermally cycled PVA cryogels exhibit compressive moduli in the range of 1–18 *MPa* and shear moduli in the range of 0.1–0.4 *MPa*.

Since cryogels are crosslinked by physical and not chemical means, there is some concern about their structural stability. The modulus of PVA in aqueous solution increases with the soak time. The increase in strength in the course of aqueous aging is the result of an increase of the supramolecular packing of the polymer chains (18). This is accompanied with the loss of soluble PVA.

Instead of freezing and thawing, the same effects can be achieved using a theta solvents or solvent mixtures (18). At the theta point, the solvent quality is such that the random Brownian motions are enough to keep the chain in an ideal Gaussian distribution. Below this critical threshold the chain segments prefer to be next to each other rather than to a solvent molecule, and the chain shrinks. The theta point is characterized by the Flory interaction parameter, χ, which is dimensionless, and depends on temperature, and pressure.

The cycling is done with different solvents with a different χ parameter. The first solvents have a low χ, while the second solvents have a high χ, with the phase transition at about χ of 0.5. A solvent with a χ of 0 corresponds to a medium, which is very similar to a monomer.

In the lattice model of a polymer chain, this occurs where the free energy comes entirely from the entropy associated with various chain patterns on the lattice. In such a case, the temperature has essentially no effect on structure, and the solvent is considered as athermal. Athermal solvents are basically good solvents. If the solvent quality is sufficiently poor, the chain will completely precipitate out of the solution. This effect can also be obtained by manipulation of the temperature of the solution, i.e., in the freeze-thaw technique. The theta temperatures for PVA in various solvents are shown in Table 2.3.

It is believed that forcing the PVA chains in solution into the close proximity of the theta point by using a theta solvent, a spinodal decomposition mechanism would occur. This results in the formation of a physical association that is resistant to dissolution. Thus, a PVA hydrogel is formed by the controlled use of solvents having a χ parameter that is sufficient to cause a gelation as it forces the PVA chains to associate by a physical mechanism.

Table 2.3 Theta temperatures for Poly(vinyl alcohol) in various solvents (18, 19)

Solvent pair	Volume Ratio	T_Θ/[°C]
tert-Butanol/Water	32.0:68.0	25
Ethanol/Water	41.5:58.5	25
Methanol/Water	41.7:58.5	25
i-Propanol/Water	39.4:60.6	25
n-Propanol/Water	35.1:64.9	25

To prevent a random crashing out of the polymer, the solvent quality must be carefully controlled. Gels that are formed in this way are addressed as *thetagels* (18).

High compression PVA thetagels can be made by placing PVA in a reverse osmosis membrane with NaCl and then making the outside concentration of NaCl quite high to compress the solution. The NaCl concentration will climb as the water leaves the reverse osmosis membrane gelling the PVA at high pressure. The concentration of PVA can be modified by the ratio of NaCl to PVA inside the reverse osmosis membrane (18). Further, instead of chemical crosslinking, ions of multivalent metals, e.g., Ca^{2+} may serve to enhance the physical crosslinking properties (20).

Still another method to introduce crosslinking by hydrogen bonds consists of the addition of multifunctional molecules, e.g. amino acids. Further, it has been shown that succinic acid or ethylenediamine is active in this way (21).

As shown in Table 2.4, some amino acids behave similarly regardless of whether the *L*- or *D*-isomer is used. However, there is a significant number of amino acids that preferentially gel PVA as either the *L*- or the *D*-isomer.

2.2.1.2 *Chemical Crosslinking*

On the other hand, crosslinking can be achieved by mixing individually prepared aqueous solutions of PVA and poly(acrylic acid) (PAA). Eventually, the hydroyxl groups in PVA and the acid groups in PAA may at lest partially esterify (22). Such a system has been proposed to enhance the drug entrapment efficiency and to improve

Table 2.4 Comparison of Amino Acids in Forming Hydrogels (21,23)

Amino acid	L	D		L	D		L	D
Ala	+	-	Thr	+	+	Met	-	+
His	++	-	Asp	-	-	Val	+	+
Pro	++	++	Leu	-	+	Gln	+	-
Arg	++	-	Trp	++	+	Nle	+	
Hyp	-		Cys	+		Gly	++	
Ser	+	++	Lys	++	++	Phe	+	+
Asn	+	-	Tyr	+	+			
Ile	-	+	Glu	-	-			

++: Gels in water, + Gels in 1 M $NaHCO_3$, - No gel

the swelling behavior of a drug delivery system.

Several applications of this type have been described (24–26). A challenging application is the photodynamic therapy of infected wounds (27). Hydrogels can be synthesized by the addition of 1,2-epoxy-5-hexene to PVA. The epoxide opens as it forms pendent vinyl groups. The reaction is shown in Figure 2.2.

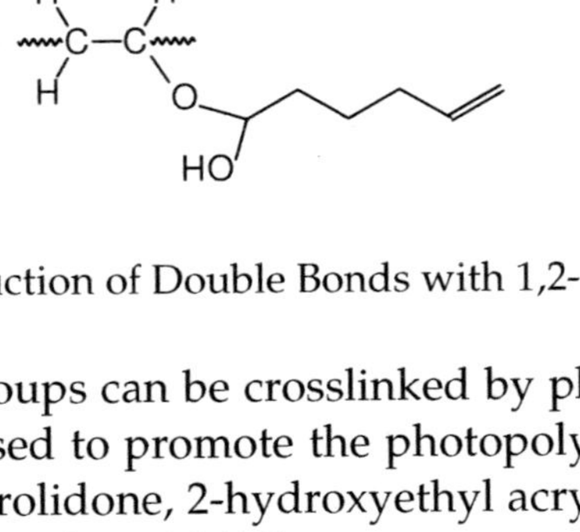

Figure 2.2 Introduction of Double Bonds with 1,2-Epoxy-5-hexene (12)

These vinyl groups can be crosslinked by photopolymerization. The monomers used to promote the photopolymerization reaction are *N*-vinyl-2-pyrrolidone, 2-hydroxyethyl acrylate, *N*,*N*-dimethylacrylamide, and acrylic acid (12).

Biodegradable and biocompatible hydrogels are on PVA that is crosslinked with a biodegradable crosslinking agent (28). Likewise, biodegradable regions are incorporated into the hydrogel during its formation. Examples of biodegradable crosslinking agents are shown in Table 2.5.

Table 2.5 Biodegradable Crosslinking Agents (28)

Crosslinker
2-(Acryloyloxy)ethyl succinate
2-(Methacryloyloxy)ethyl succinate
Carboxyethylacrylate
Vinylazlactone
2-Hydroxyethyl methacrylate glycolate
Hydroxyethyl acrylate glycolate

Aldehydes are suitable as additional crosslinking agents for hydrogels. Hardening can be achieved by treating the hydrogel with formaldehyde, acetaldehyde, glutaraldehyde, terephthalaldehyde, or hexamethylenediamine. Unfortunately, these treatments decrease the biocompatibility of the hydrogel (29).

2.3 Properties

The properties of PVA are dependent on the degree of hydrolysis. The effects of the degree of hydrolysis on the morphology and diameter of electrospun PVA fibers and the water resistance have been studied (30). In electrospun PVA nanofibers, multi-wall carbon nanotubes (CNT)s can be incorporated (31). The multi-wall CNTs are synthesized by chemical vapor deposition.

Polymers from 2-acrylamido-2-methyl-1-propane sulfonic acid sodium salt may have much lower degrees of hydrolysis than those from PVA, and reach the same properties (14).

In some cases, it is desirable to tailor PVA films of variable solubility for applications that have different dissolution temperature requirements. For example, pouches containing detergent for consumer cleaning applications are preferably soluble in cold water, at 10°C or above. At the opposite extreme, PVA laundry bags preferably only dissolve in hot water (32).

To influence solubility of PVA, functional comonomers can be incorporated during the polymerization into PVAc. For example, copolymerizing an acrylate or methacrylate ester with VA, results in a lactone functional group on the polymer chain after transesterification (32). On the other hand, a low molecular ester, e.g. of glycerol may be introduced. In the course of a transesterification, a plasticizer is introduced.

When ethylene vinyl acetate (EVA) is blended with chitosan (CS), antibacterial blends are obtained (33). Blends with a low molecular weight CS grade show an enhanced phase morphology, transparency, an enhanced water barrier properties. In contrast, blends with a high molecular weight CS grade are translucent with a clearly separated phase morphology. Moreover, into EVA antimicrobial properties can be imparted if acetic acid is incorporated into the polymer formulation before casting from solution.

2.3.1 *Swelling of Hydrogels*

The swelling properties of hydrogels are important in the development of hydrogel-based artificial muscles, bio-actuators, and sensors. The swelling of hydrogels is dependent on the pH. This property can be used for the development of pH sensors (22).

A fiber-based pH sensor contains a pH sensitive hydrogel that is coupled to an optical fiber containing a Bragg grating. The network structure of the hydrogel changes dependent on pH and ionic strength of the solution in which it is immersed. In this way, the volume changes.

These changes can be monitored. When the hydrogel is coupled with an optical fiber with Bragg gratings, the changes in swelling can be monitored optically. However, the response time of these gels are in the range of hours. This property restricts their use in systems where the pH is changing only slowly, e.g. in petroleum wells (22).

Methyl cellulose/PVA hydrogels are stable within a wide temperature range, and show a reversible thermoresponsivity (34). Materials that are themoresponsive show an irregular behavior in their differential scanning calorimetry curves.

2.4 Applications

PVA materials are widely used for paper processing agents, textile sizing agents, dispersants, adhesives, films, in the agro industries, because of their excellent film forming, surface activity and hydrogen bond-forming ability (35,36).

Copolymers from vinyl alcohol and AMPS are used in aqueous dispersions for drilling fluids, hydraulic cement compositions, coatings, and papermaking compositions (14).

2.4.1 *Papermaking*

In the field of paper processing applications, vinyl alcohol polymers are used for improving the quality of printed matter. They are used as a surface-sizing agent for printing and writing paper, as an under-sizing agent for artificial paper and coated paper, as a dispersant for fluorescent dyes, and as a filler binder for ink jet recording materials (36).

2.4.1.1 *Binder Material*

Paper coating compositions with an improved low shear viscosity at a high solids level of fine particle size calcium carbonate have been developed. This is achieved by dissolving, a partially hydrolyzed, low molecular weight PVA powder in an aqueous slurry of the pigment particles which is a predominantly fine particle size calcium carbonate. There are several advantages in the preparation of such a composition in this way (37):

- The poly(vinyl alcohol) does not need to be solubilized prior to mixing
- The poly(vinyl alcohol) can be solubilized in the calcium carbonate slurry without heating
- The low shear viscosity of the calcium carbonate slurry is significantly reduced, thus allowing greater mixing efficiency
- The solids level of the pigment slurry can be increased without increasing the shear viscosity
- Binding of the calcium carbonate to a cellulosic substrate is accomplished with a relatively small amount of PVA

- No additional binders are needed in the final coating formulation
- Excellent ink jet printability is achieved.

2.4.1.2 *Papers for Ink Jet Printing*

Papers for printing with ink jet printers are coated with an ink receiving layer or glossy layer. PVA can be used as a binder in the ink receiving layer. It may be combined with another water-soluble or water-dispersable resin. These materials include cellulosic materials, gums, and synthetic resins, such as poly(meth)acrylamide.

Also, copolymers of VA and amine functional comonomers, such as *N,N′*-dimethylaminoethyl methacrylate improve the properties of such papers. These are hydrolyzed in the final stage to get the vinyl alcohol moiety.

Amine functional poly(vinyl alcohol) is typically produced by the copolymerization of VA with amine functional monomers, such as trimethyl-(3-methacrylamido-propyl)ammonium chloride, *N*-vinylformamide, or acrylamide (AAm), followed by saponification to form the PVA derivative. However, there are disadvantages to this approach. The selection of amino comonomers is very limited due to their incompatibility with the saponification conditions to produce poly(vinyl alcohol).

Another approach for the production of amine functional PVA involves the reaction of PVA with amino moieties. For example, reacting aminobutyraldehyde dimethyl acetal with the hydroxyl groups of PVA results in the corresponding acetal of PVA.

In addition, PVA can be grafted by a free radical graft copolymerization with amine functional monomers, such as *N,N′*-dimethylaminoethyl methacrylate. The graft copolymerization has some advantages over the traditional copolymerization with VA and subsequent saponification (38). A drawback of graft polymerization reactions is the simultaneous production of homopolymer or copolymers of the monomers being grafted. This results ultimately in a blend of polymers. Copolymers from a variety of vinyl monomers with amine functional comonomers, synthesized in the presence of a large amount of PVA, offer further advantages over homogeneous aqueous graft copolymerization reactions (39).

By using large amounts of a PVA as a hydroxyl containing colloid stabilizer, more of the stabilizer is available for grafting reactions compared with traditional emulsion systems, thus affording a unique emulsion copolymer. This results in a single polymeric ink jet paper binding system which provides (39):

- Excellent image quality
- Complete fixation of ink dyes to the paper under adverse humidity conditions
- High binding power of various pigments.

The advantages of this emulsion polymerization approach to produce amine functional emulsion polymers are: The amine functional copolymer produced in emulsion form allows for the product to be prepared at a higher overall solids level compared with aqueous solution graft polymerization reactions.

The higher achievable solids levels translates into lower processing costs and a lower cost product. The emulsion polymer provides the combined properties of a dye-fixative polymer with the high binding strength of the hydroxyl containing polymer colloid stabilizer in one polymeric ink jet binder package.

The ethylenically unsaturated monomers are easily incorporated into the amine functional PVA emulsion. They serve to broaden the accessible polymer compositions and end-use performance features. The polymer morphology and property features are very different given by the nature, size, distribution, and composition of the emulsion polymer particles. Further, the comparatively high levels of colloid stabilizer and water-soluble amine functional polymers alter the particle stabilization mechanism and the ultimate properties of the emulsion polymer (39).

Comonomers for ink jet paper coatings are shown in Table 2.6. The preparation of the ink jet coating has been described in detail (39). Colloidal silica particles that are negatively charged on their surface are preferably used as fillers of these layers.

2.4.1.3 Absorbent Sheets

Fabric creping has been employed in papermaking processes that include mechanical or compactive dewatering of the paper web in

Table 2.6 Comonomers for Ink Jet Paper Coatings

Comonomers (39)	
Vinyl acetate	N,N′-Dimethylaminoethyl methacrylate
Methyl methacrylate	N,N′-Dimethylaminoethyl methacrylate
Styrene	N,N′-Dimethylaminoethyl methacrylate
Vinyl acetate	2-Trimethylammoniumethyl methacrylate chloride
Graft Comonomers (38)	
N,N′-Dimethylaminoethyl methacrylate	
4-Vinylpyridine	

order to tailor the properties of the final product. Conventional though drying processes do not take full advantage of the drying potential of Yankee dryers because it is difficult to adhere a partially dried web of intermediate consistency to a surface rotating at high speed. However, a creping adhesive has been developed that enables high speed transfer of the web of an intermediate consistency.

Adhesives based on PVA and poly(amide) can be utilized to transfer and adhere a web of intermediate consistency to a Yankee dryer sufficiently to allow for high speed operation and high jet velocity impingement (40). Suitable modifiers include quaternary ammonium complexes with a linear amide, such as 2-hydroxyethyl di-(2-alkylamido-ethyl)methyl ammonium methyl sulfate.

2.4.2 *Textile Applications*

PVA copolymers where the comonomer directly provides an acid functionality are known. The acid functionality can be introduced by alkyl acrylates, or alkyl methacrylates, respectively, or by dialkyl fumurates. In these copolymers, the acid functionality is introduced by the hydrolysis of the ester comonomer units (41).

Depending on the precise conditions, such as the catalyst, its concentration, and the solvent medium used to hydrolyze the VA ester units in the copolymer, the other ester units may or may not be hydrolyzed to the corresponding acid. Generally speaking VA ester units are far more readily hydrolyzed than alkyl ester units. If the alkyl ester units are also hydrolyzed, and if enough base is present, the resulting acid units may also be neutralized which then become ionomer units.

PVA copolymer ionomers are uniquely useful in preparing textile sizing compositions. This is because of their extraordinary ability to be desized both in water and in diluted caustic solutions. They are far more readily desized than non-ionmeric PVA copolymer compositions.

The preferred comonomer for VA is methyl acrylate (MA) since MA and VA have the same molecular weight. On the other hand, methacrylates are more reactive than acrylates, but both are far more reactive than VA, so that typically they are completely reacted, while less reactive VA has to be stripped off, and would be recycled in a commercial continuous process.

The PVA copolymer ionomers are blended with starches. Specific examples of naturally occurring starches include starches in corn, wheat, potato, sorghum, rice, etc.

In general size solutions are clear and slightly viscous if only PVA polymers are used. When starches are part of the blends, some haziness is sometimes present. This indicates that the starch is rather suspended than fully dissolved (41).

2.4.3 *Adhesive Applications*

Pressure-sensitive adhesives are extensively used in masking tapes, double-faced pressure-sensitive adhesive tapes, surface-protective films, and packaging tapes. Pressure-sensitive adhesives of the aqueous dispersion type do not contain any organic solvent. This is relevant to environmental preservation, resource saving, and safety.

In particular, rubber-based pressure-sensitive adhesives of the aqueous dispersion type are extensively used, as they have advantages such as reduced limitations on adherent selection and excellent low-temperature adhesiveness.

Aqueous dispersion type pressure-sensitive adhesive compositions can be formulated from poly(*N*-vinyl-2-pyrrolidone) (PVP), poly(ethylene glycol) (PEG), PVA, and PAA, or poly(methacrylic acid). It has been found that by the addition of PEG and PVA to an acrylic polymer, the initial pressure-sensitive adhesive force of the adhesive in application to dewy or wet surfaces can improve its constant-load peeling property (42).

In electrolytic capacitors, a protective adhesive layer must be positioned between the dielectric layer and the solid electrolyte layer. The protective adhesive layer can be a PVA composition (43). The presence of hydroxyl groups in the polymer provides adhesive characteristics to the protective adhesive layer. It helps to establish bonds between the dielectric layer and the solid electrolyte.

2.4.4 *Corrosion Inhibition*

A composite from PVA and aniline was tested for its ability in protecting mild steel against acid corrosion (44). In fact, the addition of PVA to the acid reduces the corrosion of the metal. The efficiency of inhibition increases with the increase of concentration. A maximum of efficiency is reached at 2000 *ppm*.

2.4.5 *Membranes*

Sulfonated PVA and blends with sulfonated poly(ether ether ketone) have been prepared. The wettability has been studied. In addition, the ion exchange capacity, the proton conductivity and the water sorption and desorption, respectively have been investigated (45). A good correlation has been established between the surface energy and the ion exchange capacity of the membranes.

The materials have potential applications as electrolyte membranes for fuel cells. Membranes based on PVA show good a proton conductivity together with low methanol permeability. However, they suffer from poor mechanical properties and thermal stability.

2.4.6 *Medical Applications*

Hydrogels that are based on PVA are suitable for biomaterial applications (46). Hydrogels have been used for optical lenses (47), as corneal prostheses (48), for catheters and artificial kidneys (49), thin film wound dressings, subcutaneous drug delivery devices, and coatings for catheters. In addition, the use for nucleus replacement (50) and the treatment of acne and pimples (51) has been described.

Physical crosslinks can be formed during freeze-thaw cycles without the need of potentially toxic monomers that are sometimes used as chemically crosslinked gels. PVA hydrogels have been investigated for artificial cartilage applications because they have the ability to mimic human tissue. In particular, PVA is suitable for the fabrication of medical devices for synthetic articular cartilages because of its viscoelastic nature, high water content, and biocompatibility (52). Fibrous composites with PVA as the matrix have been evaluated as potential nondegradable meniscal replacements (46).

The creep resistance of PVA hydrogels can be improved by high temperature annealing. Unfortunately, annealing collapses the pores, thus reducing the water content. As a result, the lubricity of the hydrogel surface is reduced. However, the polymers can be modified with AAm. The incorporation of AAm improves the lubricity of the gels while maintaining a high creep resistance (52).

2.4.6.1 *Drug Release*

The controlled release of drugs from hydrogels sometimes exhibits a rapid release after placement into the release media. Usually, the initial burst ends after a short period of time and the rate of release stabilizes. This effect may be advantageous in some cases, however, in general, such a behavior appears to be harmful, or even dangerous, particularly, when an extreme release rate causes an overdose of the drug.

Surface crosslinking is an effective way to minimize the burst. Moreover, in the course of the crosslinking reaction, certain amounts of the drug are removed from the near surface area (53). The burst and cumulative quantities of proxyphylline released from PVA formulations treated by surface extraction and surface-preferential crosslinking are shown in Table 2.7.

Hydrogel membranes composed from sodium alginate and PVA are suitable for controlled release for the transdermal delivery of an anti-hypertensive drug, namely, prazosin hydrochloride (54). Also, pure PVA hydrogels are effective for the delivery of atenolol, which

Table 2.7 Drug Release by Surface Treatment (53)

Surface Pretreatment	Surface Posttreatment	Total Release[a] /[$g\,g^{-1}$]	Burst release[b] /[$g\,g^{-1}$]
None	None	21.8	3.6
Extraction	1 *min*	21.2	2.4
	5 *min*	19.8	2.0
	10 *min*	17.7	1.9
Crosslinking	3% glutaraldehyde	20.9	1.7
	10% glutaraldehyde	23.1	0.05

[a] After 10 *h*, gram drug per gram polymer
[b] During the first 5 *min*, gram drug per gram polymer

is effective as antagonist for blood hypertension (55).

The preparation of thermosensitive and pH sensitive PVA microspheres suitable for drug delivery has been described (25, 27, 56). PVA microspheres can be obtained by crosslinking with glutaraldehyde. AAm polymers are grafted on the microspheres to impart thermosensitivity. Then, to establish pH sensitive properties, carboxyl groups are grafted using succinic anhydride. The microspheres exhibit a good capacity for drug loading without losing their thermosensitive properties.

The use of a hydrogel for a bioartificial pancreas has been described (57). Macroencapsulation devices were prepared using the freeze-thawing technique. Into these pancreatic islets were incorporated. In vitro and in vivo studies demonstrated the action of these devices. The hyperglycemia of diabetic mice was greatly improved to near normal levels by the transplantation of rat islets into them. A controlled release of drugs can be also achieved by an electrical field (58, 59).

Modern compositions for the oral delivery of a poorly absorbed drug contain compounds with certain functionalities (60):

- An enhancer for increasing the absorption of drugs through the intestinal mucosa
- A promoter, which alone does not increase absorption of the drug through the intestinal mucosa, but which further increases the absorption of the drug in the presence of the enhancer

- A protector for the drug from physical or chemical decomposition or inactivation in the gastrointestinal tract.

Protectors include methyl cellulose, PVA, and poly(*N*-vinyl-2-pyrrolidone) (60).

2.4.6.2 *Contact Lenses*

A highly transparent PVA hydrogel can be obtained by freeze-thawing aqueous PVA solutions repeatedly at a relatively higher temperature. There is no need to use a chemical crosslinking agent. It is believed that these materials could be superior for contact lenses than the conventionally used 2-hydroxyethyl methacrylate (61).

Aqueous solutions of PVA are obtained by dissolving PVA in water at 80 °C. The concentration of the polymers is around 5–20%. The solutions are then stored at room temperature for two weeks. Repeated freeze-thawing cycles are then performed. This procedure consists of 8 *h* freezing at 0°C. The thawing lasts 16 *h* at 37°C. A total of 15 cycles is appropriate (17).

2.4.6.3 *Spinal Prostheses*

Lower back pain is frequently caused by a degenerative disk disease. The anisotropic structure of the intervertebral disk efficiently achieves the appropriate mechanical properties required to cushion complex spinal loads. The inner viscoelastic material, i.e., the nucleus pulposus, occupies 20–40% of the total disk cross sectional area. The nucleus usually contains 70–90% water. Namely, the nucleus is composed of hydrophilic proteoglycans that attract water into the nucleus and thus generate an osmotic swelling pressure of 0.1–0.3 MPa. This pressure supports the compressive load on the spine (18).

The nucleus is constrained laterally by a highly structured outer collagen layer, the annulus fibrosus. The nucleus pulposus is always in compression, whereas the annulus fibrosus is always in tension. Although it comprises only one third of the total area of the disk cross section, the nucleus supports 70% of the total load exerted on the disk. The intervertebral disk becomes less elastic with age, reaching the elasticity of hard rubber in most middle-aged adults as the nucleus looses water content. This water loss will also cause the disk to shrink in size and jeopardizes its properties.

A successfully designed artificial disk can replace a worn out natural disk. Several artificial disk prostheses have been proposed. Many of these prostheses attempt complete replacement of the disk, including the nucleus and the annulus fibrosus.

As an alternative to the complete replacement of intervertebral disks, the nucleus pulposus alone can be replaced, leaving the annulus fibrosus intact. This approach is advantageous if the fibrosis is intact because it is less invasive and the annulus can be restored to its natural fiber length and fiber tension.

It is desirable to use materials that have similar properties as the natural nucleus. Bladders filled with air, saline, or a thixotropic gel have been designed. To prevent a leakage, the membrane material comprising the bladder must be impermeable, which inhibits the natural diffusion of body fluid into the disk cavity, preventing the access to nutrients.

More natural disk replacement materials are based on polymeric hydrogels (18). Hydrogels are good analogs for the nucleus pulposus. They typically possess good viscoelastic properties and offer a good mechanical behavior. In addition, they contain a large amount of free water which permits a prosthesis made from a hydrogel to creep under compression and handle the cyclical loading without loss of elasticity, similar to a natural nucleus pulposus. The water permeability of these materials also allows the diffusion of body fluids and nutrients into the disk space.

PVA, PVP and its copolymers yield prostheses with mechanical properties similar to natural disks. These materials have the additional advantage of having clinical success in other medical devices. As mentioned already, gels formed from PVA are prepared via a freeze-thaw process or via external crosslinking agents. In addition, the gels may contain therapeutic drugs which slowly diffuse after implantation.

The wear properties of thermally cycled, dehydrated PVA have been investigated under a variety of conditions. The wear rate found in unidirectional pin-on-disk against alumina experiments was comparable to that of ultra high molecular weight poly(ethylene). However, in reciprocating tests, the wear rate was up to 18 times larger. To improve the wear properties, PVA of higher molecular weight and additionally crosslinked by γ-radiation with doses of more than 50 kGy was tested.

Such a treatment reduces the wear rate considerably. However, in both radiation and thermally crosslinked PVA, the wear rate does not appear adequate for applications where the opposing surface has high hardness. Additionally, the irradiation would adversely affect bioactive materials loaded into the gel (18).

2.4.6.4 *Thermogelling*

Thermogelling polymers are liquids at room temperature and solids at body temperature. Methods of implanting a hydrogel into a mammal are by injecting the solution as a liquid at a temperature below the body temperature of the mammal into a selected site. Then the liquid undergoes a thermal phase transition to form a solid hydrogel *in situ* in the body as the implant warms to body temperature. These thermal gelling materials in can be used in a wide variety of applications including (62):

- Nucleus pulposus replacement
- Wound care
- Disk replacement
- Cartilage replacement
- Joint replacement
- Surgical barriers
- Gastrointestinal devices
- Cosmetic and reconstructive surgery.

Actually, this technique is a minimally invasive procedure, since in the best case, only a percutaneous injection via a needle is needed. Otherwise, a surgically invasive procedure is required for the insertion or implantation of a hydrogel (62).

Preferred hydrogels include polymer blends or copolymers of poly(N-isopropylacrylamide) (PNIPAAm) and a second polymer, e.g., PVA or PEG. Screening studies were performed to elucidate the regions of polymer concentrations that exhibit a thermogelling behavior. Aqueous PVA solutions of 5, 10, and 15% w/v, and PNIPAAm solutions of 15, 25, 35, and 45% w/v were prepared. The PVA solutions were mixed with the PNIPAAm solutions in volumetric ratios of 1:1, 1:5, and 1:10. An at room temperature miscible solution was formed. The gelation behavior at physiological temperature is shown in Table 2.8.

Table 2.8 Gelation of Blends of PVA and PNIPAAm (62)

PNIPAAM concentration	15%	25%	35%	45%
PVA Concentration 5%	1:1 (SS)	1:1 (S)	1:1 (S)	1:1 (S)
	1:5 (L)	1:5 (L)	1:5 (SS)	1:5 (SS)
	1:10(L)	1:10(L)	1:10(L)	1:10(I)
PVA Concentration 10%	1:1 (L)	1:1 (S)	1:1 (S)	1:1 (S)
	1:5 (L)	1:5 (L)	1:5 (L)	1:5 (SS)
	1:10(L)	1:10(L)	1:10(L)	1:10(L)
PVA Concentration 15%	1:1 (L)	1:1 (S)	1:1 (SS)	1:1 (S)
	1:5 (L)	1:5 (SS)	1:5 (S)	1:5 (S)
	1:10(L)	1:10(L)	1:10(L)	1:10(L)

I: immiscible, S: solid, SS: semi solid, L: liquid

The appearance of the hydrogel was then classified as solid (S), semi-solid (SS), or liquid (L). The solid (S) designation was used for materials that remain solid and did not extrude liquid upon application of pressure from a hand-held laboratory spatula. The semi-solid (SS) designation was used for systems that exhibited two-phase behavior at 37°C with one part solid and one part liquid. The liquid (L) designation was used for samples that remained liquid, i.e., the samples appear as a solution or a slurry (62).

Similar experiments as shown in Table 2.8 were performed with blends from PEG and PNIPAAm, PNIPAAm grafted PEG polymers, PNIPAAm branched PEG polymers, and PEG-PNIPAAm-PEG triblock polymers (62).

2.5 Suppliers and Commercial Grades

Suppliers and commercial grades are shown in Table 2.9.

Table 2.9 Examples for Commercially Available PVA Polymers (63)

Tradename	Producer
Celvol	Celanese Corporation
Denka Poval (Series)	Denka
Kuraray Poval® (Series)	Kuraray America, Inc.
MonoSol® (Series)	MonoSol, LLC
Mowiflex™ (Series)	Kuraray America, Inc.
Mowiol® (Series)	Kuraray America, Inc.
Mowital® (Series)	Kuraray America, Inc.
Vinylon (fiber)	Kuraray Co., Ltd. (36)

2.6 Safety

PVA as such is regraded as a potential non-toxic material. For this reason it is used in cosmetic and medical applications. However, copolymers and graft polymers may exhibit an enhanced toxicity. Health and safety factors are detailed in the literature (64, p. 189).

2.7 Environmental Impact and Recycling

PVA is susceptible to biodegradation in the presence of suitably acclimated microorganisms (35). Some PVA-based blends, composites, and copolymers can be biodegraded. If such materials are dispersed in aqueous systems they can directly interfere with the life cycle of aquatic organisms.

Most of the microorganisms that can degrade PVA are aerobic bacteria belonging to the Pseudomonas genus, the Alcaligenes genus, and the Bacillus genus. In addition, PVA may be degraded by fungi, such as Phanerochaete crysosporium. Case studies are available, on the biodegradation under composting conditions (65), in soil environments (66), and in aqueous environments (67).

Tradenames appearing in the references are shown in Table 2.10.

Table 2.10 Tradenames in References

Tradename Description	Supplier
Airvol® (38,39)	Air Products and Chemicals, Inc.
Amberlyst® 15 Ion exchange resins, heterogeneous catalysts (15,16)	Rohm & Haas
Amres® Polyamide-epichlorohydrin wet strength resin (14,40)	Georgia-Pacific Resins, Inc.
Baytron® P Complex of poly(3,4-ethylenedioxythiophene) and poly(styrenesulfonic acid) (43)	Bayer AG
Blankophor® Optical brightener (14)	Bayer
Celvol® (Series) Poly(vinyl alcohol) (14)	Celanese
CO-BOND® (Series) Modified starches (14,40)	National Starch and Chemical Comp.
Crepeccel® (Series) Creping agents (40)	Calgon Corp.
Ecocite® Poly(vinyl butyral) copolymer (32)	DuPont
Elvaloy® (Series) *n*-Butyl acrylate copolymers (32)	DuPont
Elvanol® (Series) Poly(vinyl alcohol) (28,32)	DuPont
Eudragit® Coating Lacquers for use on medicinal tablets (60)	Evonik Roehm GMBH
Fluorad® (Series) Surfactant (60)	3M Comp.
Gelucire® (Series) Fatty acid esters (60)	Gattefosse S. A.
Gelvatol® Poly(vinyl alcohol) (28)	Monsanto
Hostalux® Optical Brightener (14)	Hoechst
Leucophor® Optical brightener (14)	Clariant

Table 2.10 (cont.)

Tradename Description	Supplier
Lomar® D	Geo Specialty Chemicals, Inc. (Henkel)
Sodium salt of the formaldehyde condensation product of naphthalene sulfonic acid (14)	
Luviskol® VA 73 W	BASF AG
50% Solution of a copolymer of vinylpyrrolidone and vinylacetate (70:30) in water (28)	
Mowiol®	Kuraray Europe GmbH
Poly(vinyl alcohol) (28,43)	
Polyviol®	Wacker
Poly(vinyl alcohol) (28)	
Poval®	Kuraray Co. Ltd.
Cationic poly(vinyl alcohol) (38,39)	
Tinopal®	Ciba-Geigy
Optical brightener (14)	
Tween® (Series)	Uniqema
Ethoxylated fatty acid ester surfactants (60)	
Varisoft® (Series)	Goldschmidt Chemical Corp.
Fatty amide amides (creping agents) (40)	
Veova® (Series)	Resolution Performance Products LLC Corp. (Shell)
Vinyl ester of VERSATIC® acid 9 (15)	
Vinol® 107	Air Products
Hydrolyized poly(vinyl alcohol) (28)	
Zonyl® (Series)	DuPont
Fluorinated nonionic surfactant (60)	

References

1. W. Haehnel and W.O. Herrmann, Verfahren zur Darstellung von polymerem Vinylalkohol, DE Patent 450 286, assigned to Consortium Elektrochem Ind., October 05, 1927.
2. W.O. Herrmann and H. Wolfram, Process for the preparation of polymerized vinyl alcohol and its derivatives, US Patent 1 672 156, assigned to Consortium Elektrochem Ind., June 05, 1928.
3. W.O. Herrmann and W. Haehnel, Über den Poly-vinylalkohol, *Berichte der deutschen chemischen Gesellschaft*, 60(7):1658–1663, July 1927.

4. I. Sakurada, *Polyvinyl alcohol fibers*, Vol. 6 of *International Fiber Science and Technology Series*, M. Dekker, New York, 1985.
5. C.A. Finch, ed., *Polyvinyl alcohol: Developments*, Wiley, Chichester, 1992.
6. Y. Anufriyeva, R. Gromova, M. Krakovyak, V. Kuznetsova, V. Lushchik, T. Nekrasova, A. Sorokin, and T. Sheveleva, Chemical and intramolecular structure of water-soluble copolymers of vinyl alcohol and vinyl acetate, *Polymer Sci. (USSR)*, 26(6):1427–1434, 1984.
7. R.M. Nowak, J.T.K. Woo, and D.H. Heinert, Copolymers of vinyl amine and vinyl alcohol as flocculants, US Patent 3 715 336, assigned to The Dow Chemical Company (Midland, MI), February 6, 1973.
8. R. Vicari, Production of vinyl alcohol copolymers, US Patent 6 818 709, assigned to Celanese International Corporation (Dallas, TX), November 16, 2004.
9. L. Trukhmanova, S. Ushakov, and T. Markelova, Synthesis of water-soluble copolymers of vinyl alcohol and its diacetyl-*p*-aminosalicylic ester, *Polymer Sci. (USSR)*, 6(7):1488–1492, 1964.
10. F. Unger, M. Wittmar, and T. Kissel, Branched polyesters based on poly-[vinyl-3-(dialkylamino)alkylcarbamate-*co*-vinyl acetate-*co*-vinyl alcohol]-graft-poly(d,l-lactide-*co*-glycolide): Effects of polymer structure on cytotoxicity, *Biomaterials*, 28(9):1610–1619, March 2007.
11. C.-H. Huang, H.-M. Wu, C.-C. Chen, C.-W. Wang, and P.-L. Kuo, Preparation, characterization and methanol permeability of proton conducting membranes based on sulfonated ethylene-vinyl alcohol copolymer, *J. Membr. Sci.*, 353(1-2):1–9, May 2010.
12. R.A. Bader, Synthesis and viscoelastic characterization of novel hydrogels generated via photopolymerization of 1,2-epoxy-5-hexene modified poly(vinyl alcohol) for use in tissue replacement, *Acta Biomaterialia*, 4(4):967–975, July 2008.
13. L.M. Robeson and T.L. Pickering, Amine functional poly(vinyl alcohol) for improving properties of recycled paper, US Patent 5 380 403, assigned to Air Products and Chemicals, Inc. (Allentown, PA), January 10, 1995.
14. R. Vicari, Vinyl alcohol copolymers for use in aqueous dispersions and melt extruded articles, US Patent 7 790 815, assigned to Serisui Specialty Chemicals America, LLC (Dallas, TX), September 7, 2010.
15. R. Vicari, F. Barsan, and B.F. Hann, Method to purify poly(vinyl alcohol), US Patent 7 388 069, assigned to Celanese International Corporation (Dallas, TX), June 17, 2008.
16. B.F. Hann, Method to purify poly(vinyl alcohol), US Patent 7 524 924, assigned to Celanese International Corporation (Dallas, TX), April 28, 2009.

17. S. Gupta, S. Sinha, and A. Sinha, Composition dependent mechanical response of transparent poly(vinyl alcohol) hydrogels, *Colloids Surf., B*, 78(1):115–119, June 2010.
18. J.W. Ruberti and G.J.C. Braithwaite, Systems and methods for controlling and forming polymer gels, US Patent 7 745 532, assigned to Cambridge Polymer Group, Inc. (Boston, MA), June 29, 2010.
19. J. Brandrup and E.H. Immergut, eds., *Polymer Handbook*, John Wiley & Sons, New York, 3rd edition, 1989.
20. S. Hua, H. Ma, X. Li, H. Yang, and A. Wang, pH-sensitive sodium alginate/poly(vinyl alcohol) hydrogel beads prepared by combined ca2+ crosslinking and freeze-thawing cycles for controlled release of diclofenac sodium, *Int. J. Biol. Macromol.*, 46(5):517–523, June 2010.
21. B.D. Ratner, P.D. Nair, M.S. Boeckl, and E.R. Leber, Hydrogels formed by non-covalent linkages, US Patent 7 300 962, assigned to University of Washington (Seattle, WA), November 27, 2007.
22. S. Quintero, R. Ponce F, M. Cremona, A. Triques, A. d'Almeida, and A. Braga, Swelling and morphological properties of poly(vinyl alcohol) (PVA) and poly(acrylic acid) (paa) hydrogels in solution with high salt concentration, *Polymer*, 51(4):953–958, February 2010.
23. B.D. Ratner, P.D. Nair, M.S. Boeckl, and E.R. Leber, Hydrogels formed by non-covalent linkages, US Patent 6 949 590, assigned to University of Washington (Seattle, WA), September 27, 2005.
24. M.J. Nugent and C.L. Higginbotham, Preparation of a novel freeze thawed poly(vinyl alcohol) composite hydrogel for drug delivery applications, *Eur. J. Pharm. Biopharm.*, 67(2):377–386, September 2007.
25. M.J. Mc Gann, C.L. Higginbotham, L.M. Geever, and M.J. Nugent, The synthesis of novel pH-sensitive poly(vinyl alcohol) composite hydrogels using a freeze/thaw process for biomedical applications, *Int. J. Pharm.*, 372(1-2):154–161, May 2009.
26. F.A. Sheikh, N.A. Barakat, B.-S. Kim, S. Aryal, M.-S. Khil, and H.-Y. Kim, Self-assembled amphiphilic polyhedral oligosilsesquioxane (POSS) grafted poly(vinyl alcohol) (PVA) nanoparticles, *Mater. Sci. Eng., C*, 29(3):869–876, April 2009.
27. R.F. Donnelly, C.M. Cassidy, R.G. Loughlin, A. Brown, M.M. Tunney, M.G. Jenkins, and P.A. McCarron, Delivery of methylene blue and meso-tetra (n-methyl-4-pyridyl) porphine tetra tosylate from cross-linked poly(vinyl alcohol) hydrogels: A potential means of photodynamic therapy of infected wounds, *J. Photochem. Photobiol. B Biol.*, 96 (3):223–231, September 2009.
28. T. Hirt, T. Holland, V. Francis, and H. Chaouk, Degradable poly(vinyl alcohol) hydrogels, US Patent 6 710 126, assigned to Bio Cure, Inc. (Norcross, GA), March 23, 2004.

29. D.N. Ku, Poly(vinyl alcohol) hydrogel, US Patent 6 231 605, assigned to Restore Therapeutics (Atlanta, GA), May 15, 2001.
30. J.-C. Park, T. Ito, K.-O. Kim, K.-W. Kim, B.-S. Kim, M.-S. Khil, H.-Y. Kim, and I.-S. Kim, Electrospun poly(vinyl alcohol) nanofibers: effects of degree of hydrolysis and enhanced water stability, *Polym. J.*, 42(3): 273–276, March 2010.
31. M.J. Kim, J. Lee, D. Jung, and S.E. Shim, Electrospun poly(vinyl alcohol) nanofibers incorporating pegylated multi-wall carbon nanotube, *Synth. Met.*, 160(13-14):1410–1414, July 2010.
32. D.C. Urian, P.A. Morken, and D.L. Visioli, Poly(vinyl alcohol) composition comprising a polyol, US Patent 7 781 506, assigned to E.I. du Pont de Nemours and Company (Wilmington, DE), August 24, 2010.
33. P. Fernandez-Saiz, M. Ocio, and J. Lagaron, Antibacterial chitosan-based blends with ethylene-vinyl alcohol copolymer, *Carbohydr. Polym.*, 80(3):874–884, May 2010.
34. C. Xiao and N. Geng, Tailored preparation of dual phase concomitant methylcellulose/poly(vinyl alcohol) physical hydrogel with tunable thermosensivity, *Eur. Polym. J.*, 45(4):1086–1091, April 2009.
35. E. Chiellini, A. Corti, S. D'Antone, and R. Solaro, Biodegradation of poly(vinyl alcohol) based materials, *Progress in Polymer Science*, 28(6): 963–1014, June 2003.
36. A. Jikihara and N. Fujiwara, Vinyl alcohol polymer, US Patent 7 141 638, assigned to Kuraray Co., Ltd. (Kurashiki, JP), November 28, 2006.
37. J.R. Boylan, Multifunctional poly(vinyl alcohol) binder for fine particle size calcium carbonate pigment, US Patent 6 441 076, assigned to Celanese International Corporation (Dallas, TX), August 27, 2002.
38. J.J. Rabasco, E.H. Klingenberg, and J.R. Boylan, Ink jet paper coatings containing amine functional monomer grafted poly(vinyl alcohol), US Patent 6 348 256, assigned to Celanese International Corporation (Dallas, TX), February 19, 2002.
39. J.J. Rabasco, Ink jet media comprising a coating containing amine functional emulsion polymers, US Patent 6 455 134, assigned to Air Products Polymers, L.P. (Allentown, PA), September 24, 2002.
40. S.L. Edwards, G.H. Super, S.J. McCullough, D.J. Baumgartner, R.W. Eggen, D.P. Duggan, J.E. Krueger, D.W. Lomax, and C.A. Jones, Fabric crepe process for making absorbent sheet, US Patent 7 704 349, assigned to Georgia-Pacific Consumer Products LP (Atlanta, GA), April 27, 2010.
41. R.A. Hayes, Poly(vinyl alcohol) copolymer ionomers, their preparation and use in textile sizes, US Patent 6 387 991, assigned to E. I. du Pont de Nemours & Company (Wilmington, DE), May 14, 2002.

42. Y. Tosaki, H. Nagatsu, S. Kouno, and T. Yatagai, Aqueous dispersion type pressure-sensitive adhesive composition and pressure-sensitive adhesive product, US Patent 7 396 868, assigned to Nitto Denko Corporation (Osaka, JP), July 8, 2008.
43. M. Biler, Solid electrolytic capacitor containing a protective adhesive layer, US Patent 7 460 358, assigned to AVX Corporation (Myrtle Beach, SC), December 2, 2008.
44. R. Karthikaiselvi, S. Subhashini, and R. Rajalakshmi, Poly (vinyl alcohol - aniline) water soluble composite as corrosion inhibitor for mild steel in 1 m hcl, *Arabian Journal of Chemistry*, In Press, 2010.
45. P. Kanakasabai, P. Vijay, A.P. Deshpande, and S. Varughese, Cross-linked poly(vinyl alcohol)/sulfonated poly(ether ether ketone) blend membranes for fuel cell applications–surface energy characteristics and proton conductivity, *J. Power Sources*, 196(3), February 2010.
46. J.L. Holloway, A.M. Lowman, and G.R. Palmese, Mechanical evaluation of poly(vinyl alcohol)-based fibrous composites as biomaterials for meniscal tissue replacement, *Acta Biomaterialia*, 6(12):4716–4724, December 2010.
47. K.F. Mueller, Dimethylacrylamide-copolymer hydrogels with high oxygen permeability, US Patent 4 954 587, assigned to Ciba-Geigy Corporation (Ardsley, NY), September 4, 1990.
48. S. Mori, E. Tabei, and H. Umehara, Chloro-terminated polysilane and process for making, US Patent 5 292 415, assigned to Shin-Etsu Chemical Company, Limited (Tokyo, JP), March 8, 1994.
49. N. Shimoyama, M. Yokota, and T. Uemura, Surface-treated plastic article and method of surface treatment, US Patent Application 20 020 006 521, January 17, 2002.
50. Q.-B. Bao and P.A. Higham, Hydrogel intervertebral disc nucleus, US Patent 5 047 055, assigned to Pfizer Hospital Products Group, Inc. (New York, NY), September 10, 1991.
51. A.C. Hymes, Therapeutic method for treating acne or isolated pimples and adhesive patch therefor, US Patent 6 455 065, assigned to LecTec Corporation (Minnetonka, MN), September 24, 2002.
52. H. Bodugoz-Senturk, C.E. Macias, J.H. Kung, and O.K. Muratoglu, Poly(vinyl alcohol)-acrylamide hydrogels as load-bearing cartilage substitute, *Biomaterials*, 30(4):589–596, February 2009.
53. X. Huang, B.L. Chestang, and C.S. Brazel, Minimization of initial burst in poly(vinyl alcohol) hydrogels by surface extraction and surface-preferential crosslinking, *Int. J. Pharm.*, 248(1-2):183–192, November 2002.
54. R.V. Kulkarni, V. Sreedhar, S. Mutalik, C.M. Setty, and B. Sa, Interpenetrating network hydrogel membranes of sodium alginate and

poly(vinyl alcohol) for controlled release of prazosin hydrochloride through skin, *Int. J. Biol. Macromol.*, 47(4):520–527, November 2010.

55. E.-R. Kenawy, M.H. El-Newehy, and S.S. Al-Deyab, Controlled release of atenolol from freeze/thawed poly(vinyl alcohol) hydrogel, *J. Saudi Chem. Soc.*, 14(2):237–240, April 2010.
56. G. Fundueanu, M. Constantin, and P. Ascenzi, Poly(vinyl alcohol) microspheres with pH- and thermosensitive properties as temperature-controlled drug delivery, *Acta Biomaterialia*, 6(10):3899–3907, October 2010.
57. M. Qi, Y. Gu, N. Sakata, D. Kim, Y. Shirouzu, C. Yamamoto, A. Hiura, S. Sumi, and K. Inoue, PVA hydrogel sheet macroencapsulation for the bioartificial pancreas, *Biomaterials*, 25(27):5885–5892, December 2004.
58. S. Murdan, Electro-responsive drug delivery from hydrogels, *J. Controlled Release*, 92(1-2):1–17, September 2003.
59. K. Juntanon, S. Niamlang, R. Rujiravanit, and A. Sirivat, Electrically controlled release of sulfosalicylic acid from crosslinked poly(vinyl alcohol) hydrogel, *Int. J. Pharm.*, 356(1-2):1–11, May 2008.
60. S.-H. Choi and S.-W. Cho, Oral formulation for delivery of poorly absorbed drugs, US Patent 7 666 446, assigned to ProCarrier, Inc. (Park City, UT), February 23, 2010.
61. R.-Y. Ma and D.-S. Xiong, Synthesis and properties of physically cross-linked poly(vinyl alcohol) hydrogels, *J. China Univ. Min. Technol.*, 18 (2):271–274, June 2008.
62. A.M. Lowman, M.S. Marcolongo, and A.J.T. Clemow, Thermogelling polymer blends for biomaterial applications, US Patent 7 708 979, assigned to Synthes USA, LLC (West Chester, PA) Drexel University (Philadelphia, PA), May 4, 2010.
63. IDES Integrated Design Engineering Systems, The Plastics Web®, IDES Inc. 209 Grand Avenue Laramie, WY 82070 USA [electronic:] http://www.ides.com/prospector/, 2006.
64. F.L. Marten, "Vinyl alcohol polymers," in H.F. Mark, N. Bikales, C.G. Overberger, and G. Menges, eds., *Encyclopedia of Polymer Science and Engineering*, Vol. 17, pp. 167–198. Wiley Interscience, New York, 2nd edition, 1988.
65. C. David, C.D. Kesel, F. Lefebvre, and M. Weiland, The biodegradation of polymers: Recent results, *Angew. Makromol. Chem.*, 216(1):21–35, March 1994.
66. R. Solaro, A. Corti, and E. Chiellini, A new respirometric test simulating soil burial conditions for the evaluation of polymer biodegradation, *J. Environ. Polym. Degrad.*, 6(4):203–208, October 1998.
67. J.J. Porter and E.H. Snider, Long term biodegradability of textile chemicals, *J. Water Pollut. Control Fed.*, 48(9):2198–2210, September 1976.

3

Polysaccharides

This chapter summarizes the industrial applications of water-soluble polysaccharides. This chapter is arranged in a slightly different way than the other chapters since the polysaccharides are the basic materials as such. Actually the preparation does not start from the monomers and there are no polymerization reactions to be discussed, but rather a modification mostly of the side chains. Because of the heterogeneity of the material, each polysaccharide is dealt with completely in a separate section.

3.1 Polymers

Polysaccharide types are summarized in Table 3.1.

Table 3.1 Polysaccharide Types

Type	References
Starch	
Chitosan	
Guar gum	(1)
Hydroxyethyl guar	
Hydroxypropyl guar	
Hydroxybutyl guar	(2)
Hydroxyethyl cellulose	(1)
Carboxymethyl cellulose	(1)
Carboxymethyl hydroxyethyl cellulose	(1)
Xanthan gum	(1)

3.2 Starch

Starch is mostly important as a food as it is found in corn, wheat, maize, rice, potatoes, and other plants. Starch is a polysaccharide from glucose and consists of linear amylose and branched amylopectin. The latter is not soluble in water. Chemically modified starch is used as a food additive, but some types of modified starches are also used for applications besides of food. Industrial applications apart from food include:

- Papermaking
- Adhesives for wallpaper and bookbinding
- Laundry and textile applications
- Biodegradable polymers
- Oil field applications.

True solutions of starch in water are difficult to prepare using conventional cooking techniques and require the application of specialized techniques, such as autoclaving at elevated temperatures and steam pressures (3). Steam jet cooking is another technique for preparing starch solutions, which is simpler and more economical than autoclaving, and is suitable for continuous processing.

Because of these processing advantages, jet cooking has been used to prepare starch solutions for commercial applications. The method of steam jet cooking involves pumping a water slurry of starch through an orifice located in a heating chamber, i.e., the hydroheater, where the starch slurry contacts a jet of steam with high temperature, and pressure.

There are two basic steam jet cooker designs that are commercially used. The first of these designs is referred to as thermal jet cooking. Here, the amount of steam is carefully controlled to achieve complete steam condensation during the cooking process. This means that little or no excess steam passes through the cooker.

The second of these designs is referred to as excess steam jet cooking. Here, the steam which enters the hydroheater exceeds the amount required to achieve the required cooking temperature and pressure, thus allowing considerable amounts of excess steam to pass through the cooker along with the cooked starch solution.

The intense turbulence caused by the passage of this excess steam through the hydroheater promotes mechanical shearing and degradation of starch molecules, especially those having the highest molecular weight, and it also produces starch solutions with reduced viscosity. The high degree of turbulence and mechanical shear of the excess steam jet cooking process also can convert a water immiscible phase into an aqueous dispersion of micrometer-sized droplets.

An inherent property of starch pastes and solutions is their tendency to form gels on cooling, and this property is commonly referred to as retrogradation (3). Retrogradation is caused by aggregation of starch molecules through hydrogen bonding and crystallization.

The tendency of starch solutions to retrograde and form gels increases with the amylose content of the starch because amylose is a straight chain polymer with little or no branching. Although retrogradation has also been observed in amylopectin solutions, retrogradation is much slower with amylopectin, and is generally observed only after solutions have been allowed to stand for prolonged periods of time.

3.2.1 *Modified Starch Types*

3.2.1.1 *Starch based Polymers*

Biodegradable polymers, the structure uses and methods of processing have been extensively reviewed (4). Polymers based on starch cannot be thermally processed without a plasticizer or a gelatinization agent. This results from the fact that its decomposition temperature is lower than its melting temperature before gelatinization. For this reason, plasticizers and additives have been developed which gelatinize starch in the course of thermal processing.

Starch can be modified with cationic polymers. The modified starches can be used as a dry strength agent in paper industry (5). For example, *N*-vinyl formamide (NVF) can be hydrolyzed to a degree of 95%, which the amount vinylamine units present in the polymer. The hydrolysis of NVF is performed in aqueous sodium hydroxide solution. For the modification, oxidized maize starch has been used.

Oxidized starch bears carboxyl groups as anionic groups. Oxidizing agents include ammonium persulfate, hydrogen peroxide, sodium hypochlorite, ozone, or *tert*-butyl hydroperoxide (6).

A process has been described where an aqueous slurry of oxidized starch is digested together with a cationic polymer in a continuous cooker (7).

3.2.1.2 *Starch Esters*

Starch half esters not being crosslinked have been reported to be biodegradable detergent builders (8). A liquid fat or oil can be gelled by adding 1% or more of a starch ester of a fatty acid (9). It is advantageous to use rather oligomeric dextrin esters. Such esters base are excellent to give thixotropic properties to such compositions.

3.2.1.3 *Crosslinked Starch*

Crosslinked cationic and anionic starches are suitable for heavy metal control (10). In the same way, crosslinked amphoteric starch is active (11). Cationic starch maleates with quaternary ammonium moieties are accessible via the reaction of starch and epichlorohydrin followed by the reaction with 2,3-epoxypropyltrimethylammonium chloride and finally by esterification with maleic anhydride. The first two steps of the reaction are shown in Figure 3.1.

3.2.1.4 *Carboxymethyl starch*

Carboxymethyl starch can be crosslinked with glycol dichlorides (12). 1,5-Dichloro-3-oxapentane, i.e., diglycol dichloride is prepared from diethylene glycol by the reaction with thionyl chloride in benzene and pyridine solution (12). Longer chain oligoether glycol compounds with end chain chloride moieties can be prepared in a similar manner (12).

3.2.2 *Uses of Starch Compositions*

3.2.2.1 *Textile Sizes*

Unmodified starches can be used as textile sizes since they are inexpensive, but they often flake off the yarn when used as sizes. In

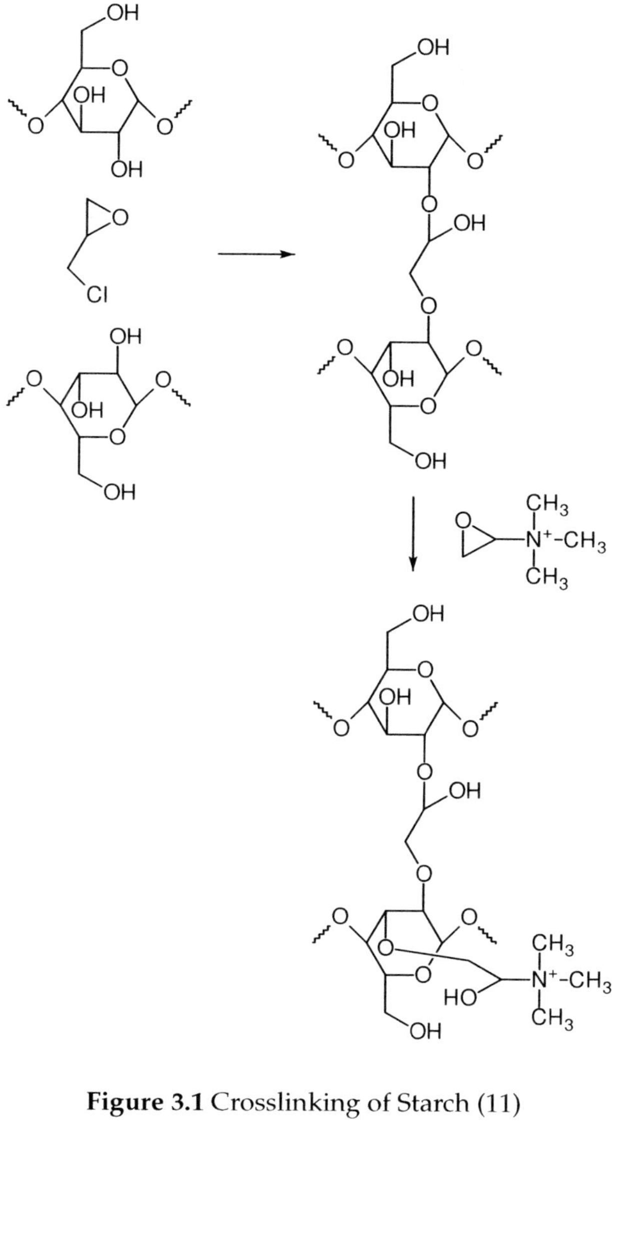

Figure 3.1 Crosslinking of Starch (11)

addition, they do not give stable solutions, and often desizing requires use of enzymes (13). Many modified starches are known which are improvements in various ways over simple starches but may be considerably more expensive.

Blending of readily desizable polymers such as poly(vinyl alcohol) (PVA) polymers is advantageous (14). Examples of modified starches include α-starch, fractionated amylose, moist heat treated starch, enzymatically modified starches acid treated starch, dialdehyde starch, etherified or esterified starches.

3.2.2.2 Oil Field Applications

Lubricating additives for drilling fluids can be prepared from poly-(butene) as the lubricant by steam jet cooking of a mixture of starch, water, and the lubricant. Steam jet cooking aids to uniformly suspend the lubricant in droplets in the aqueous starch matrix with sizes in the μm range. These types of additives avoid the need for toxic emulsifiers, surfactants, or short chain hydrocarbon solvents for dispersing the lubricants. They impart lubricity to drilling muds and inhibit the fluid loss in geological formations by enhancing the filtration control properties of the mud (3).

Selectively crosslinked starches are useful as fluid loss control additives in subterranean treatment fluids (15). The crosslinking agent is selected from epichlorohydrin, phosphorus oxychloride, adipic-acetic anhydride, or sodium trimetaphosphate. Blends of crosslinked starches may be used. For example, a blend of epichlorohydrin crosslinked starch and phosphorus oxychloride crosslinked starch may be used.

Modified starch is used in combination with ceramic particulate bridging agents to provide fluid loss control (16). Generally, these starches may be a crosslinked ether derivative of a partially depolymerized starch or a partially depolymerized crosslinked ether derivative of a starch. The composition deposits filter cakes that can readily be removed without the use of strong acids or other hazardous chemicals that may create problems on the well site, e.g., corrosion of the equipment (16).

The common technique is to use aqueous acids to break a filter cake. The acid can degrade the starch, which acts as a bonding material for the bridging particles (17). In particular, an acid foam is effective to remove the filter cake containing calcium carbonate particulates and starch.

Dispersants are often used in subterranean well cement compositions to facilitate mixing the cement composition (18). Such dispersants are extensively used, *inter alia*, to reduce the apparent viscosities of the cement compositions in which they are utilized to allow the cement composition to be pumped with less friction pressure and less horsepower.

In addition, the lower viscosity often allows the cement composition to be pumped in turbulent flow. Turbulent flow characteristics are desirable, for instance, when pumping cement compositions into subterranean wells to more efficiently remove drilling fluid from surfaces in the well bore as the drilling fluid is displaced by the cement composition being pumped.

The inclusion of dispersants in cement compositions is also desirable in that the presence of the dispersants may facilitate the mixing of the cement compositions and reduce the water required. This may be desirable because cement compositions having reduced water content are often characterized by improved compressive strength development.

A low molecular weight starch with anionic groups can be used as a cement dispersant for cementing in subterranean formations (18). This dispersant is biodegradable. The starch in the dispersant is oxidized.

When the starch contains aldehyde groups it can be further reacted with a sulfite salt to provide a sulfite adduct of an oxidized starch. Sulfite adducts of oxidized starches are especially suitable. Another method of forming a sulfite adduct of an oxidized starch includes reacting an acetone formaldehyde condensate with starch under alkaline conditions, followed by addition of a sulfite salt (19). Examples of cement dispersants are shown in Table 3.2.

3.2.2.3 Water Treatment

Water treating agents may be prepared by the grafting poly(acrylamide) (PAAm) on to starch. Due to the of hydroxyl and amide groups and the branched structure, acrylamide (AAm) grafted

Table 3.2 Examples of Cement Dispersants (18)

Type →	%	% Bisulfite	% Sulfite	T/[°C]	t /[h]
Dialdehyde	1.7	1.9	–	56	16
Dialdehyde	3.8	–	5.2	63	8.5
Dialdehyde	5.3	–	3.6	72	5
Type →	%	% Hypochlorite	% Sulfite	T/[°C]	t /[h]
Amylose	8.2	0.8	4.7	53	3
Amylose	8.1	0.8	4.6	53	24
Type →	%	% Periodate	% Sulfite	T/[°C]	t /[h]
Amylose	7.3	0.8	0.5	72	24
Amylose	7.3	0.8	0.5	72	24

starch has a better flocculating effect than PAAm under certain conditions (20) .

A PAAm with single cations in its structure is limited in its application as a flocculating agent. Sometimes it must be used in combination with an anionic type polymeric flocculant, thereby increasing troubles in operation. Owing to viscoelasticity, at increased concentrations, the mechanical load of stirring increases rapidly, which results in that the reaction is hard to control, and the product is also difficult to be diluted in use.

3.3 Chitosan

Chitosan (CS) is produced by partial or complete deacetylation of chitin. Chitin is a naturally occurring polysaccharide, which is the second most abundant natural product on earth preceded only by cellulose (21). Structurally, chitin is a polysaccharide consisting of 2-acetamido-2-deoxy-β-*D*-glucopyranose units, some of which are deacetylated. Chitin is not a single polymer type with a fixed stoichiometry, but a class of polymers of *N*-acetylglucosamine with different crystal structures and degrees of deacetylation and with a fairly large variability from species to species.

If in the molecule are increasingly deacetylated moieties, the molecule is addressed as CS. Typically, CS has a degree of deacetylation that is between 50% and 100%. The degree of deacetylation in the commercially available CS is usually in the range of 70–78%. CS is prepared by the deacylation of chitin, as shown in Figure 3.2.

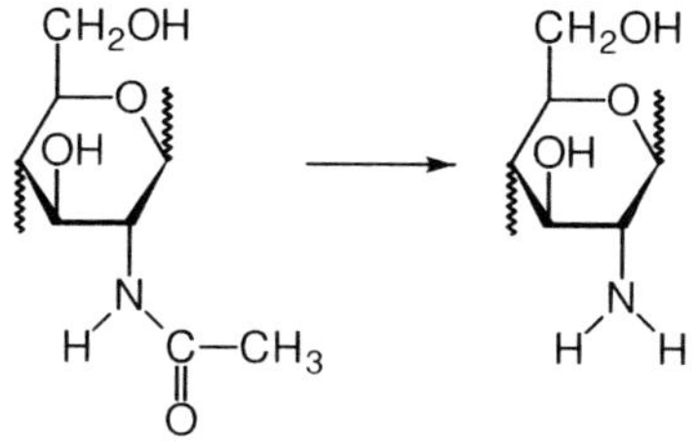

Figure 3.2 Deacylation of Chitin

Commonly, the deacetylation of chitin is achieved by the treatment with concentrated sodium hydroxide (22,23). The large number of free amine groups causes CS to be a weak base. However, because CS is a polysaccharide containing many primary amine groups, it forms water-soluble salts with many organic and inorganic acids. For example, CS is somewhat more soluble in dilute aqueous acids, usually carboxylic acids, as the chitosonium salt.

Nevertheless, the solubility of CS in acidified water, for example in acetic or hydrochloric acid, is still only in the range of 1–2% . If the pH of the solution is increased above 6, the polymer precipitates, thus limiting its solubility. The viscosity of the aqueous CS solution depends on the molecular weight of the polymer (21).

The solubility of CS can be increased by partial oxidation. To oxidize a CS based polymer, oxidizers can be used, such as sodium hypochlorite, sodium periodate, hydrogen peroxide, or peracetic acid. The selection of the oxidizer and the concentration of oxidizer should be sufficient to oxidize or degrade the CS based polymer to a desired solubility. The increased solubility of the oxidized CS may be explained by the degradation into shorter chain segments and the introduction of carboxyl groups.

In addition, water-soluble CS can be prepared by hydrolysis with hydrochloric acid (24). This method, however, requires excessive amounts of HCl and an overly long period of time to hydrolyze the CS (24). In the enzymatic method, water-soluble CS is prepared by enzymatic treatment. First, the CS is dissolved in a poor acid solution. Afterwards the hydrolysis takes place by an enzymatic treatment of about for 24 *h*. Finally the product is freeze dried (24). CS has a wide variety of applications. These are summarized in Table 3.3.

Table 3.3 Fields of Application of Chitosan (25)

Specific Application
Food and dietary products
Fining agent to clarify beverages
Antibacterial and deodorant textile fibers
Cosmetic products
Hair shampoos
Skin creams and lotions
Agricultural biodegradable films
Water treatment and filtration processes
Wet strength additive for paper

3.3.1 *Nanoparticles*

Crosslinked core and core-shell nanoparticle polymers can be prepared from CS. These products may be used as detergents and as additives for pharmaceutical compositions including drug delivery (26, 27).

The crosslinking is initiated by the reaction with acids, such as tartaric acid, citric acid, or salicylic acid in the presence of a water-soluble carbodiimide, such as 1-[3-(dimethylamino)propyl]-3-ethylcarbodiimide hydrochloride.

It is believed that the amine groups (not the nitrogen in the amide groups) in the CS react with the carboxylic groups to form an amide linkage so that the poly acids form an intramolecular bridge. The carbodiimide reacts with the water eliminated during formation of amide linkage between CS and the carboxylic acids. Due to this intramolecular bridging the initially coiled CS structure is transformed into a globular spherical nanoparticle (27).

This method allows the formation of polycations, polyanions, and polyampholyte nanoparticles. According to the same principle, hyaluronan nanoparticles can be prepared (28, 29). Hyaluronic acid is a linear polysaccharide consisting of alternating units of α-1,4-*D*-glucuronic acid and α-1,3-*N*-acetyl-*D*-glucosamine, c.f. Figure 3.3.

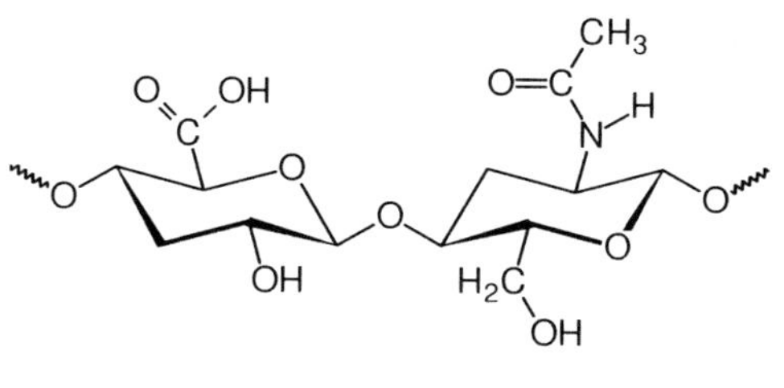

Figure 3.3 Hyaluronic acid

3.3.2 *Deodorizing Preparations*

The application of nanoscale CS leads to a long lasting deodorizing effect (30). The absorption of CS derivatives by the Stratum corneum of the skin can be increased. In addition, the production of the preparations and their compatibility with anionic surfactants are considerably facilitated by the use of the nanoscale chitosans.

The formulations have a pleasant feeling on the skin and show high stability. The moisturizing effect of nanoscale CS counteracts a possible drying out of the skin, particularly in the case of alcohol-containing aerosol formulations.

3.3.3 *Contact Lens Solutions*

CS is a non-toxic biopolymer with a weak antimicrobial activity. However, the use of CS to preserve pharmaceutical compositions has been hampered by its insolubility at pH above 6 (31).

The water solubility at near neutral pH can be improved by derivatization with hydrophilic functional groups, such as carboxymethyl or glycol substituents, or by selective *N*-acetylation.

The acetylation can be done in aqueous acidic solution by the reaction with an acetylating agent in the presence of a phase transfer reagent. Phase transfer reagents include quaternary ammonium salts, quaternary phosphonium salts, crown ethers, and pyridinium salts. Several procedures of the preparation have been described in detail (31).

CS derivatives with borate or phosphate buffers have higher antimicrobial activity in comparison to citrate, tris buffers, and in water. Surfactant additives may help in the cleaning of the lenses. Poly(ethylene oxide) (PEO) polymers are suitable surfactants.

A series of compositions based on glycol CS have been prepared and tested with respect to their antimicrobial activity, biocompatibility, an other required properties (31).

3.3.4 *Intranasal Protein Drug Delivery*

The possibility of the synthesis of well-defined, highly purified peptides and proteins on a large scale has revolutionized many areas of medicine. The nasal route has been successfully used for the administration of a number of peptide drugs. However, the bioavailability of peptides from these formulations is usually low.

Commonly, higher molecular proteins must be administered by injection because they are inadequately absorbed by the body when administered by other routes. It has been found that the intranasal administration of proteins having a molecular weight of 10 *k Dalton* or greater can be achieved using a powder formulation comprising the protein and a CS derivative. The effective absorption of the protein can be achieved using such a formulation (32).

Also vaccine compositions including CS have been investigated for intranasal administration (33). It has been found that CS enhances the immune response of antigens and thus provides an adjuvant effect.

3.4 Carboxymethyl cellulose

Cellulose and starch are shown in Figure 3.4. These molecules differ just in the stereochemistry of the ether linkage that forms the polymer. Examples of cellulose ethers are summarized in Table 3.4.

Cellulose ethers find widespread applications in (34):

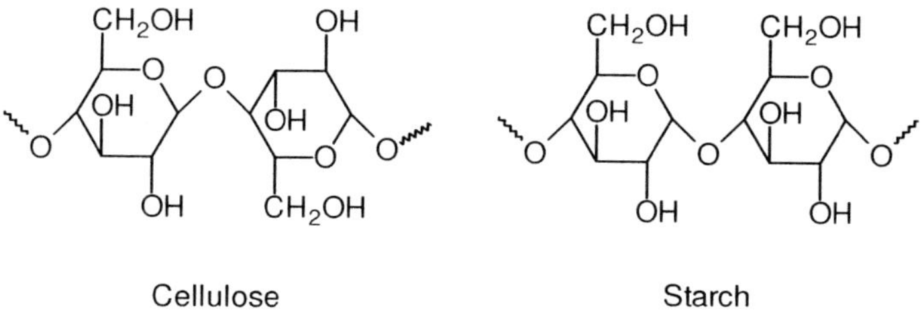

Figure 3.4 Cellulose and Starch

Table 3.4 Cellulose Ethers (34)

Ether type	Acronym
Carboxymethyl cellulose	CMC
Hydroxyethyl cellulose	HEC
Hydroxypropyl cellulose	HPC
Methyl hydroxyethyl cellulose	MHEC
Ethyl hydroxyethyl cellulose	EHEC
Methyl cellulose	MC
Methyl hydroxypropyl cellulose	MHPC

- Adhesives
- Emulsion stabilizers
- Film forming agents
- Food additives
- Lacquers and paints
- Paper coating compositions
- Petroleum production fluids
- Plastic sheets in packaging
- Printing pastes
- Protective colloids
- Suspension stabilizing agents
- Textile finishing compositions
- Thermoplastic materials
- Thickening agents.

Cellulose ethers are typically produced by alkaline treatment of cellulose, e.g., with sodium hydroxide, to form an alkali cellulose and subsequent etherification (34). For example, hydroxyethyl cellulose is used in well cement compositions (35).

Mercerization is an alkaline treatment of cellulose (36). The fibers become that somewhat lustrous. In general the reaction is an important initial step in the production of cellulose derivatives as the cellulose is activated by the mercerization reaction.

During this process the cellulose converts from the cellulose I, the native form, to cellulose II form. I and II designate the different crystal form evidenced by X-ray spectroscopy. The latter form is thermodynamically more favorable. The mercerized cellulose pulp may be converted into cellulose ether by converting the mercerized cellulose pulp into a cellulose floc.

The final step is the etherification. Alkali cellulose is prepared as intermediate. Eventually, to get carboxymethyl cellulose (CMC), the alkali cellulose is treated with sodium chloroacetate (34).

The solution viscosity of CMC produced from mercerized and recovered cellulose pulp is significantly greater than that produced from non-mercerized cellulose pulp (34).

An efficient quality and process control through the manufacturing of CMC is desirable. In particular, the molecular weight should be identical with those of the material pulp. The molecular weight is generally determined by gel permeation chromatography. Special solvents have been developed, containing complex forming agents, such as cadmium ethylenediamine or iron tartrate (37).

CMC can be further modified by the esterification of the free pendant hydroxy group using a specific anhydride or a combination of anhydrides. Examples have been given with propionic anhydride and also with a mixture of acetic anhydride and butyric anhydride (38).

These polymers are easily formulated into either lacquer or enamel type coatings where they are used as rheology modifiers and binder components providing improved aluminum flake orientation and improved hardness. They can be applied to a substrate in the form of an organic solvent solution, an amine neutralized waterborne dispersion, a fully neutralized aqueous and organic colloidal dispersion, or as a dispersion in aqueous ammonia.

They can provide a water clear, high gloss, protective coating for a variety of substrates, especially for metal and wood (38). In analogy to CMC, methyl cellulose can be prepared using methylchloride instead of sodium chloroacetate (34).

3.4.1 *Thickeners*

CMC is used as thickener in various formulations for a wide range of applications. Special derivates are used for this purpose.

3.4.1.1 *Carboxymethyl cellulose esters*

Carboxymethyl cellulose esters, i.e., CMC propionate, CMC acetate propionate, CMC butyrate, and CMC acetate butyrate, are useful as rheology modifiers (38). At low concentrations of less than 5%, amine neutralized waterborne dispersions have shown exponential viscosity changes with changes of as little as 0.5% concentration of the C_2–C_4 esters. This rapid viscosity build is especially useful in the reduction of runs and sags in waterborne spray applications.

3.4.1.2 *Fixing of Hazardous Dust*

Liquid compositions for fixing or sterilizing dust, e.g., asbestos dust, ash containing dioxin, or dust containing microorganisms such as viruses have been developed (39). Potential applications are in the course of demolition of building materials containing asbestos or by exchange or removal of filters used for hazardous microorganisms. A typical formulation is reproduced in Table 3.5.

Table 3.5 Dust Fixing Formulation (39)

Ingredient	Function	Amount/[%]
Gum arabic powder	Adhesive	5
Poly(ethylene glycol)	Surfactant	1
Carboxymethyl cellulose	Thickener	1
Water		93

3.4.1.3 *Food Additive Slurry*

A food additive slurry composition and a similar powder composition has been described, which are useful for enriching magnesium by adding the compositions to foods, such as yogurt, cow milk, juice, milk powder, etc. (40).

Magnesium has the actions of relaxing and dilating muscle and blood vessels, and is an indispensable mineral to our wellness. When magnesium is deficient, it is considered that a human being easily suffers from hypertension, angina pectoris, and hyperlipemia. In addition, magnesium is greatly involved in calcium metabolism and, when magnesium is deficient, various symptoms accompanied with calcium metabolism abnormality are manifested. In the suggested slurries, CMC is used as emulsion stabilizer. An amount of around 2.5% of CMC is suitable (40).

The substitution degree of the carboxymethyl groups is preferably 0.6–1.0. When the substitution degree is smaller than 0.3, there is a tendency that the resistance to acids, alkalies, and salts is deficient. As a result, the stability of a magnesium ingredient in foods becomes deficient. In contrast, when the substitution degree exceeds 2.0, the viscosity of the aqueous solution becomes too high.

3.4.2 *Superabsorbent Polymers*

Superabsorbent polysaccharide based polymers may be obtained by the grafting of acrylonitrile, acrylic acid, or AAm on to starch or cellulose. Despite their very high water absorption, these grafted polysaccharides, prepared by radical polymerization are not known to be biodegradable (12).

3.4.3 *Papermaking*

The strength properties of the pulp can be increased by adding CMC during alkali cooking or during the delignification process (41). A high ionic strength is needed, which occurs naturally by the presence of the cooking liquor, basically from sodium hydroxide, and from calcium which is released from the wood raw material. A cooking temperature of 130–160°C is established in the course of cooking. This high temperature is also advantageous for the deposition of CMC on to the fibers.

It is possible to deposit CMC permanently on to cellulose fibers. The resulting manufactured paper receives a substantial increase in its strength properties. No additional process stages are required so the operating costs can be kept low (41).

3.4.4 *Textile Printing*

CMC is used as a thickener and flow improver for reactive textile printing. For those applications, conventional sodium CMC must have a degree of substitution of at least 2.

Conventional CMC with a lower degree of etherification cannot be used because the residual hydroxyl groups in CMC are likely to react with the dyes (42). In contrast, alginate, where only the C_6 position is a carboxyl function so that it has a degree of substitution of carboxyl functions of 1, generally does not react with reactive dyes.

The preparation of a CMC with a high degree of substitution is cost intensive. However, it has been found that sodium CMC that is prepared by a special method can be used as an additive in textile printing without the need of high degrees of substitution (42). The key of this behavior seems to be the introduction of a grinding step. The size reduction is carried out as a dry grinding or wet grinding. For improving the ease of dissolution, it is advisable to grind the additive just before use.

3.4.5 *Laundry Compositions*

It has been found that a mixture of cyclic amine based polymers, oligomers or copolymers and hydrophobically modified cellulosic based polymers or oligomers, imparts fabric appearance and integrity benefits that are greater than the benefits achieved by a corresponding amount of either component by itself (43).

Hydrophobically modified CMC polymers are those where longer chain ethers are attached to the main chain. For example, the hexylether of CMC can be synthesized directly from cellulose in ethanolic aqueous sodium hydroxide suspension with monochloroacetic acid and hexylchoride as grafting reagent (43). Basically, a chloroalkane is added to the conventional process of manufacturing CMC.

Several heavy duty detergent compositions have been prepared containing a mixture of cyclic amine based polymers and hydrophobically modified CMC. Such a composition is listed in Table 3.6. More examples can be found elsewhere (43).

Table 3.6 Heavy Duty Detergent Compositions (43)

Component	Amount/[%]
C_{12} linear alkyl benzene sulfonate	9.31
Oligo ethylene oxide sulfate	12.74
Zeolite Builder	27.79
Sodium carbonate	27.31
Poly(ethylene glycol) 4000	1.60
Dispersant	2.26
C_{12}–C_{13} alcohol ethylene oxide	1.5
Sodium Perborate	1.03
Soil Release Polymer	0.41
Enzymes	0.59
Cyclic Amine Based Polymers or Oligomers	3.0
Hydrophobically Modified CMC	1.0
Others (perfume, brightener, suds suppressor, etc.)	11.46

3.4.6 *Shaped Activated Carbon*

Activated carbons are extensively used (44):

- To purify, decolorize, deodorize, dechlorinate, and detoxicate potable waters
- For solvent recovery and air purification in inhabited spaces
- In the purification of many chemical and foodstuff products.

Shaped carbon bodies are generally produced from powdered carbon particles with organic or inorganic binders. Carbon powder, binder, water, and other ingredients are mixed to form a material that is subsequently shaped. Conventionally, a thermal treatment at high temperatures is necessary to increase the product strength and the water stability.

However, a binder for shaped activated carbon and a method for its manufacture have been developed that do not require a costly high temperature heat treatment step (44). Such a binder is CMC and its derivative salts.

In addition, crosslinking agents based on urea or epoxides are added. The crosslinking reaction occurs at temperatures at around 200°C. The properties of shaped activated carbon with a CMC binder and with other conventional binders are compared in Table 3.7.

Table 3.7 Properties of Shaped Activated Carbon (44)

Binder	$\rho_a/[g\ ml^{-1}]$	Hardness
Clay	0.36	81
CMC	0.33	94
Phenolic	0.40	99

ρ_a: Apparent density

3.4.7 *Cosmetics and Medical*

Augmentation of the skin can be an important factor in recovering from injury or for cosmetic purposes (45). There are situations in which loss of tissue can leave an indentation in the skin. For example, surgical removal of a dermal cyst, lipoatrophy or solid tumor can result in loss of tissue volume. In other cases, injuries, such as gunshot wounds, knife wounds, or other excavating injures may leave an indentation in the skin. Regardless of the cause, it is desirable to provide a dermal filler that can increase the volume of the tissue.

Collagen is often used as an injectable material for soft tissue augmentation. Additionally, numerous other materials, including proteins, fats, hyaluronic acid, polyalcohols, and other polymers have been used as injectable dermal fillers.

Compositions of CMC have unique properties that allow such compositions to be injected into the skin to fill spaces and to provide support where support is desired. One example for needed support is dermal augmentation in the face where dermal and subdermal volume is lost due to aging. CMC has the unique property of being an elastic gel with unique physical properties such as dynamic, plastic and zero shear viscosity, tissue adhesiveness, cohesiveness and flow characteristics. In addition, it can achieve these properties without the requirement of covalent crosslinking.

CMC is particularly unique because chemical modifications of CMC expand the number of physical properties that make it an ideal injectable polymer for human treatment. For example, a change in the degree of substitution has a dramatic effect on thixotropy and on viscosity of the gel. Its biocompatibility and viscoelastic properties makes it uniquely useful for injection into human skin where it becomes a space filling, biocompatible polymer (45).

In detail, a composition has been developed based on CMC, PEO, and calcium ions for the preparation of an ionically crosslinked dermal filler. Ionically crosslinked gels can be made by simply mixing appropriate amounts of CMC, PEO, and calcium ions together in a solution. Additionally, the solution may be acidified to promote the crosslinking of the polyacid and polyether molecules through hydrogen bonds.

Moreover, crosslinked gels can be fabricated which incorporate drugs to be delivered to the surgical site. Drugs that are anti inflammatory, such as aspirin, ibuprofen, and ketoprofen can be useful. Further, it can be desirable to use drugs that increase the formation of new tissues at the site of application.

The materials perform well for dermal augmentation and behave as elastic gels at frequencies from 0.01 *Hz* to 100 *Hz*. The elastic modulus is higher than the viscous modulus, therefore, the material remains a gel at all rates of deformation (45).

These properties are not shown by hyaluronic acid based compositions. There, the elastic and viscous moduli crossover at some frequency and stress. At frequencies lower than this transition point, these materials are predominately viscous fluids that do not act as space filling gels.

3.4.8 *Enzyme Activity*

The activity of cellulase enzymes is generally measured using traditional biochemical activity tests based on the ability of the cellulase enzyme in question to hydrolyze soluble cellulose derivatives such as CMC. In the course of the degradation the viscosity of the aqueous solutions is reduced. This serves as a measure for the activity (46).

For example, to measure the activity of endoglucanase, an aqueous CMC substrate solution is prepared, using a tris buffer at pH 9.0. The enzyme sample to be analyzed is dissolved in the same buffer. Both solutions are mixed together and transferred to a viscosimeter, thermostated at 40°C.

Viscosity readings are taken as soon as possible after mixing and again 30 *min* later. The amount of enzyme that reduces the viscosity to one half under these conditions is defined as *I unit* of CMC-endoase activity (47).

Apart from soluble cellulose derivatives, studies on the viscosity of the enzyme itself are useful to give information about the effects of temperature, denaturant concentration and pH on the protein denaturation. In addition, the conformational change of cellulase enzymes can be studied by using viscosity measurements (48).

3.5 Guar

3.5.1 *Phase Separated Solutions*

It has been found that aqueous solutions of guar and hydroxypropyl cellulose form phase separated solutions over a range of polymer concentrations (49). A phase separated mixture can be formed by simultaneously dissolving 2% dry guar and 2% dry hydroxypropyl cellulose in water by stirring. After stirring for approximately 1 *h*, the sample is allowed to rest for 1 *h* to confirm the phase separation.

The phase separated solution can then be gently stirred by hand to remix the guar-rich and HPC-rich phases as emulsion.

The rheology of this mixed two-phase polymer mixture shows drastic effects. The fluid is of sufficiently low viscosity to be easily pourable and pumpable. Its viscosity is substantially less than a 2% guar solution. Thus, the hydroxypropyl cellulose phase dramatically reduces the rheology of the guar polymer.

By itself, the 2% guar would be too viscous to pump. The addition of 2% hydroxypropyl cellulose to this polymer in solution, however, reduces the complex viscosity by more than an order of magnitude at low frequencies, i.e., at low shear rates (49).

3.5.2 *Fracturing Fluids*

An aqueous fracturing fluid is prepared by blending a hydratable polymer into an aqueous fluid. The aqueous fluid may be water, brine, aqueous based foams, or water-alcohol mixtures.

The hydratable polymer serves as the gelling agent. Examples include galactomannan gums, guars and derivatized guars. Preferred gelling agents are guar gum, hydroxypropyl guar and carboxymethyl hydroxypropyl guar (50,51). Guar and derivatized guars can be crosslinked by the addition of borate compounds (50,52).

In general, organic polyhydroxy compounds having hydroxyl groups positioned in the *cis*-form on adjacent carbon atoms or on carbon atoms in a 1,3-relationship react with borates to form five or six member ring complexes. At alkaline pH above about 8.0 these complexes form didiol crosslinked complexes (53). This leads to a valuable reaction with dissociated borate ions in the presence of polymers having the required hydroxyl groups in a *cis*-relationship. The reaction is fully reversible with changes in pH.

An aqueous solution of the polymer will gel in the presence of borate when the solution is made alkaline, and will liquify again when the pH is lowered below about 8. If a dry powdered polymer is added to an alkaline borate solution it will not hydrate and thicken until the pH is dropped below about 8. The critical pH at which gelation occurs is modified by the concentration of dissolved salts. The effect of the dissolved salts is to change the pH at which a sufficient quantity of dissociated borate ions exists in solution to cause gelation.

The addition of an alkali metal base such as sodium hydroxide enhances the effect of condensed borates such as borax by converting the borax to the dissociated metaborate. Polymers which contain an appreciable content of such *cis*-hydroxyl groups besides of guar gum, include locust bean gum, dextrin, and PVA (53).

In the course of service of fracturing fluids the gels must be broken, otherwise no free flow is possible after fracturing. Therefore, gel beakers are added. The breakers should become active only when the fracturing task has been completed. However, it is preferable that the compositions should be prepared in one stroke at the surface.

3.5.2.1 Gel Breakers

Bromine can be used as gel breaker (54). However, sulfamate stabilized bromine-based breakers show a better performance. Instead of bromine, bromine chloride, chlorine, or a mixture, can be used. The sulfamate used in the production of such breaker products is effective in stabilizing the active bromine species over long periods of time, especially at ultimately high pH. In particular, sulfamic acid or sodium sulfamate is used (55) .

Unlike hypobromites, sulfamate stabilized bromine breakers do not oxidize or otherwise destroy organic phosphonates which are typically used as corrosion and scale inhibitors. In fact, these breakers are compatible with PAAm containing slickwater fracturing fluids as long as they are not contacted with hydrogen sulfide.

Further, these breakers can provide a controlled rate of viscosity decay, allowing the breaker to be mixed with the fracturing fluid, avoiding a second treatment in the downhole region to break the fracturing fluid polymer. The rate of viscosity decay can be controlled by pH adjustments.

Tetrasodium propylenediaminetetraacetic acid, ethylenediamine tetraacetic acid, trisodium hydroxyethylenediaminetetraacetic acid and other aminocarboxylic acids and their salts can be used to directly break crosslinked guar in gelled fracturing fluids, particularly at elevated temperatures of 50–140°C (56).

The aminocarboxylic acids may be provided in an extended release form such as encapsulation by a polymer or otherwise, by pelletization with binder compounds, or absorbed on a porous substrate. Encapsulation permits a slow or timed release.

Enzyme breakers are the preferred gel breakers because they are not themselves consumed in the breaking process. Suitable enzyme breakers include galactosidase and mannosidase hydrolases.

They have an activity in the pH range of 5–10 and are effective to attack the 1,4-β-*D*-mannosidic linkages or the 1,6-α-*D*-glactomannosidic linkages in guar derivatives (50).

3.5.2.2 *Stabilizing Crosslinked Guars*

Viscoelastic treating fluids that are gelled with a crosslinked guar or guar derivative can be stabilized by the addition of ethylene glycol (EG) (57). These fluids are more stable in that viscosity is maintained, particularly at elevated temperatures. The additive may also increase the viscosity.

In particular, these compositions are more stable at high temperatures, up to 180°C. This discovery will allow the guar based systems to be used at a higher temperature, and will help minimize formation damage after hydraulic fracturing operations when less of the guar polymer is used, but the same viscosity is achieved through the use of a glycol. In other words, the introduction of these additives to the guar systems could possibly lower the amount of guar polymer needed to obtain the fluid viscosity necessary to perform gelled fluid applications or treatments (57).

3.6 Carrageenan

Carrageenans are a series of mono sulfate esters of poly(galactoside)s. The varieties are designed in front by Greek letters, e.g., κ-carrageenan. These polysaccharides contain as basic structure two repeating galactose units, one being sulfated, the other non sulfated. The units are connected via alternating α 1-3 and β 1-4 ether linkages.

In a typical process for making pure carrageenan, crude seaweed is first washed with cold water to remove sand and other particulates that may be present after the seaweed has been harvested. Carrageenan typically does not swell during the cold wash, primarily because carrageenan in seaweed is associated with the structural components of the seaweed, generally cellulose.

Depending on the seaweed species, following the cold wash a hot water extraction procedure is typically performed in which the extracted carrageenan is treated with aqueous base at high temperature. Generally, the base used is an alkali or alkaline earth metal hydroxide, for example, NaOH, $Ca(OH)_2$, or KOH. Establishing a high temperature in alkaline medium results in the formation of 3,6-anhydro linkages in the galactose units of carrageenan.

The hot extract is then filtered to remove insoluble material such as cellulose, hemicellulose and other particulates, and acid is added. The filtrate can then be concentrated to about 4% carrageenan for further processing. Optional procedural steps after extraction may include centrifugation and bleaching. Pure carrageenan is typically obtained by precipitation of the extract from the aqueous solution with KCl or an alcohol such as isopropanol. On the other hand, improved methods of preparation using a shear stress treatment have been disclosed (58).

Carrageenans are used as thickeners in food, e.g. ice cream or condensed milks. Another food applications is as clarifier to remove haze causing proteins. Other uses are in fire fighting foams as thickener to cause the foam to become sticky, and in general as thickeners in cosmetic applications. Carrageenans are soluble in hot water. The sodium salt forms soluble in cold water. In cold water, only λ-carrageenan is soluble.

3.6.1 *Medical Applications*

3.6.1.1 *Drug Encapsulation*

Hard capsule for pharmaceutical drugs are made from polysaccharides. Traditionally, hard capsules for pharmaceutical drugs are molded from film compositions. Gelatin is generally used as the base material. Further, a plasticizer such as glycerin and sorbitol is added. Such capsules generally contain 10–15% of water in the film. If the water content in the capsule film decreases below 10%, the plasticity of the film is lost, resulting in the distinctive deterioration of the impact resistance (59).

Alternative formulations are made up of κ-carrageenan and ι-carrageenan as the gelatinizing agent. Auxillary agents to enhance the gelation are potassium chloride or calcium chloride, among others.

The hard capsule for pharmaceutical drugs can be produced according to the conventional immersion molding method as in the case of gelatin hard capsules. A water-soluble cellulose derivative, a gelatinizing agent and an auxiliary agent for gelation, optionally together with a coloring agent, an opaquer, a flavor, etc., are compounded together with water to prepare an aqueous solution, in which an immersion molding pin is immersed to obtain a hard capsule.

The hard capsule does not exhibit fragility under the condition of low humidity. The drugs filled therein can be prevented from deteriorating because of the lower water content. The water content in the capsule film is usually in a range 4–6%.

Table 3.8 Mechanical Tests (59)

	Drop weight impact test		Finger impact test	
Gelling agent	Failure per 50	Water /[%]	Failure per 10	Water /[%]
Carrageenan	0	1.1	0	0
Gelatin	46	8.8	10	0

3.6.1.2 Controlled Drug Release

In controlled drug release formulations gelling polymers are an essential ingredient. Cellulose derivatives such as hydroxypropyl methyl cellulose have been described that may used in combination with ι-carrageenan (60).

3.6.1.3 Toothpastes

Toothpastes based on carrageenan as binder exhibit good properties such as stability, low stringiness, and good rheology. Moreover, these formulations have an appealing taste, a good cleansing effect and are easy to rinse. However, the wider use of carrageenan has been limited by its high cost in comparison other binders, in particular CMC.

Carrageenan, when used without other additional binders is typically present in a concentration of 0.6–1.2%. Carrageenan can sometimes be used in lesser amounts when mixed with natural or synthetic gums and other thickeners such as CMC or xanthan. The amount of carrageenan can be further reduced by special methods of preparation of the formulation.

It has found that viscosity enhancements of 100% can be obtained when toothpaste formulations prepared from carrageenans are allowed to quiescently cool (61). Low levels of carrageenan are required, down to 0.05%. Table 3.9 shows the properties representative formulations of ι-carrageenan.

Table 3.9 Cuban Values of Carrageenan Formulations (61)

	ι-Carrageenan/[%]						
	0.05	0.75	0.10	0.15	0.20	0.30	0.40
T/[°C]	Cuban Values						
35	–	–	–	1	1	3	4
40	–	–	–	1	1	3	5
45	–	–	–	1	1	4	6
50	–	–	–	1	2	4	7
55	0	–	0	1	2	6	8
60	0	–	1	2	4	9	–
65	0	–	1	5	6	11	–
70	0	–	2	7	10	≥12	–
75	0	2	3	11	≥12	≥12	≥12
80	1	3	6	11	≥12	≥12	≥12
85	1	4	9	–	–	–	–
90	1	4	11	–	–	–	–

The temperatures in Table 3.9 correspond to the maximum temperature of the formulation prior to quiescent cooling. The Cuban test is described below in detail.

The quality of a toothpaste is described by a practical test, the Cuban or rack test. Cuban test values are directly related to the viscosity of the toothpaste. In the Cuban test, the paste is squeezed from a tube through a fixed orifice across a grid of parallel rods, increasingly spaced apart. The test results are expressed as the greatest space number, from 1-12, which represents the longest distance between rods that support the dentrifice ribbon without having it break.

The rack is 300 *mm* long and 100 *mm* wide. The stainless steel rods are spaced at increasing distances apart starting at 3 *mm* between rods 1 and 2 with space number 1 and the distance between rods increases by 3 *mm* from rod to rod. Thus the distance between rods 2 and 3 is 6 *mm*, and the distance between the twelfth and thirteenth rod (space number 12) is 39 *mm*. Ratings of 1-2 and 9-12 are not acceptable, 3 and 8 are acceptable, 4-7 are good (62).

Thus, in Table 3.9 an optimum in the Cuban values can be easily discovered at low additions of carrageenan and at a high temperature of preprocessing.

3.6.2 *Other Applications*

3.6.2.1 *Fire Fighting Foams*

In fire fighting foams, carrageenan has been suggested among other polysaccharides, as a water-soluble polymeric film former (63). It may be used for the formulation of alcohol resistant agents which are used to fight both polar (water-soluble) and nonpolar solvent fires and fuel fires.

These polymeric film formers, dissolved in alcohol resistant agents, precipitate from solution when the bubbles contact polar solvents and fuel, and form a vapor repelling polymer film at the solvent interface or foam interface, preventing a further foam collapse (64).

3.6.2.2 *Anti-icing Compositions*

An anti-icing agent composition which is useful for deicing and anti-icing aircraft is shown in Table 3.10. Besides the usual EG carrageenan is used a thickener. The gel-forming carrageenan gums employed as thickeners exhibit the desired shear thinning charac-

Table 3.10 Anti-icing Agent Composition (65)

Component	Amount/[%]
EG	49.875
Water	49.875
ι-Carrageenan	0.25

teristics described above, yet are resistant to pump shear-induced degradation.

This particular characteristic is important since anti-icing fluids are typically applied using conventional ground-based deicing equipment which incorporates a pump driven spraying system. The carrageenan-thickened aqueous glycol based anti-icing fluids exhibit sufficient shear thinning to be readily pumpable in a conventional aircraft ground deicing equipment (65).

3.7 Suppliers and Commercial Grades

Suppliers and commercial grades are shown in Table 3.11.

Table 3.11 Examples for Commercially Available Polysaccharide Polymers

Tradename	Producer	Remarks
Hawaii Chitopure®	Synedgen, Inc.	Ultra-pure, medical grade chitosan
Opticel®	Hercules Inc.	Water-soluble cellulose ether
Cellit®	Dow Chemical Comp.	Organic cellulose esters
AquaPAC®	Aqualon Corp.	Polyanionic cellulose
Carbotron.TM.	Dow	Cellulose derivative
Celpol®(Series)	Noviant, Nijmegen	Polyanionic cellulose
FILTER-CHEK®	Halliburton Energy Services, Inc.	Modified Cellulose
LIQUI-VIS	Baroid	Hydroxyethyl cellulose
Mil-Pac LV	Baker Hughes	Low viscosity polyamine cellulose
Natrosol® 250LR	Aqualon Corp.	Hydroxyethyl cellulose
PAC	Halliburton Energy Services, Inc.	Polyanionic cellulose
Polyquaternium® 10	Nalco Chemical Comp.	Cationic cellulose derivative
Hydagen® HCMF	Cognis	Chitosan lactate

Tradenames appearing in the references are shown in Table 3.12.

Table 3.12 Tradenames in References

Tradename Description	Supplier
Aculyn™ (Series) hydrophobically-modified poly(acrylate) (2)	Rohm and Haas
Admul® WOL 1403 Polyglyceryl polyricinoleate (30)	Kerry Group Services Ltd.
Aerosil® Fumed Silica (30)	Degussa AG
Alcalase® Proteolytic enzyme, detergent (46)	Novo Industries A/S
Ampholak™ 7TX Amphoteric surfactant (46)	Kenobel AB
AquaPAC® Polyanionic cellulose (15)	Aqualon Corp.
Araldite® (Series) Epoxy resins (38)	Ciba
Broma™ FLA Starch (16)	TBC Brinadd
Calgon® T Sodium hexametaphosphate (42)	Calgon Corp.
Carbolite™ Sized ceramic proppant (51)	Carbo Corp.
Carbopol® (Sseries) Poly(acrylate) (30, 60, 62)	Lubrizol Advanced Materials, Inc.
Cartaretine® Copolymers of adipic acid and dimethylamino-hydroxypropyl diethylenetriamine (30)	Sandoz
Celite® 545 Diatomaceous earth (60)	Celite Corp.
Celluzyme® Detergent enzymes (43)	Novozymes A/S
Cera Bellina® Modified beeswax (30)	Koster Keunen Holland B.V.
Ceramicrete Magnesium-based ceramic particulate bridging agent (16)	Argon National Labs.
Chimexane® Polyglyceryl-3 cetyl ether (30)	Societe Chimex Corp. France

Table 3.12 (cont.)

Tradename Description	Supplier
Cremophor® GS 32 Polyglyceryl-3 Distearate (30)	BASF
Dacron® Poly(ethylene terephtthalate) (51)	DuPont
Dehymuls® PGPH Polyglyceryl-2 dipolyhydroxystearate (30)	Cognis IP Management GmbH
Drewplus® Antifoaming agent (38)	Ashland Aqualon
Dymed® poly(aminopropyl) biguanide (31)	Bausch & Lomb, Inc.
Epon® Epoxy resin (38)	Shell Chemical Comp.
Esperase® Proteolytic enzyme, detergent (46)	Novozymes A/S Corp.
Finsolv® C_{12}-C_{15} Alkyl Benzoate (30)	Finetex Co.
Gelcarin® GP 379 Calcium iota carrageenan (65)	FMC Corp. Marine Colloids Division
Hercoflat® PP (38)	Hercules Inc.
Hostamer® V2825 AMPS terpolymer (35)	Clariant GmbH
Hydagen® HCMF Chitosan lactate (21,30)	Cognis
Irgasan® 5-Chloro-2-(2,4-dichlorophenoxy)-phenol, Bacteriostatic agent (30)	Ciba-Geigy
Isolan® GI 34 Polyglyceryl-4-isostearate (30)	Evonik Goldschmidt GmbH
Isolan® PDI Diisostearoyl polyglyceryl-3-diisostearate (30)	Evonik Goldschmidt GmbH
Jaguar® (Series) Cationic guar gum (30)	Rhodia Inc. Corp.
Keltrol™ Xanthan gum (60)	CP Kelco U.S., Inc.
Lameform® TGI Poly(glycerin-3-diisostearate), emulsifier for cosmetics and pharmaceuticals (30)	Cognis IP Management GmbH

Table 3.12 (cont.)

Tradename Description	Supplier
Lamequat® L	Cognis IP Management GmbH
Hydroxypropyl hydrolyzed collagen, cationic protein (30)	
Maxacal®	Gist-Brocades N.V
Proteolytic enzyme (46)	
Maxatase®	Gist-Brocades N.V
Proteolytic enzyme (46)	
Merquat® (Series)	Calgon Inc.
Copolymers of acrylic acid with dimethyl diallyl ammonium chloride (30)	
Methocel®	Dow
Methylcellulose (49,60)	
Microcel® C	Blanver Farmoquimica LTDA
Microcrystalline cellulose (46)	
Mirapol® (Series)	Miranol
Polyquaternium cosmetics (30)	
Oxiplex®	FzioMed, Inc.
CMC and PEO polymers, surgical implants (45)	
Pluronic® (Series)	BASF AG
Ethylene oxide/propylene oxide block copolymer, defoamers (31)	
Polybor®	U.S. Borax of Valencia
Polymeric borate (2)	
Polymin® SK	BASF AG
Ethyleneimine-grafted water-soluble poly(amidoamine) formed from adipic acid and a triamine and crosslinked with a bischlorohydrin ether (5)	
Polyquad®	Alcon Research, Ltd.
Polyquaternium 1, $(C_6H_{12}N)_nC_{16}H_{36}N_2O_6 \times 3Cl$ (31)	
Protasan™ UPG213	NovaMatrix
Chitosan glutamate (32)	
Retsch® ZM-1	Retsch GmbH
Grinding mill (3)	
Satiaxane®	Cargill France SAS
Xanthan gums (60)	
Savinase®	Novo Nordisk A/S
Proteolytic enzyme for detergent usage (46)	
Shale Guard™ NCL100	Weatherford Int.
Shale anti-swelling agent (51)	
Syloid®	Davison
Synthetic silica (38)	

Table 3.12 (cont.)

Tradename Description	Supplier
Tego Care® 450 Poly-glyceryl-3-methylglucose distearate (30)	EVONIK Goldschmidt GmbH
Tetronics® Modified poly alkylene oxide (31)	BASF
Thermalock™ Cement for corrosive environments (18)	Halliburton Energy Services, Inc.
Troykyd® Defoamer (38)	Troy Chemical Corp.
Troysol® Antifoaming agent (38)	Troy Chemical Corp.
Wellguard™ 7137 Interhalogen gel breaker (1,55)	Albemarle Corp.
XAN-PLEX™ D Polysaccharide viscosifying polymer (3)	Baker Hughes INTEQ
Xantural™ Xanthan gum (60)	CP Kelco U.S. Inc.
Zeolex® Silicic acid, aluminum sodium salt (38)	J. M. Huber Corp.
Zyderm™ Bovine collagen (45)	Allergan, Inc. Corp.
Zyplast™ Collagen fibers crosslinked with glutaraldehyde (45)	Allergan, Inc. Corp.

References

1. J.F. Carpenter, Bromine-based sulfamate stabilized breaker composition and process, US Patent 7 576 041, assigned to Albemarle Corporation (Baton Rouge, LA), August 18, 2009.
2. X. Wang, Q. Qu, J.C. Dawson, and D.V.S. Gupta, Thermal insulation compositions containing organic solvent and gelling agent and methods of using the same, US Patent 7 713 917, assigned to BJ Services Company (Houston, TX), May 11, 2010.
3. G.F. Fanta, H.M. Muijs, K. Eskins, F.C. Felker, and S.M. Erhan, Starch-containing lubricant systems for oil field applications, US Patent 6 461 999, assigned to The United States of America as represented by the Secretary of Agriculture (Washington, DC) Shrieve Chemical Products (The Woodlands, TX), October 8, 2002.

4. H. Liu, F. Xie, L. Yu, L. Chen, and L. Li, Thermal processing of starch-based polymers, *Prog. Polym. Sci.*, 34(12):1348–1368, December 2009.
5. P. Lorencak, A. Stange, K. Diehl, and N. Mahr, Modifying starch with cationic polymers and use of the modified starches as dry-strength agent, US Patent 6 746 542, assigned to BASF Aktiengesellschaft (Ludwigshafen, DE), June 8, 2004.
6. T. Aitken and W.D. Pote, Cationization of starch utilizing alkali metal hydroxide, cationic water-soluble polymer and oxidant for improved wet end strength, US Patent 4 097 427, assigned to Nalco Chemical Company (Oak Brook, IL), June 27, 1978.
7. P.D. Buikema and T. Aitken, Making a lightly oxidized starch additive by adding a cationic polymer to starch slurry prior to heating the slurry, US Patent 4 146 515, assigned to Nalco Chemical Company (Oak Brook, IL), March 27, 1979.
8. J.H. Finley, Dextrin carboxylates and their use as detergent builders, US Patent 4 029 590, assigned to FMC Corporation (Philadelphia, PA), June 14, 1977.
9. T. Suzuki, I. Amano, K. Chiba, and R. Tofukuji, Dextrin ester of fatty acids and use thereof, US Patent 5 840 883, assigned to Chiba Flour Milling Co., Ltd. (Chiba, JP) Kose Corporation (Tokyo, JP), November 24, 1998.
10. W.E. Rayford and R.E. Wing, Crosslinked cationic and anionic starches: Preparation and use in heavy metal removal, *Starch-Stärke*, 31(11):361–365, 1979.
11. G.-X. Xing, S.-F. Zhang, B.-Z. Ju, and J.-Z. Yang, Study on adsorption behavior of crosslinked cationic starch maleate for chromium(VI), *Carbohydr. Polym.*, 66(2):246–251, October 2006.
12. C. Couture, D. Bergeron, and F. Picard, Crosslinked polysaccharide, obtained by crosslinking with substituted polyethylene glycol, as superabsorbent, US Patent 7 365 190, assigned to Archer-Daniels-Midland Company (Decatur, IL), April 29, 2008.
13. R.A. Hayes, Poly(vinyl alcohol) copolymer ionomers, their preparation and use in textile sizes, US Patent 6 387 991, assigned to E. I. du Pont de Nemours & Company (Wilmington, DE), May 14, 2002.
14. R.A. Hayes and G.D. Robinson, Poly(vinyl alcohol)copolymer sizes having high capacity to be desized, US Patent 5 362 515, assigned to E. I. Du Pont de Nemours and Company (Wilmington, DE), November 8, 1994.
15. T.R. Sifferman, J.M. Swazey, C.B. Skaggs, N. Nguyen, and D.B. Solarek, Fluid loss control additives and subterranean treatment fluids containing the same, US Patent 6 180 571, assigned to Monsanto Company (St. Louis, MO) National Starch and Chemical Investment Holding Corporation (Wilmington, DE), January 30, 2001.

16. T. Munoz, Jr. and B.L. Todd, Treatment fluids comprising starch and ceramic particulate bridging agents and methods of using these fluids to provide fluid loss control, US Patent 7 462 581, assigned to Halliburton Energy Services, Inc. (Duncan, OK), December 9, 2008.
17. A.F. Chan, Method and composition for removing filter cake from a horizontal wellbore using a stable acid foam, US Patent 7 514 391, assigned to Conocophillips Company (Houston, TX), April 7, 2009.
18. B.R. Reddy and L.S. Eoff, Biodegradable dispersants for cement compositions and methods of cementing in subterranean formations, US Patent 7 273 100, assigned to Halliburton Energy Services, Inc. (Duncan, OK), September 25, 2007.
19. W.T. Cain, E. Ruso, D.M. Blum, and I. Cutting, Process for the synthesis of 5-[(substituted amino]-8-[phenyl or substituted-phenyl]-3h,6h-1,4,5a,8a-tetraazaace-naphthylen-3-ones and intermediates therefor, US Patent 5 247 086, assigned to American Cyanamid Company (Wayne, NJ), September 21, 1993.
20. Z. Cai, H. Yan, and Y. Tao, Method for manufacturing grafted polyacrylamide flocculant of cationic/ampholytic ions, US Patent 5 990 216, assigned to Guangzhou Institute of Environmental Protection Sciences (Guangzhou, CN), November 23, 1999.
21. B.R. Reddy, L.S. Eoff, and E.D. Dalrymple, Well treatment fluid and methods with oxidized polysaccharide-based polymers, US Patent 7 007 752, assigned to Halliburton Energy Services, Inc. (Duncan, OK), March 7, 2006.
22. J.R. Trinkle, K.-O. Hwang, and W. Fan, Chitosan production, US Patent 7 488 812, assigned to Cargill, Incorporated (Wayzata, MN), February 10, 2009.
23. J.C. Cowan, A.N. Blanchard, C.G. Benoit, and T.L. Rodrigue, Process for the treatment for chitinaceous materials and for the deacetylation of chitin, US Patent 7 544 785, assigned to Venture Chemicals, Inc. (Lafayette, LA), June 9, 2009.
24. M.K. Jang, C.Y. Choi, W.S. Kim, B.G. Kong, Y.I. Jeong, H.P. Yang, and J.T. Jang, Method for preparing water-soluble free amine chitosan, US Patent 7 345 165, assigned to Jae Woon Nah (KR), March 18, 2008.
25. W. Fan, J.A. Bohlmann, J.R. Trinkle, J.D. Steinke, K.-O. Hwang, and J.P. Henning, Chitosan and method of preparing chitosan, US Patent 7 413 881, assigned to Cargill, Incorporated (Minneapolis, MN), August 19, 2008.
26. M. Bodnar, J. Hartmann, and J. Borbely, Preparation and characterization of chitosan-based nanoparticles, *Biomacromolecules*, 6(5):2521–2527, 2005.
27. J. Borbély and M. Bodnár, Nanoparticles from chitosan, US Patent 7 740 883, assigned to University of Debrecen (HU), June 22, 2010.

28. M. Bodnár, L. Daróczi, G. Batta, J. Bakó, J. Hartmann, and J. Borbély, Preparation and characterization of cross-linked hyaluronan nanoparticles, *Progress in Colloid & Polymer Science*, 287(8):991–1000, 2009.
29. M. Maroda, M. Bodnár, S. Berkó, J. Bakó, G. Eros, E. Csányi, P. Szabó-Révész, J.F. Hartmann, L. Kemény, and J. Borbély, Preparation and investigation of a cross-linked hyaluronan nanoparticles system, *Carbohydr. Polym.*, In Press, 2010.
30. C. Panzer and R. Wachter, Deodorizing preparations containing nanosacle chitosans and/or chitosan derivatives, US Patent 6 916 465, assigned to Cognis Deutschland GmbH & Co. KG (Duesseldorf, DE), July 12, 2005.
31. W.M. Hung, K.L. Bergbauer, K.C. Su, and G. Wang, Water soluble, randomly substituted partial N-, partial O-acetylated chitosan, preserving compositions containing chitosan, and processes for making thereof, US Patent 7 683 039, assigned to Adjuvant Pharmaceuticals, LLC (Alpharetta, GA), March 23, 2010.
32. A.M. Dyer, P.J. Watts, Y.-H. Cheng, and A. Smith, Pharmaceutical formulations for intranasal administration of protein comprising a chitosan or a derivative thereof, US Patent 7 662 403, assigned to Archimedes Development Limited (Nottingham, GB), February 16, 2010.
33. L. Illum and S.N. Chatfield, Vaccine compositions including chitosan for intranasal administration and use thereof, US Patent 7 323 183, assigned to Archimedes Development Limited (Nottingham, GB), January 29, 2008.
34. R.B. Harding, S.L.H. Crenshaw, P.E. Gregory, and D.H. Broughton, Cellulose ethers and method of preparing the same, US Patent 7 022 837, assigned to BKI Holding Corporation (Wilmington, DE), April 4, 2006.
35. G.F. DiLullo Arias, P.J. Rae, and D.T. Mueller, Multi-functional additive for use in well cementing, US Patent 6 235 809, assigned to BJ Services Company (Houston, TX), May 22, 2001.
36. P. Mansikkamäki, M. Lahtinen, and K. Rissanen, The conversion from cellulose I to cellulose II in NaOH mercerization performed in alcohol-water systems: An X-ray powder diffraction study, *Carbohydr. Polym.*, 68(1):35–43, March 2007.
37. Y. Uda and S. Matsumoto, Method of determining the molecular weight distribution of carboxymethylcellulose or a salt thereof, US Patent 5 521 100, assigned to Dai-Ichi Kogyo Seiyaku Co., Ltd. (Kyoto, JP), May 28, 1996.
38. J.M. Allen, A.K. Wilson, P.L. Lucas, and L.G. Curtis, Carboxyalkyl cellulose esters, US Patent 5 668 273, assigned to Eastman Chemical Company (Kingsport, TN), September 16, 1997.

39. A. Furuya, Liquid for preventing dispersion of or for fixing dispersible dust, US Patent 7 799 241, September 21, 2010.
40. H. Hojo and N. Kubota, Food-additive slurry composition and powder composition, and food composition containing these, US Patent 7 264 834, assigned to Maruo Calcium Company Limited (Akashi-Shi, JP), September 4, 2007.
41. C. Gustavsson, V. Snekkenes, and K. Olsson, Method for the modification of cellulose fibres, US Patent 7 214 291, assigned to Kvaerner Pulping AB (Karlstad, SE), May 8, 2007.
42. R. Kniewske, R. Kiesewetter, E. Reinhardt, and K. Szablikowski, Carboxymethylcellulose and its use in textile printing, US Patent 5 463 036, assigned to Wolff Walsrode Aktiengesellschaft (Walsrode, DE), October 31, 1995.
43. R.K. Panandiker, J.A. Leupin, and W.C. Wertz, Laundry detergent compositions with a combination of cyclic amine based polymers and hydrophobically modified carboxy methyl cellulose, US Patent 6 835 707, assigned to The Procter & Gamble Company (Cincinnati, OH), December 28, 2004.
44. P.D.A. McCrae, T. Zhang, and D.R.B. Walker, Method of making shaped activated carbon, US Patent 6 696 384, assigned to MeadWestvaco Corporation (Stamford, CT), February 24, 2004.
45. R. Berg, S. Falcone, W.G. Oppelt, and S.M. Cortese, Compositions of polyacids and polyethers and methods for their use as dermal fillers, US Patent 7 192 984, assigned to Fziomed, Inc. (San Luis Obispo, CA), March 20, 2007.
46. J.R. Knorr, B. Hepler, and M. Holmgren, Laundry detergent composition containing level protease enzyme, US Patent 6 235 697, assigned to Colgate-Palmolive Co. (New York, NY), May 22, 2001.
47. M. Schulein and K.B. Levring, Fungal cellulase composition containing alkaline cmc-endoglucanase and essentially no cellobiohydrolase, US Patent 5 691 178, assigned to Novo Nordisk A/S (Bagsvaerd, DK), November 25, 1997.
48. N. Ghaouar, A. Aschi, L. Belbahri, S. Trabelsi, and A. Gharbi, Study of thermal and chemical effects on cellulase enzymes: Viscosity measurements, *Physica B:*, 404(21):4246–4252, November 2009.
49. P.F. Sullivan, G.J. Tustin, Y. Christanti, G. Kubala, B. Drochon, and T.L. Hughes, Aqueous two-phase emulsion gel systems for zone isolation, US Patent 7 703 527, assigned to Schlumberger Technology Corporation (Sugar Land, TX), April 27, 2010.
50. J.B. Crews, Polyols for breaking of fracturing fluid, US Patent 7 160 842, assigned to Baker Hughes Incorporated (Houston, TX), January 9, 2007.

51. D.P. Kippie and L.W. Gatlin, Shale inhibition additive for oil/gas down hole fluids and methods for making and using same, US Patent 7 566 686, assigned to Clearwater International, LLC (Houston, TX), July 28, 2009.
52. J.W. Dobson, Jr., S.L. Hayden, and B.E. Hinojosa, Borate crosslinker suspensions with more consistent crosslink times, US Patent 6 936 575, assigned to Texas United Chemical Company, LLC. (Houston, TX), August 30, 2005.
53. T.C. Mondshine, Crosslinked fracturing fluids, US Patent 4 619 776, assigned to Texas United Chemical Corp. (Houston, TX), October 28, 1986.
54. R.D. Goodenough, J. Place, and C.F. Parks, Stable bromo-sulfamate composition, US Patent 3 558 503, assigned to Dow Chemical Co., January 26, 1971.
55. J.F. Carpenter, Breaker composition and process, US Patent 7 223 719, assigned to Albemarle Corporation (Richmond, VA), May 29, 2007.
56. J.B. Crews, Aminocarboxylic acid breaker compositions for fracturing fluids, US Patent 7 208 529, assigned to Baker Hughes Incorporated (Houston, TX), April 24, 2007.
57. P.A. Kelly, A.D. Gabrysch, and D.N. Horner, Stabilizing crosslinked polymer guars and modified guar derivatives, US Patent 7 195 065, assigned to Baker Hughes Incorporated (Houston, TX), March 27, 2007.
58. A.G. Tsai, L.K. Ledwith, R. Kopesky, M.G. Lynch, W.R. Blakemore, and P.J. Riley, Production of carrageenan and carrageenan products, US Patent 6 479 649, assigned to FMC Corporation (Philadelphia, PA), November 12, 2002.
59. T. Yamamoto, K. Abe, and S. Matsuura, Hard capsule for pharmaceutical drugs and method for producing the same, US Patent 5 431 917, assigned to Japan Elanco Company, Ltd. (Osaka, JP), July 11, 1995.
60. A. Magnusson and M. Thune, Modified release pharmaceutical formulation, US Patent 7 781 424, assigned to AstraZeneca AB (Sodertalje, SE), August 24, 2010.
61. A.D. Ballard, Process for making toothpaste using low levels of carrageenan, US Patent 6 447 755, assigned to FMC Corporation (Philadelphia, PA), September 10, 2002.
62. V.B. Randive and V.K. Gadkari, High moisture toothpaste, US Patent 6 159 446, assigned to FMC Corporation (Philadelphia, PA), December 12, 2000.
63. M. Beck, S. Champ, M. Tonnessen, A. Ziemer, G. Goebel, and M. Pfeiffer, Fire extinguishing and/or fire retarding compositions, US Patent Application 20 070 289 752, assigned to BASF AG, Ludwigshafen, DE, December 20, 2007.

64. K.P. Clark, Use of fluorine-free fire fighting agents, US Patent 7 172 709, assigned to Chemguard, Inc. (Mansfield, TX), February 6, 2007.
65. R.J. Tye, G.E. Lauterbach, and P.R. Standel, Aircraft anti-icing fluid containing carrageenan, US Patent 4 698 172, assigned to FMC Corporation (Philadelphia, PA), October 6, 1987.

4

Poly((meth)acrylic acid)

Although we are mostly dealing with poly(acrylic acid) (PAA) in this volume we have to discuss the related methacrylic compounds. In some applications copolymers are used that are not water soluble. I will only discuss here applications in which the solubility in water is essential.

Around 1900, Otto Röhm was engaged in the polymerization of acrylic acid (AA) and esters (1, 2). He did not continue to focus on PAA directly, and became highly famous for the invention of poly(methyl methacrylate).

Poly(methacrylic acid) was first described in 1880 (3). The difficulty to purify methacrylic acid (MA) by distillation without spontaneous polymerization was already recognized at this time.

4.1 Monomers

Monomers and comonomers are summarized in Table 4.1. The chemical structures of these monomers are listed in Figure 4.1.

4.1.1 Acrylic acid

An early method of the manufacture of AA is the carbonylation reaction of acetylene, i.e., the reaction with carbon monoxide and water.

AA can be produced by the method of vapor-phase catalytic oxidation of propylene in two stages with the use of air. In the first stage propylene is mixed with air and steam. The mixed gas propylene is then converted into acrolein and acrylic acid as a by-product. In the second stage, the acrolein is basically converted into AA.

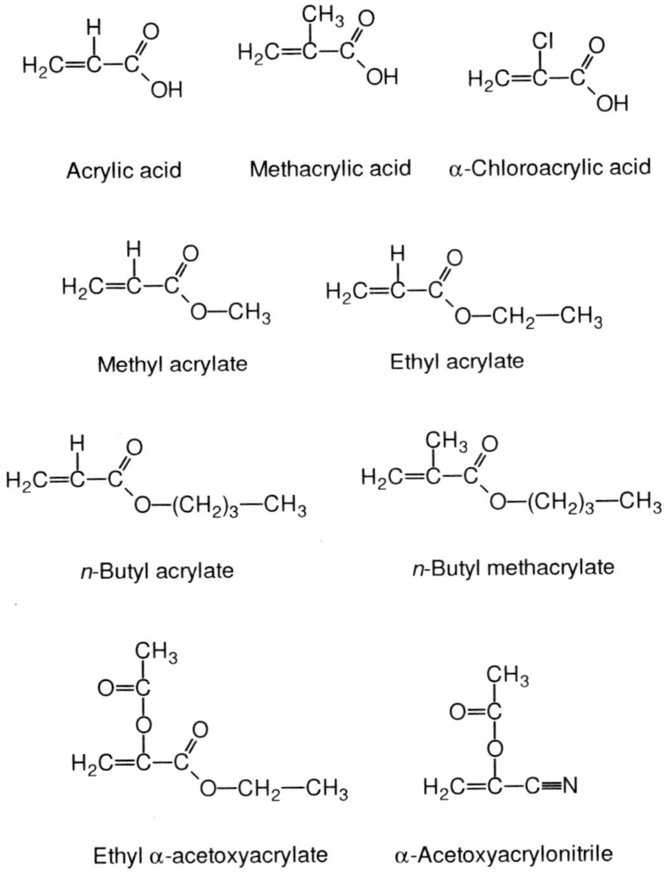

Figure 4.1 Monomers used for PAA

Table 4.1 Monomers and Comonomers for PAA Polymers

Compound	References
Acrylic acid	
Methacrylic acid	
Methyl acrylate	
Ethyl acrylate	
n-Butyl acrylate	(4)
n-Butyl methacrylate	(4)
α-Chloroacrylic acid	(5)
Ethyl α-acetoxyacrylate	
α-Acetoxyacrylonitrile	

The AA and the steam are cooled and recovered as an aqueous solution. In a further purification step, AA is isolated by extraction or distillation (6). A multi component catalyst containing molybdenum and vanadium effects a conversion of acrolein of 90% or more at a reaction temperature of about 220–340°C with a contact time of about 0.5–6 s.

The conversion of propylene into AA is highly exothermic. The reaction gas should contain in addition to the starting materials and products, an inert diluent gas, e.g., atmospheric nitrogen, or saturated hydrocarbons, such as methane or propane. Nevertheless, the gas can absorb only a small part of the heat of reaction. Therefore, tube-bundle heat exchangers which are filled with the oxidation catalyst are generally used, since in them, the predominant part of the heat evolved during the reaction can be removed by convection and radiation to the cooled tube walls (7).

Typical by-products of the gas phase oxidation, are furfural, and benzaldehyde. Higher oxidized by-products are acetic acid and propionic acid (8). Further, AA oligomers are formed. In case of purification by distillation, these oligomers in the bottom liquid. From the heterogeneously catalyzed gas phase oxidation of propylene at a temperature of 270°C a gaseous mixture of products was obtained and had a composition as given in Table 4.2.

However, the AA oligomers formed can be dissociated and the resulting dissociation gas is subjected to a countercurrent rectification before it is recycled (8). An improvement of the yield of AA can

Table 4.2 Composition of a Raw Output of Acrylic acid (8)

Compound	[%]
Acrylic acid	11.8201
Acetic acid	0.2685
Water	5.2103
Formic acid	0.0279
Formaldehyde	0.0989
Acrolein	0.1472
Propionic acid	0.0028
Furfurals	0.0033
Allyl acrylate	0.0014
Allyl formate	0.0005
Benzaldehyde	0.0038
Maleic anhydride	0.1352
Benzoic acid	0.0113
Phthalic anhydride	0.0147
CO_2	1.8042
CO	0.5898
Propane	0.5519
Propylene	0.2693
O_2	4.0253
N_2	75.0145

be obtained by direct cooling with a quench liquid.

The purification of AA with respect to water occurs by azeotropic distillation. However, there are methods that do not require an azeotropic solvent (9). This consists of a step of crystallization of the crude AA.

A more recent method for the manufacture of AA consists of the direct oxidation of propane in the gas phase (10). The reaction proceeds in the presence of multimetal oxide catalysts at 300–440°C.

4.1.2 *Methacrylic acid*

MA can be obtained from the reaction of acetone cyanohydrin and sulfuric acid to form methacrylamide as an intermediate which is in turn further reacted with water to form MA (11). The reaction is shown in Figure 4.2.

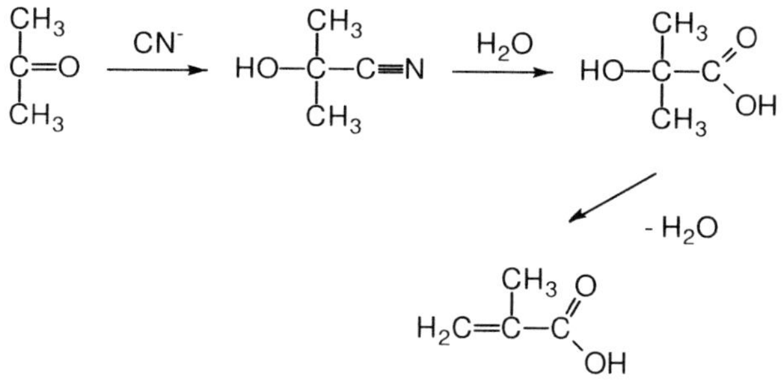

Figure 4.2 Formation of Methacrylic acid via the Cyanhydrin

This method is also addressed as ACH method and is widely used. A disadvantage of this method is that large amounts of waste sulfuric acid and ammonium bisulfate are produced (12).

MA can be also synthesized by the catalytic gas phase oxidation of *i*-butane (13), *i*-butene, *tert*-butyl alcohol, or methacrolein (14).

The catalysts suitable for the oxidation of methacrolein are multimetal catalysts based on Mo, V, P, Cu, Cs and NH_4 as the essential, active components. The preparation has been described in detail (15).

In the process with *i*-butane this starting material is subjected to a partial catalytic dehydrogenation into *i*-butene. This is eventually oxidized. As an intermediate in the oxidation process, methacrolein is formed (13).

Unpurified MA generally has impurity levels of about 5% that include water, acetic acid, AA, acetone, methacrolein, acrolein, isobutyric acid, 2-hydroxyisobutyric acid, mesityl oxide, 3-methacryloxy-2-methylpropionic acid, methyl methacrylate, propionic acid, and methacrylamide. The presence of specific impurities in the unpurified crude MA is dependent at least in part on the production process employed to produce the original crude MA.

High purity glacial MA can be obtained by purifying the crude MA in a series of successive distillation steps. It is desirable to add a polymerization inhibitor to inhibit the polymerization of MA. Suitable inhibitors are hydroquinone and 4-methoxyphenol. When such phenolic inhibitors are used, oxygen containing gas may be added to the distillation column to enhance the effectiveness of the inhibitor (11).

4.2 Polymerization and Fabrication

4.2.1 *Copolymers*

Copolymerization of vinyl acetate (VA) and AA is not practical because the reactivity of AA is significantly faster than that of VA. So the initial polymer formed in a typical free radical polymerization is very high in AA content. After AA is depleted, poly(vinyl acetate) (PVAc) is produced (4).

The polymers of AA produced during the early stages of the polymerization are not miscible with the PVAc produced during the later stages of the polymerization. The resulting blends are phase separated and exhibit inferior mechanical properties and will appear to be inhomogeneous.

While the random copolymerization of VA and AA can be achieved to some degree of success by the controlled and continuous addition of the more reactive monomer, this procedure is difficult to control and reproduce (4). However, copolymers from styrene and AA have been prepared via bulk reaction in a corotating twin screw extruder. Afterwards, the polymer products so prepared were feeded via gear pumps to a compounding extruder for the devolatilization of residual monomers.

4.2.1.1 *Reactive Extrusion*

Mixtures of styrene and AA monomers, in the desired feed ratios were combined with toluene as inert diluent. *tert*-Butyl perbenzoate is used as radical initiator. The temperature in the reactor extruder was maintained at 140°C, whereas the temperature in the compounding extruder was kept at 190°C. The properties of the resulting polymers are summarized in Table 4.3.

Table 4.3 Properties of Acrylic acid Styrene Copolymers (4)

Feed Ratio St/AA	M_w /[*Dalton*]	T_g /[°C]	MF10 /[$dg\ min^{-1}$]	MF30 /[$dg\ min^{-1}$]
92/8	86,000	125	6.74	7.27
86/14	83,000	137	2.93	2.97
80/20	71,000	142	5.03	5.02
82/18	138,000	139	0.68	0.66

T_g: Glass transition temperature
MF10: Melt flow index, preheating 10 *min* at 200°C
MF30: Melt flow index, preheating 30 *min* at 200°C

The melt flow values were determined after 10 *min* (MF10) and 30 *min* (MF30) preheating at 200°C and 0.3 *M Pa* (44 *psi*). In a similar way, by bulk reaction in an extruder, other copolymers from AA and acrylic esters have been prepared. These comonomers include methyl acrylate, ethyl acrylate, acrylic acid. methacrylic acid, *n*-butyl acrylate, and *n*-butyl methacrylate.

4.2.1.2 *Emulsion Polymerization*

Highly functionalized ethylene vinyl acetate (EVA) copolymers can be prepared by emulsion polymerization. The preparation of EVA polymers using emulsion polymerization techniques with medium pressure equipment is a standard technique. However, the incorporation of more than 5% of a functionalized comonomer, e.g., AA is difficult in emulsion polymerization because of many factors (16):

- It is important to control the pH of the emulsion because the reactivity of the functionalized monomer and the stability of the emulsion vary with pH.
- The polymerization of functionalized monomers in the aqueous phase can lead to high viscosity emulsions and grit formation. Differences in the monomer reactivity ratios often result in non homogeneous copolymers
- Levels of AA higher than 1.5% inhibit the batch copolymerization reaction of ethylene and VA
- The incorporation of large amounts of functionalized monomer in the growing polymer creates a polar micellar environment that discourages the migration of gaseous ethylene to the micelles.

Similar problems arise in related polymerization techniques, c.f., the arguments given above in Section 4.2.1 (4). Improved methods for emulsion polymerization have been developed that allow the incorporation of 20–60% of a functionalized monomer (16).

There is generally a limit as to how much AA can be incorporated into these systems. This limit can be greatly extended by use of 2-hydroxyethyl acrylate as an additional comonomer. It was found that using a 1:1 ratio of acrylic acid and 2-hydroxyethyl acrylate, it is possible to obtain up to about 20% AA in the polymer.

Only small amounts of 2-hydroxyethyl acrylate with AA are needed to effect polymerization at a reasonable rate. Accordingly, the amount of AA can be raised to 40% or more. Apparently, the 2-hydroxyethyl acrylate and AA pair and the ethylene and VA pair react very well together, independent of the reactivity of the other pair.

To prepare the functionalized polymers, emulsion polymerization is superior to solution polymerization. Solution polymerization methods tend to yield polymers that are low in molecular weight. The solution tends to have high viscosity, a low conversion rate at the end of the reaction with high residual monomers, and requires the removal of the solvent at the end of the reaction.

Functionalized monomers may be styrene sulfonic acid, 2-acrylamido-2-methyl-1-propane sulfonic acid (AMPS), and sodium vinyl sulfonate. These monomers are shown in Figure 4.3.

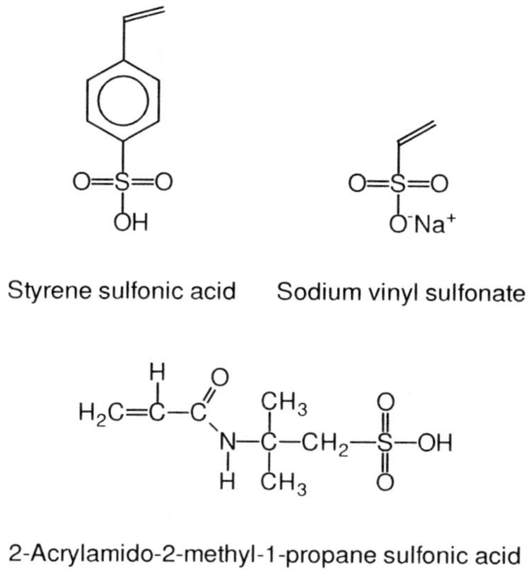

Figure 4.3 Monomers for Functionalization

A series of acrylate monomers also belong to this group, i.e., 2-hydroxyethyl acrylate, 2-hydroxyethyl methacrylate, hydroxypropyl acrylate, hydroxypropyl methacrylate, hydroxybutyl acrylate, and hydroxybutyl methacrylate. The preferred hydroxyl containing monomer is 2-hydroxyethyl acrylate.

A special technique in charging the monomers has been found to be advantageous. A large initial charge containing smaller amounts of AA and 2-hydroxyethyl acrylate allows the reaction to reach faster conversion rates before being flooded with the higher amounts of AA present in the slow add portion. The reaction can therefore continue at a reasonable rate throughout the course of the slow add portion and the inhibitory effect of the AA is largely overcome.

Minimizing the amount of AA and of 2-hydroxyethyl acrylate in the initial charge allows a substantial uptake of ethylene during the batch polymerization of the initial charge which then carries over into the slow add portion of the reaction.

The amount of ethylene taken up during the slow add is controlled by the faster conversion rates built up by polymerization of the initial charge, by the rate of emulsified monomer addition, as well as by the amount of buffer in the monomer emulsion.

Preferably, the emulsion polymerizing is initiated by a redox initiation system. As polymerization initiators, *tert*-butyl hydroperoxide and hydrogen peroxide have been used. Reducing agents include ascorbic acid, sodium formaldehyde sulfoxylate, sodium meta bisulfite, and 2-hydroxy-2-sulfinicacetic acid. Ascorbic acid seems to be the preferred reagent.

The aqueous emulsion copolymers are useful in personal care products and pharmacological products which are available in a variety of types and forms. A classification according to personal product types would consist of hair care products, bath products, cleansing products, skin care products, shaving products, deodorant products, antiperspirant products, and oral hygiene products (16).

4.2.2 *Hydrolysis of Poly(acrylamide)*

Instead of polymerizing AA (optionally with comonomers) directly, PAA can be obtained by the controlled hydrolysis of poly(acrylamide) (PAAm). In his way, hydrogels have been prepared (17). Such copolymers have been investigated for their metal retention properties. These properties are useful in environmental procedures for purification of wastewaters as well as in the analysis of metal ions in water.

4.2.3 *Slightly Crosslinked Polymers*

Slightly crosslinked PAA is usually produced by a precipitation polymerization method (18). A solvent in which AA can be dissolved but PAA is insoluble is used. Such solvents may be benzene, ethylacetate, cyclohexane, or toluene.

As crosslinking agent a multifunctional vinyl monomer is used such as divinylbenzene, diallyl phthalate, ethylene glycol dimethacrylate or sugar derivatives such as allyl saccharose. The crosslinking agent is added in very small amounts of 0.05%.

Suitable peroxide polymerization initiators are dibenzoyl peroxide, methyl ethyl ketone peroxide, cumene hydroperoxide, *tert*-butyl hydroperoxide, or cyclohexanone peroxide. Azo polymerization initiators include 2,2′-azobisisobutyronitrile, etc.

When AA is polymerized in such a solvent, a slightly crosslinked PAA precipitates out, which is collected and dried. The resulting powder has a low crosslinking density. Therefore, the water solubility is maintained. However, when the crosslinking density becomes higher, the polymer does not dissolve in the water but it swells in water.

4.3 Properties

Blends of polymers of VA and copolymers derived from AA exhibit thermodynamic miscibility. These polymer blends exhibit excellent mechanical compatibility and provide enhanced toughness and strength (4). They are useful films and in compounded formulations, such as coatings, emulsions, and adhesives.

Thermodynamically miscible blends of PVAc and AA containing binary copolymers are shown in Table 4.4. Miscible blends of

Table 4.4 Copolymers for Miscible Poly(vinyl acetate) Blends (4)

Monomer 1	Amount / [%]	Monomer 2	Amount / [%]
Acrylic acid	5 – 30	Methyl acrylate	70 – 95
Acrylic acid	5 – 40	Ethyl acrylate	60 – 95
Acrylic acid	16 – 40	Styrene	60 – 84

PVAc AA based terpolymers can also be prepared. The terpolymers include further ethyl acrylate, *n*-butyl acrylate, 2-ethylhexyl acrylate, or styrene as the third comonomer.

4.4 Applications

4.4.1 Superabsorbent Polymers

PAA in its free acid form and does not function as an superabsorbent polymer (SAP) for neutral to acidic aqueous media. The polymer has a low charge density. Therefore, a major driving force for absorption and retention, i.e., electrostatic repulsion, is missing.

In contrast, partially neutralized PAA has a sufficient charge density and can be used as a SAP by itself (19). SAP can be also built from an admixture of a poly(*N*-vinylamine) and PAA (19). Such binary SAP materials are an improved class of SAPs.

4.4.2 *Viscosifier for Aqueous Compositions*

PAA polymers severely increase the viscosity of aqueous solutions. This property is of interest in laundry detergents, cosmetic compositions, pharmaceutical compositions, and oil field applications.

4.4.2.1 *Toilet Bowl Cleaner*

PAA polymers are used as viscosity modifiers in compositions for a toilet bowl cleaner (20). The aqueous composition contains a nonionic surfactant, a disinfecting agent, a polymeric viscosity modifier, and a perfume. A composition for a toilet bowl cleaner is shown in Table 4.5.

Table 4.5 Toilet Bowl Cleaner Composition (20)

Compound	Amount/[%]
Ethoxylated non-ionic surfactant	3.5
Quaternary AA homopolymer	0.9
Alkyl dimethyl benzyl ammonium chloride	0.5
Perfume	0.4
EDTA tetrasodium salt	0.12
Formaldehyde	0.075
Water	94.51

The composition exhibits a pH in the neutral range. The PAA viscosity modifier permits the toilet bowl cleaning composition to have a Brookfield viscosity of about 175–250 $mPa\,s$. Recall that pure water has a viscosity of 1 $mPa\,s$ at room temperature.

4.4.2.2 *Hot Melts*

Poly(sodium acrylate) can be used as a thickener for an aqueous dispersion of a hot melt adhesive powder (18). Nevertheless, since poly(sodium acrylate) has a big moisture absorption property, so when poly(sodium acrylate) is dissolved in the water, clod is apt to be produced, and as a result, a long time is necessary to dissolve poly(sodium acrylate) in the water.

However, an aqueous dispersion of poly(sodium acrylate) does not show plastic flow and thus cannot prevent the sedimentation of a dispersed hot melt adhesive powder completely (18). When slightly crosslinked PAA is added to water, it dissolves therein, but increases the viscosity of the water only slightly. However, when an alkali agent is added to neutralize the PAA, the viscosity is remarkably increased because the neutralized PAA expands its chain due to the reciprocal electrostatic repelling force of each carbanion (18).

It is preferable that ammonia or organic amines such as triethanolamine, diisopropanolamine, aminomethylpropanol, trimethylol aminomethane, and tetrahydroxyethlenediamine may be used for neutralization. However, sodium hydroxide, potassium hydroxide and the like can be also used (18).

A dispersion of a hot melt adhesive powder with a slightly crosslinked PAA as thickener shows a plastic flow. This property prevents substantially the sedimentation of the hot melt adhesive powder.

4.4.3 *Laundry Detergents*

Multifunctional copolymers from AA use at least one of the following comonomers: α-Chloroacrylic acid, α-acetoxy AA, ethyl α-acetoxyacrylate, maleic anhydride, and VA. For example, the preparation of a terpolymer of ethyl α-acetoxyacrylate, AA and VA has been described. The solution polymerization technique is used, with isopropanol as solvent (5). *tert*-Butyl peroxide is used as radical initiator.

The desirable properties are obtained by hydrolysis after polymerization. For example, an α-acyloxyacrylic ester will result in α-hydroxyacrylic acid moieties after hydrolysis. The same is true for α-chloroacrylic acid. VA moieties will give vinyl alcohol moieties.

The multifunctional polymers are used in laundry detergents and cleaners, as a stabilizer in textile bleaching, as well as the bleaching of fiber pulps for papermaking. When used in laundry detergents, the polymers provide not only an incrustation inhibition, but also an augmentation of the primary detergency while stabilizing the bleaches such as perborates or percarbonates at the same time. Further, these copolymers are also excellent complexing agents for polyvalent metal ions such as calcium, iron, manganese and copper ions (5).

4.4.4 *Emulsifier Compositions*

Emulsifiers can be composed predominantly of both water-soluble surfactant and polyelectrolytes, thereby enabling the surfactant and polyelectrolyte to interact in the water before they reach an interface (21).

However, the stabilization of emulsions can be also achieved by using an oil soluble surfactant and a polyelectrolyte, i.e., PAA of opposite charge. For example, hexadecyl amine has a very low solubility in water.

It orients, to some degree, at the oil/water interface. On the other hand, PAA adsorbs at the oil/water interface. It seems the polymer then helps to anchor the oil soluble surfactant at the interface and vice versa (22) .

4.4.5 *Pulps*

There are three major types of pulping methods known in the pulp and paper industry. The first method is chemical, the second is mechanical, and the third is a combination of chemical and mechanical (23).

In chemical pulps, sufficient lignin is dissolved to allow the fibers to separate with little mechanical action. However, a portion of the lignin remains with the fiber and an attempt to remove this during digestion would result in excess degradation of the pulp. The degradation is a depolymerization of the cellulose.

For this reason 3–4% lignin is normally left in hardwood chemical pulps and 4–10% lignin is left in softwood chemical pulps after the cook or digestion. The lignin is subsequently removed by bleaching in separate pulp mill operations if completely delignified and whitened pulps are being produced.

Chemical wood pulping processes are the sulfate process or the sulfite process. The sulfite process has several advantages over the sulfate process. These advantages include improved yield, lower cost cooking chemicals, higher brightness pulps and more easily bleached pulps. However, the sulfite method also has two distinct disadvantages: Only a limited number of species can be pulped and the pulps produced are distinctly weaker than those made using the kraft or sulfate process.

The bleaching of the pulp is a standard method of removing color from pulp. It is current state-of-the-art technology for all chemical and mechanical pulps. The bleaching of a pulp and the subsequent delignification of pulp is usually performed by several chemical stages, with each stage being referred to by a letter designation. These stages are summarized in Table 4.6.

Table 4.6 Stages of Bleaching (23)

Code	Description
C	Chlorination
E	Extraction with sodium hydroxide
H	Hypochlorite reaction with sodium hypochlorite
Y	Hydrosulfite reaction with sodium hydrosulfite
D	Chlorine Dioxide reaction
P	Peroxide Reaction with peroxides
O	Oxygen Reaction

Five or six stages are needed to produce a full bleach brightness level corresponding to the color of 89–91% MgO. Most commonly, these stages in order are CEDED, CEHDED and OCEDED (23).

Several auxiliary chemicals are needed to provide an adequate performance. These auxiliary chemicals include sodium silicate for stability and chelation, sodium hydroxide for alkalinity, chelating agents such as ethylenediamine tetraacetic acid (EDTA) for controlling of transition metals, and magnesium sulfate for cellulose stability. Magnesium sulfate does not influence the delignification, but provides a small measure of protection for the pulp viscosity.

Diethylenetriaminepentakis(methylphosphonic acid) (DTMPA) is a known chelant and is in use to enhance the brightness in both mechanical and chemical pulp production. The structure is shown in Figure 4.4.

There is a synergism between DTMPA and PAA. Thus, a certain portion of the DTMPA can be replaced by PAA. This is extremely

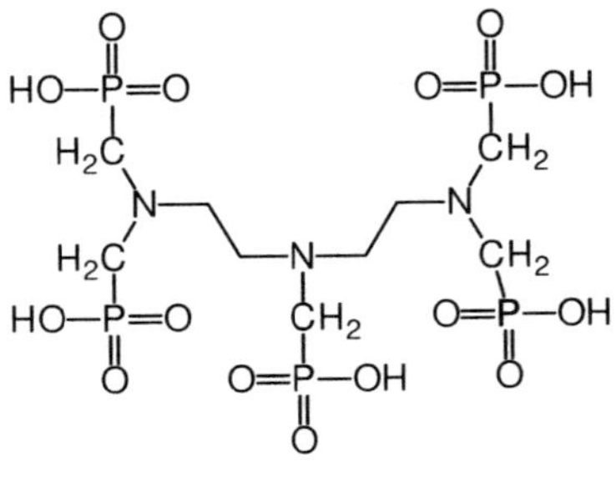

Diethylenetriaminepentakis(methylphosphonic acid)

Figure 4.4 Chelant for Bleaching

valuable because PAA is much less expensive than DTMPA (23). Usually an alkaline solution is prepared.

4.4.6 *Surface Coating*

PAA can be grafted on a gold substrate by using a diazonium-induced anchoring reaction (24,25). This process is an efficient technique to impart covalent adhesion of polymeric coatings on to surfaces without requiring electrical conductivity. In this process, aryl diazonium salts are reduced with iron powder to give surface active aryl radicals that form a grafted poly(phenylene) type film on the surface of the substrate. This film may be considered as a primer layer. In addition, the initiation of the radical polymerization of the vinyl monomer occurs, which is also present in the solution.

The growing polymeric radicals react with the poly(phenylene) primer layer to form a very homogeneous thin organic film on the surface. The final coating is strongly grafted on the surface. For example, it has a strong persistence after a long ultrasonic treatment, even in a in a good polymer solvent. The thickness of the polymer film can be controlled by the time of reaction, which is typically a few minutes.

In detail, *p*-phenylenediamine is treated with sodium nitrite in aqueous solution, containing also acrylic acid. The azo moieties are coupled to the surface of the substrate by adding iron powder.

This very simple and efficient grafting method provides a powerful tool for the covalent coating of organic or inorganic surfaces possessing complex geometrical shapes (24). At the same time, polymerization starts.

The interfacial electrochemistry of the PAA coated gold electrodes was studied with impedance spectroscopy and cyclic voltametry. PAA grafted on to a gold electrode can reversibly bind metal ions.

Since PAA is considered as a general purpose chelating material, it is able to capture different heavy metal ions at low concentrations. The release of the metal ions from the grafted PAA films is achieved with electro induced acidification by applying an anodic potential at the electrode. In this way, a localized water electrolysis process is promoted. The application of electrochemical pH switchable PAA films has potential applications in the field of waste treatments of heavy metals (25).

4.4.7 *Polishing Integrated Circuits*

Polishing slurries for metal commonly contain an oxidizing agent, solid abrasive particles or powders, an oxidized-metal dissolving agent, and a protective-film forming agent. The basic mechanism is that the metal surface is first oxidized and then this oxide layer scraped off by the solid abrasive particles.

The metal surface is not perfectly planar but has valleys and hills in the microscopic range. In the polishing process the material in the hills should be removed in order to effect smoothing of the structure. If, however, the oxide layer in the valleys of the metal film surface is also removed, the oxide layer in the valleys may unwantedly be etched further, resulting in a loss of the effect of smoothing. In order to prevent this result, a protective film forming agent can be added to the polishing slurry. It is important to balance the effects of the oxidant and the protective film forming agent.

The oxidizing agent is selected from hydrogen peroxide, nitric acid, potassium periodate, or hypochlorous acid. In the polishing slurries a combination of two protective film forming agents are used.

As the first protective film forming agent, compounds that are forming chelates with copper are suitable, e.g., EDTA or benzotriazole (26). The second protective film forming agent is selected from PAA and their ammonium salts, the corresponding methacrylic compounds, as well as PAAm.

It has been discovered that the use of the first protective film forming agent in combination with a second protective film forming agent of the classes explained above enables a sufficiently low etching rate in the valleys of the material to be etched. The use of the combination of two protective film forming agent allows to carry out the polishing process at a practical rate even without adding any solid abrasive grains in the polishing solution (26).

4.4.8 *Anti Reflective Coatings in Semiconductor Technology*

In the field of manufacturing semiconductor elements there has been applied a lithography technique in which a photoresist coating is formed on a substrate such as silicon wafer. After selectively exposing this element with actinic rays, a resist pattern on the substrate is formed (27).

Basically, anti reflective coatings prevent the reflection of incident light and substrate reflected light at the surface of a resist upon the formation of a pattern by a photolithography technique. The use of short and single wavelength radiation upon exposure is popular. This technology causes mutual interference of the incident light, reflected light from the photoresist or substrate interface, and multiple reflected light within the photoresist layer. The mutual interference causes a problem of standing waves. This adversely affects the pattern.

This problem can be solved by an anti reflective coating on a photoresist layer. An anti reflective coating composition has been proposed that contains PAA, poly(*N*-vinyl-2-pyrrolidone), perfluoro octanoic acid, and tetramethylammonium hydroxide (27).

Various additives may be included. Specific examples of surfactants are poly(oxyethylene) alkyl ethers, ammonium salts of alkylbenzenesulfonic acid, or lauric acid amide propylhydroxy sulfone betaine (27).

4.4.9 *Crosslinked Cellulose*

Cellulose products such as absorbent sheets and other structures are composed of cellulose fibers which, in turn, are composed of individual cellulose chains. Cellulose fibers are usually crosslinked to impart advantageous properties such as increased absorbent capacity, bulk, and resilience to structures containing cellulose fibers. High-bulk fibers are generally highly crosslinked fibers characterized by high absorbent capacity and high resilience (28).

Common cellulose crosslinking agents include aldehyde and urea-based formaldehyde addition products. While these crosslinking agents have been widely used in some environments, their applicability to absorbent products that contact human skin is limited by safety concerns. These crosslinkers are known to cause irritation to human skin.

Alternatively, cellulose can be crosslinked by carboxylic acids. Suitable acids are PAA polymers, poly(maleic acid) polymers, and the respective copolymers. Crosslinking catalysts include alkali metal salts of phosphorous containing acids such as sodium hexametaphosphate and sodium hypophosphite. The crosslinking catalyst is typically present in an amount of about 10%.

To achieve crosslinking, the fibers are treated with PAA to provide fibers with 2% PAA on the fiber. Sodium hypophosphite is used as a catalyst present at a ratio of 1:10 relative to PAA. The treated fibers are cured at temperatures from 165–195°C. Also a mixture of PAA and maleic acid as a second crosslinking agent can be used.

The effect of the absorbent capacity and the wet bulk of fibers on the method of treatment is summarized in Table 4.7. In general, the bulk and absorbent capacity of cellulose fibers crosslinked with blends of polycarboxylic acids is greater than fibers that are crosslinked with either only one polycarboxylic acid.

Furthermore, the problem associated with discoloration to crosslinked fibers crosslinked with only citric acid is improved by using a polymeric polycarboxylic acid blend without sacrificing the beneficial aspects of the absorbent capacity of the fibers. Cellulosic fibers crosslinked with a blend of a polymeric polycarboxylic acid and citric acid have been found to have improved brightness relative to fibers crosslinked with only citric acid.

In addition, the resistance of polymeric polycarboxylic acid crosslinked fibers to reversion, i.e., the loss of crosslinks, imparts stability to fibers crosslinked with polymeric polycarboxylic acid blends. For example, while cellulose fibers crosslinked with citric acid alone

Table 4.7 Absorbent capacity and Wet Bulk of Fibers Treated with PAA and Maleic acid Compared with PAA Alone Crosslinked Fibers at Various Cure Temperatures

	Blend 1		Blend 2		Poly(acrylic acid)	
T_c	AC	WB	AC	WB	AC	WB
/[°C]	/[$g\,g^{-1}$]	/[$cm^3 g^{-1}$]	/[$g\,g^{-1}$]	/[$cm^3 g^{-1}$]	/[$g\,g^{-1}$]	/[$cm^3 g^{-1}$]
165	14.9	14.8	15.0	14.7	13.6	13.2
170	15.3	15.1	15.4	15.1	14.2	13.8
180	15.9	15.7	16.2	16.0	15.2	14.9
195	16.5	16.2	16.8	16.6	15.9	15.6

Blend 1	PAA 3 *k Dalton* + maleic acid
Blend 2	PAA 2 *k Dalton* + maleic acid
AC	Absorbent capacity
WB	Wet bulk

are highly susceptible to crosslink reversion, fibers crosslinked with blends of citric acid and a polymeric polycarboxylic acid exhibit advantageous absorbent properties and stable crosslinks.

4.4.10 *Teeth Bleaching Gel*

Aqueous bleaching gels for bleaching teeth have been described (29). These gels contain a very high percentage of hydrogen peroxide. Hydrogen peroxide has become the bleaching agent of choice for use in dental bleaching gels. Hydrogen peroxide is a powerful oxidizer which serves to bleach the colored materials in the teeth, thereby producing a whiter appearance.

Unfortunately, at room temperature, hydrogen peroxide will attack the gelling agents used to make the dental bleaching gels. As a result of this attack, the gelling agents break down over the course of time. The degradation eventually results in the viscosity of the gel becoming too low to be useful.

Viscosity is very important to the effectiveness of dental bleaching gels. If the viscosity is too low the gel will flow uncontrollably from the dispensing tube and become difficult to manipulate for the purposes of varying or equalizing the bleaching treatment. Furthermore, if the viscosity is too low, the gel is more likely to flow away from the teeth resulting in a reduced residence time. Residence time is the time the dental bleaching gel actually contacts the tooth enamel. The effectiveness of a dental bleaching gel is directly proportional to its residence time.

To circumvent these drawbacks, dental bleaching gels containing high concentrations of hydrogen peroxide are generally refrigerated until immediately prior to use. Refrigeration slows down the hydrogen peroxide attack on the gelling agent and also slows down hydrogen peroxide decomposition. However, refrigeration is both expensive and inconvenient.

Thickening agents can be used to circumvent these problems. The thickening agent is a PAA type in an amount of 0.25–3%. Preferred thickeners include the crosslinked PAA resins (29). These polymers are homopolymers of copolymers of AA crosslinked with allyl sucrose, poly(alkyl ether)s of divinyl glycol, or allyl pentaerythritol. These polymers swell in water up to 1000 times their original volume to form a gel when exposed to a pH environment above 4.0–6.0.

Various stabilizing agents have been investigated in an attempt to develop hydrogen peroxide containing dental bleaching gels that are stable at room temperature. The stabilizing agent in the aqueous gel is present in an amount of around 0.15%. The stabilizing agent is selected from *trans*-1,2-cyclohexylene dinitrilo tetraacetic acid (CDTA), ethylenediamine tetraacetic acid (EDTA), *N*-(2-hydroxyethyl) ethylenediamine triacetic acid (HEDTA), nitrilotriacetic acid (NTA), diethylene triamine pentaacetic acid (DTPA), triethylene tetraamine hexaacetic acid (TTHA), and ethylene glycol bis (2-aminoethylether) tetraacetic acid (GEDTA). The most preferred stabilizers include CDTA, $CaNa_2$ EDTA, Na_2 EDTA, Na_4 EDTA, HEDTA, and Na_3 HEDTA.

The combination of a PAA thickener and an aminocarboxylic acid (salt) stabilizer results in a gel that can be loaded with hydrogen peroxide and is stable as it maintains the suitable gel properties for 4–12 weeks at room temperature.

In addition, the stabilizer prevents substantial hydrogen peroxide decomposition at room temperature. The loss of hydrogen peroxide in gels utilizing PAA thickeners and aminocarboxylic acid (salt) stabilizers is less than 0.05% per day.

Bleaching compositions are shown in Table 4.8. The results of

Table 4.8 Bleaching Compositions (29)

No	TEA /[*g*]	GA /[*g*]	Stabilizer /[*g*]	35% H_2O_2 /[*g*]	Composition
1	0.102	–	20.00	0.050	PAA
2	0.102	0.030	20.00	0.050	PAA $CaNa_2$ EDTA
3	0.412	0.030	20.00	–	XG $CaNa_2$ EDTA
4	0.514	0.031	20.00	–	XG $CaNa_2$ EDTA
5	0.612	0.030	20.00	–	CMC $CaNa_2$ EDTA
6	1.262	0.030	20.00	–	SC $CaNa_2$ EDTA
7	0.102	0.0294	20.00	0.150	PAA CDTA
8	0.102	0.0310	20.00	0.100	PAA Na_2 EDTA
9	0.102	0.0309	20.00	0.100	PAA Na_4 EDTA
10	0.102	0.0972	20.00	0.100	PAA Na_3 HEDTA
11	0.102	0.10	20.00	0.100	PAA VE $CaNa_2$ EDTA
12	0.1025	0.0312	20.0189	0.250	PAA Na_3 NTA
13	0.1021	0.0318	20.0366	0.200	PAA Na_2 NTA
14	0.1032	0.0309	20.0029	0.050	PAA NaCa HEDTA
15	0.1022	0.0370	20.0305	0.200	PAA $NaCa_2$ DTPA
16	0.1020	0.0300	20.0000	0.300	PAA NaCa NTA
17	0.1026	0.0311	20.0126	0.200	PAA Ca_3 TTHA
18	0.1019	0.0311	20.0339	0.100	PAA Na_2Ca_2 TTHA
19	0.1024	0.0313	20.0032	0.100	PAA Na_2 CaEDTA

TEA	Triethanolamine
GA	Gelling agent
PAA	Poly(acrylic acid)
XG	Xanthan gum
CMC	Carboxymethyl cellulose
SC	Sodium carrageenan
EDTA	Ethylenediamine tetraacetic acid
CDTA	*trans*-1,2-Cyclohexylene dinitrilo tetraacetic acid
HEDTA	*N*-(2-Hydroxyethyl)ethylenediamine triacetic acid
VE	*D,L*-α-Tocopherol
NTA	Nitrilotriacetic acid
DTPA	Diethylene triamine pentaacetic acid
TTHA	Triethylene tetramine hexaacetic acid
GEDTA	Ethylene glycol bis (2-aminoethylether) tetraacetic acid

these stability studies continue in Table 4.9, wherein each example corresponds to a like numbered example in Table 4.7.

Table 4.9 Stability Studies (29)

No	Stability /[*d*]	H_2O_2 degrad. %/[*d*]	Self-expelling
1	7	0.04	Yes
2	84	0.004	No
3	28	0.0002	No
4	28	0.01	No
5	7	0.01	Yes
6	7	0.05	No
7	48	0.004	No
8	33	0.05	Yes
9	57	0.01	No
10	44	0.01	Yes
11	18	0.01	No
12	8	0.001	?
13	8	0.01	?
14	<5	0.03	No
15	3	–	No
16	<3	–	Yes
17	<7	0.01	No
18	<7	0.02	No
19	<3	–	No

In a typical treatment process, the soft tissues surrounding the teeth are first covered with a protecting device, e.g., a ligated rubber dam. This is important because the more hydrogen peroxide a dental bleaching gel contains, the more likely it is to burn the soft tissue upon contact. Dental bleaching gels containing more than 30% by weight of hydrogen peroxide will immediately burn any soft tissue they contact, quickly turning the tissue white.

Next a brush, or a needle, is utilized to place the dental bleaching gel described above in contact with the teeth one wishes to bleach. The dental bleaching gel is then allowed to remain in contact with the teeth for a period of time of 20–30 *min*. The bleaching effect of any dental bleaching gel is directly proportional to its residence time.

The bleaching effect can be amplified by applying a heat lamp or laser light to the applied dental bleaching gel. Heat or light increase the rate of bleaching of the hydrogen peroxide, providing a shorter period of time for whitening the teeth (29).

After the treatment the gel is removed with a gauze. The patient's mouth is then thoroughly cleaned with water and suction. When the dental bleaching gel comprises at least 20% by weight of hydrogen peroxide, only one or two such treatments are necessary.

4.4.11 Oil Field Applications

4.4.11.1 Permeability Modification

Methods that have been used to control or restrict the water production include the gellation of poly(vinyl alcohol), PAA and the condensation polymerization of phenol and formaldehyde within the pore channels of the formation.

Improved compositions for selective permeability modification of subterranean formations include copolymers with *N*-vinyl formamide, *N,N*-diallylacetamide as anchoring groups in combination of a hydrophillic monomer, such as AMPS. Other optional anchoring groups are sodium acrylate (30).

These compositions are relatively non-damaging to the oil permeability, while exhibiting the ability to decrease the water permeability substantially in water saturated zones. Therefore, the compositions may be applied successfully to a productive zone without the necessity of mechanical isolation in the wellbore. An example of a feed of such a composition is given in Table 4.10.

The chemicals are added in several steps. Before adding cumene hydroperoxide as radical initiator, the mixture is throughly degassed. The polymerization is carried out at 50°C. Detail can be found in the literature (30).

AA is also used for grafting of humic acid compositions. These compositions are utilized as fluid loss control additives in cementing compositions. The backbone of such a graft copolymer comprises a humic acid salt. The grafted monomers comprise 2-acrylamido-2-methyl-1-propane sulfonic acid sodium salt, acrylamide, AA salt, and diallyldimethylammonium chloride (31).

Table 4.10 Permeability Reducing Polymer Feed (30)

Component	Weight/[*g*]
Acrylamidomethylpropane sulfonic acid	37.53
Distilled water	100.06
Disodium EDTA	0.25
Ethyl acetoacetate	0.27
Acrylic acid	5.03
Ice	103.41
50% Potassium hydroxide	31.56
Acrylamide	77.50
N-Vinyl formamide	30.00
Distilled water	110.52
Cumene hydroperoxide in *tert*-butyl alcohol	0.50

4.4.11.2 *Surfactant Flooding*

Lignosulfonate AA graft copolymers are used in surfactant flooding. These copolymers reduce the loss to the formation during hydrocarbon recovery. The solution of sacrificial agents is injected into the formation to decrease the loss of more costly surfactants.

Lignosulfonate AA graft copolymers may be prepared when in a free radical reaction AA is polymerized with the lignosulfonate in the presence of a free radical initiator, such as hydrogen peroxide and iron sulfate. It is believed that sacrificial agents generally work by several chemical mechanisms. Possible mechanisms are (32):

1. The complexing of the sacrificial agent with polyvalent cations
2. The electrostatic attraction of the matrix and the sacrificial agent for each other
3. The functional groups of the agent attach themselves to the rock surface.

4.4.11.3 *Reverse Mud Emulsions*

Reverse emulsion types of mud have a multitude of advantages. However, these have to be weighed against environmental problems, in particular for offshore drilling. The mud itself is always recycled but the cuttings have to be removed after separating them on the surface using mechanical separator means for separating out solids (33).

Under the strictest of regulations, it is permitted to discharge cuttings into the sea only when the cuttings contain less than 1% of organic substances. This amount is greatly exceeded with an reverse emulsion type mud because the film of mud contaminates the cuttings and cannot be eliminated using the mechanical means employed.

Proposals have therefore been made to wash the cuttings before discharging them to the sea. However, the surfactants added to stabilize the reverse emulsion are so effective that the washing water itself is emulsified in the mud, such that the oil is dispersed very little in the washing water while both the volume and the viscosity of the mud increase. Adding detergents to destabilize such emulsions has also proved to be largely ineffective. Further, the detergents themselves cause environmental problems.

Water in oil emulsions have been described that can be reversed when the salinity of the aqueous phase is reduced. This is effected by adding freshwater or even seawater.

This remarkable property is achieved by using combinations of ethoxylate type non-ionic surfactants and sulfonate anionic surfactants as the emulsifying agent. However, such combinations of surfactants cannot produce all of the properties simultaneously, namely endowing the emulsion with high stability, even at high temperatures, while using additives that are biodegradable and of low toxicity.

Amphiphilic polymers have been proposed as emulsifying agents (34). These are hydrophobic modified polyacrylates, with a hydrophilic backbone formed from a crosslinked high molecular PAA. Comonomers are small amounts of long chain alkyl acrylates or alkyl methacrylates. These emulsions with high concentrations hydrophobic modified polyacrylates are destabilized by adding an electrolyte.

Polyelectrolytes can be modified to render them hydrophobic by amidification of a hydrophilic backbone through didodecylamine (33). The preferred hydrophilic backbone is poly(sodium acrylate) or a statistical copolymer of an acrylate and AMPS.

These polymers are effective as stabilizers for direct or reverse emulsions. The emulsion can be destabilized or reversed by reducing the salinity of the aqueous phase or by neutralizing the acid. Destabilizing the emulsion enables the organic oil phase to be recovered for recycling, and enables the mineral waste, such as drilling debris, to be eliminated, since it is no longer wetted by the oil.

The modification of PAA proceeds in *N*-methyl-2-pyrrolidone (NMP) solution. NMP is an aprotic solvent. Didodecylamine is coupled with the acid groups of PAA by means of dicyclohexylcarbodiimide. The derivatives are obtained in the basic form and can be converted into their acid form by treatment with 0.1 M hydrochloric acid.

Terpolymers based on AMPS are prepared in two steps. First, copolymers from AA and AMPS are prepared by radical polymerization. As an initiator, ammonium peroxodisulfate and tetramethylenediamine is used. These copolymers are then hydrophobically modified, again with didodecylamine and dicyclohexylcarbodiimide. In this way terpolymers are obtained. These polymers can be advantageously used as emulsifiers for breakable emulsions (33).

4.5 Suppliers and Commercial Grades

Suppliers and commercial grades are shown in Table 4.11.

Table 4.11 Examples for Commercially Available PAA Polymers

Tradename	Producer
Acrysol®	Rohm and Haas
Alcosperse®	Akzo Nobel
Aqualic®	Nippon Shokubai Co., Ltd
Aquatreat®	Akzo Nobel
Carbopol®	Lubrizol Advanced Materials, Inc.
Good-rite®	Noveon
Junlon®	Green Leaf Chemical Ltd.
Jurymer®	Toagosei Co., Ltd.
Sokalan®	BASF

PAA is delivered as such, as well as in its salt form, e.g. the potassium or sodium salt. The sodium salt is applied to sanitary articles. There is also food grade PAA available. The appearance is mostly a white powder or granules. The salts may appear faint yellow.

Tradenames appearing in the references are shown in Table 4.12.

Table 4.12 Tradenames in References

Tradename Description	Supplier
Amberlyst® 15	Rohm & Haas
Ion exchange resins, heterogeneous catalysts (11)	
Bruggolite® FF 6	Brueggemann
Reducing agent (16)	
Carbopol® (Sseries)	Lubrizol Advanced Materials, Inc.
Poly(acrylate) (29)	
Halad® (Series)	Halliburton Energy Services, Inc.
Fluid loss control additive (31)	
Halpasols®	Haltermann
Paraffin oils (13)	
Isopar® (Series)	Exxon
Isoparaffinic solvent (30)	
Kelig 4000	Daishowa Chemical Co.
Lignosulfonate-acrylic acid graft copolymer (32)	
Lignosite® 100	Georgia-Pacific Corp.
Lignosulfonate monomer (32)	
Ludox® (Series)	DuPont
Silicon colloid (13)	
Marasperse 92 ZCAA	Marasperse Chemical Co., Lignotech USA
Oxidized lignosulfonate (32)	
Microbond™	Halliburton Energy Services, Inc.
Cement expanding additive (31)	
NES-25	Henkel
Ethoxysulfonate surfactant (32)	
Primid® XL-552	Ems Chemie AG
β-Hydroxyalkylamide (19)	
Silicalite®	Halliburton Energy Services, Inc.
High surface area amorphous silica (31)	
Teflon®	DuPont
Tetrafluoro polymer (11, 12)	
Ultrez™ 10	B.F. Goodrich Co.
Crosslinked acrylic acid-(meth)acrylate ester copolymer (29)	

References

1. O. Röhm, *Über Polymerisationsprodukte der Akrylsäure*. Ph.D thesis, Universität Tübingen, Tübingen, DE, 1901.
2. O. Röhm, Über die Darstellung von Acrylsäuremethylester, *Berichte der deutschen chemischen Gesellschaft*, 34(1):573–574, January-April 1901.
3. R. Fittig, Ueber polymerisirte ungesättigte Säuren, *Berichte der deutschen chemischen Gesellschaft*, 12(2):1739–1744, July-December 1879.
4. J.A. Kuphal, L.M. Robeson, and D. Sagl, Miscible blends of poly(vinyl acetate) and polymers of acrylic acid, US Patent 5 171 777, assigned to Air Products and Chemicals, Inc. (Allentown, PA), December 15, 1992.
5. I. Rau, J. Perner, C. Ott, and R. Baur, Multifunctional polymers, method for the production and use thereof, US Patent 6 921 746, assigned to BASF Aktiengesellschaft (Ludwigshafen, DE), July 26, 2005.
6. K. Kadowaki, K. Sarumaru, and T. Shibano, Production of acrylic acid, US Patent 4 365 087, assigned to Mitsubishi Petrochemical Company Limited (JP), December 21, 1982.
7. E. Bastiaensen, B. Eck, and J. Thiel, Method for purifying acrylic acid or methacrylic acid by crystallization and distillation, US Patent 6 541 665, assigned to BASF Aktiengesellschaft (Ludwigshafen, DE), April 1, 2003.
8. J. Thiel, U. Hammon, D. Baumann, J. Heilek, J. Schroder, and K.J. Muller-Engel, Preparation of acrylic acid, US Patent 6 939 991, assigned to BASF Aktiengesellschaft (Ludwigshafen, DE), September 6, 2005.
9. K. Ueno, H. Hirao, N. Serata, and T. Yokogoshiya, Method for production of acrylic acid, US Patent 7 151 194, assigned to Nippon Shokubai Co., Ltd. (Osaka, JP), December 19, 2006.
10. M. Dieterle, J. Heilek, and K.J. Mueller-Engel, Method for the heterogeneously catalyzed partial direct oxidation of n-propane to acrylic acid, US Patent 7 795 470, assigned to BASF AG (Ludwigshafen, DE), September 14, 2010.
11. M.S. DeCourcy, J.E. Elder, and J.J.J. Juliette, Process for manufacturing high purity methacrylic acid, US Patent 7 723 541, assigned to Rohm and Haas Company (Philadelphia, PA), May 25, 2010.
12. C.I. Carlson, Jr., M.S. DeCourcy, and J.J.J. Juliette, Process for production of methacrylic acid, US Patent 7 253 307, assigned to Rohm and Haas Company (Philadelphia, PA), August 7, 2007.
13. S. Berndt, K.J. Muller-Engel, G.-P. Schindler, F. Rosowski, and J. Petzoldt, Method for producing methacrylic acid from isobutane, US Patent 6 933 407, assigned to BASF Aktiengesellschaft (Ludwigshafen, DE), August 23, 2005.

14. T. Hino and A. Ogawa, Method for producing methacrylic acid, US Patent 7 304 179, assigned to Mitsubishi Rayon Co., Ltd. (Tokyo, JP), December 4, 2007.
15. A. Sudo, Y. Seo, and H. Sugi, Catalyst for producing methacrylic acid and preparation method thereof, US Patent 7 825 061, assigned to Nippon Kayaku Kabushiki Kaisha (Tokyo, JP), November 2, 2010.
16. D.R. Williams, Highly functionalized ethylene-vinyl acetate emulsion copolymers, US Patent 6 762 239, assigned to National Starch and Chemical Investment Holding Corporation (New Castle, DE), July 13, 2004.
17. W. Li, H. Zhao, P.R. Teasdale, R. John, and S. Zhang, Synthesis and characterisation of a polyacrylamide-polyacrylic acid copolymer hydrogel for environmental analysis of Cu and Cd, *React. Funct. Polym.*, 52(1):31–41, July 2002.
18. M. Ogawa, N. Kioka, and K. Ito, Hot-melt adhesive powder dispersed in water with alkali thickener, US Patent 6 316 088, assigned to Nagoya Oilchemical Co., Ltd. (Aichi, JP), November 13, 2001.
19. M.A. Mitchell, T.W. Beihoffer, and R.S. Sultana, Poly(vinylamine) - base superabsorbent gels and method of manufacturing the same, US Patent 6 603 055, assigned to BASF Aktiengesellschaft (Ludwigshafen, DE), August 5, 2003.
20. H. Aszman, A. Kugler, and C. Blanvalet, Toilet bowl cleaning compositions containing a polymeric viscosity modifier, US Patent 6 372 701, assigned to Colgate Palmolive Company (New York, NY), April 16, 2002.
21. S. Mun, E.A. Decker, and D.J. McClements, Effect of molecular weight and degree of deacetylation of chitosan on the formation of oil-in-water emulsions stabilized by surfactant-chitosan membranes, *J. Colloid Interface Sci.*, 296(2):581–590, April 2006.
22. N.S. Stamkulov, K.B. Mussabekov, S.B. Aidarova, and P.F. Luckham, Stabilisation of emulsions by using a combination of an oil soluble ionic surfactant and water soluble polyelectrolytes. I: Emulsion stabilisation and interfacial tension measurements, *Colloids Surf., A*, 335(1-3):103–106, March 2009.
23. S.M. Shevchenko and P.Y. Duggirala, Methods to enhance brightness of pulp and optimize use of bleaching chemicals, US Patent 7 351 764, assigned to Nalco Company (Naperville, IL), April 1, 2008.
24. V. Mévellec, S. Roussel, L. Tessier, J. Chancolon, M. Mayne-L'Hermite, G. Deniau, P. Viel, and S. Palacin, Grafting polymers on surfaces: A new powerful and versatile diazonium salt-based one-step process in aqueous media, *Chem. Mater.*, 19(25):6323–6330, 2007.
25. X.T. Le, P. Viel, A. Sorin, P. Jegou, and S. Palacin, Electrochemical behaviour of polyacrylic acid coated gold electrodes: An application

to remove heavy metal ions from wastewater, *Electrochim. Acta*, 54(25): 6089–6093, October 2009.
26. T. Uchida, J. Matsuzawa, T. Hoshino, Y. Kamigata, H. Terazaki, Y. Honma, and S. Kondoh, Abrasive liquid for metal and method for polishing, US Patent 6 899 821, assigned to Hitachi Chemical Company, Ltd. (Tokyo, JP) Hitachi, Ltd. (Tokyo, JP), May 31, 2005.
27. Y. Takano, H. Tanaka, and D.H. Lee, Composition for antireflection coating, US Patent 6 692 892, assigned to Clariant Finance (BVI) Limited (VG), February 17, 2004.
28. J.A. Westland, R.A. Jewell, and A.N. Neogi, Polycarboxylic acid crosslinked cellulosic fibers, US Patent 6 620 865, assigned to Weyerhaeuser Company (Federal Way, WA), September 16, 2003.
29. T.C. Chadwick and H.L. Hunt, Stable tooth whitening gels containing high percentages of hydrogen peroxide, US Patent 6 824 704, assigned to The Gillette Company (Boston, MA), November 30, 2004.
30. J.C. Dawson, H.V. Le, and S. Kesavan, Compositions and methods for selective modification of subterranean formation permeability, US Patent 6 228 812, assigned to BJ Services Company (Houston, TX), May 8, 2001.
31. S. Lewis, J. Chatterji, B. King, and D.C. Brenneis, Cement compositions comprising humic acid grafted fluid loss control additives, US Patent 7 576 040, assigned to Halliburton Energy Services, Inc. (Duncan, OK), August 18, 2009.
32. G. Kalfoglou and G.S. Paulett, Method of using lignosulfonate-acrylic acid graft copolymers as sacrificial agents for surfactant flooding, US Patent 5 251 698, assigned to Texaco Inc. (White Plains, NY), October 12, 1993.
33. N. Monfreux-Gaillard, P. Perrin, and F. LaFuma, Invertible emulsions stabilised by amphiphilic polymers and application to bore fluids, US Patent 6 822 039, assigned to M-I L.L.C. (Houston, TX), November 23, 2004.
34. R.Y. Lochhead, "Electrosteric stabilization of oil-in-water emulsions by hydrophobically modified poly(acrylic acid) thickeners," in D.N. Schulz and J.E. Glass, eds., *Polymers as Rheology Modifiers*, Vol. 462 of *ACS Symposium Series*, chapter 6, pp. 101–120. American Chemical Society, Washington, DC, 1991.

5

Poly(acrylamide)

The first industrial process for the polymerization of acrylamide (AAm) and related compounds was launched in 1936 (1). The authors used the solution polymerization in water, using potassium persulfate as initiator. Another early process was that poly(acrylamide) (PAAm) was produced from poly(acrylic acid chloride) by the reaction with ammonia (2). Concomitantly, the potential use of this class of polymers in the textile industry was appreciated (3).

5.1 Monomers

The basic monomer for PAAm is AAm. Several related monomers are used as comonomers to get copolymers containing AAm. These are shown in Table 5.1.

AAm is prepared by the reaction of acrylonitrile (AN) with water. In addition to the conventional method of synthesizing AAm by the hydrolysis of AN, it is also known to effect the hydrolysis in the presence of various other catalysts, such as copper compounds (4). A multi element type Raney copper catalyst has been described, which is extremely active (5).

5.2 Polymerization and Fabrication

Typically, PAAm is made by an inverse emulsion polymerization technique. In this the AAm monomer, together with comonomers, are dissolved in water. As a free radical initiator, an oil soluble initiator is used, e.g., 2,2′-azobisisobutyronitrile. From this the term

Table 5.1 Comonomers for Poly(acrylamide) (6,7)

Amide Comonomers
Acrylamide
N-Methylol methacrylamide
N-Methylol acrylamide
Cationic Comonomers
Dimethylaminoethyl acrylate
Diethylaminoethyl acrylate
Dimethyaminoethyl methacrylate
Dimethyldiallylammonium chloride
Anionic Comonomers
Acrylic acid
Methacrylic acid
Maleic acid
Fumaric acid
2-Acrylamido-2-methyl-1-propane sulfonic acid
Vinylsulfonic acid

inverse emulsion polymerization originates. This solution is emulsified in an oil and the polymerization is initiated by raising the temperature. The water-in-oil emulsion is stabilized with a surfactant system. The polymer is found dissolved in the oil phase.

In this way the viscosity will not become too high (7). Although the viscosity of the aqueous PAAm solution is high, the effective viscosity of the emulsion is determined primarily by that of the oil and this is chosen to be suitably low. To introduce the PAAm effectively into the aqueous systems in which it should be actually used, the emulsion has to be broken. Mostly the emulsion undergoes an inversion by dilution into additional water.

Conventionally, the oil used in the emulsion polymerization is mineral oil. The limited biodegradability of mineral oils introduces drawbacks from the environmental perspective. However, processes using water-in-oil emulsions with oils from vegetable sources have been successfully developed (7).

Hydroxamated PAAms can be produced by the reaction of hydroxylamine with PAAm in aqueous solution at 50–85°C at a pH of at least 8 (8). The hydroxamated polymers are useful as chelating agents, which in turn are useful in the formation of iron complexes in drilling muds.

For some biological applications it is important to achieve very low levels of free AAm. The amount of unreacted monomer can be reduced by (9):

- Applying an ultraviolet initiator to the surface of the polymer
- Irradiating the polymer with ultraviolet light.

A liquid ultraviolet initiator composition can be prepared by dissolving 2-hydroxy-2-methylpropiophenone into a liquid poly(ethylene glycol) type. The irradiation may occur simultaneously with the drying process after polymerization.

Polymers based on AAm can be gelled by a crosslinking system containing hexamethylenetetramine, *p*-aminobenzoic acid and phenol (10). Such a composition is useful in oil field operations.

5.3 Properties

PAAm is appreciated because of its hydrophilic properties. Therefore it is not used as a construction material in a direct way, but it is dissolved or swollen in water.

5.3.1 *Mechanical Properties*

PAAm hydrogels are used for cell-based studies since their hydrophilic and semipermeable nature is advantageous for mimicking the extracellular matrix that is surrounding cells and tissue (11). Unfortunately, the use in these fields is limited due to its viscoelastic properties.

PAAm has a comparatively low shear elastic modulus value of 10 kPa (12). The modulus can be tailored by the concentration of the crosslinking agent, by the chemical composition, and by the conditions of processing (11,12).

Partially hydrolyzed aqueous solutions of PAAm can be degraded by mechanical shear. This property is of importance when used in subterraneous wells. Based on a theoretical viscoelastic fluid model the extent of degradation and the permeability of the rock can be predicted (13).

5.3.2 *Acoustic Properties*

A hydrogel acoustic coupling medium has been proposed as an alternative to water for clinical applications of focused ultrasound therapy (14). In the course of this research, detailed studies on the acoustic properties, such as sound speed, acoustic impedance, and others have been performed and reported (14). PAAm hydrogels exhibit favorable acoustic properties that vary linearly with the concentration of AAm.

5.3.3 *Thermal Properties*

In hydrogels, the thermal conductivity is around 0.8 $W\,m^{-1}K^{-1}$ and the heat capacity is around 6.5 $k\,J\,kg^{-1}K^{-1}$ (14). These values do not vary extensively with the water content.

Pyrolysis experiments using thermogravimetry indicate a two stage mechanism of degradation (15). Metal ions act as stabilizer for PAAm (16). This is not true for water-soluble polymers in general (17).

The thermal stability and the UV stability of PAAm hydrogels has been investigated (18). The hydrogels are stable at room temperature. In hot water at 95°C the hydrolysis of the pendent amide moieties to acids occurs. Low amounts of monomer are formed under UV irradiation.

5.4 Special Additives

PAAm can be used as a component of additive formulations for poly(ethylene terephthalate) (PET). These additives allow the extrusion of PET at lower temperatures which results in less formation of acetaldehyde (19). Acetaldehyde imparts an undesirable taste or flavor to bottled water stored in PET bottles. Therefore, the reduction in the amount of acetaldehyde in PET is highly beneficial in this respect.

5.5 Applications

Applications of PAAm are listed in Table 5.2. Subsequently, the recent developments in the art are detailed.

Table 5.2 Applications of Poly-(acrylamide)

Use	Remark
Adhesive	Paper industry
Flocculation	Water cleaning
Thickener	Oil field applications

5.5.1 *Membranes*

Flat sheet, tubular, and hollow fiber PAAm membranes are used in a variety of applications, including reverse osmosis and various filtration separation techniques (20).

One class of such membranes are composite membranes which may comprise a microporous support with a thin film. The properties of such membranes may be modified by the addition of various additives, coatings and posttreatments. Polyether based modifiers have been combined with PAAm as block copolymers.

A layer of PAAm is preferably prepared by interfacial polycondensation. A polyfunctional amine monomer and a polyfunctional acyl halide monomer are contacted at the surface of the microporous support.

m-Phenylenediamine can be used as amine monomer. A solution is coated on to a microporous polysulfone support. Subsequently this area is coated with a solution of trimesoyl chloride (1,3,5-benzenetricarboxylic chloride) (20).

Once they are coated, the poly(amide) (PA) membranes are passed through a convection air dryer at approximately 93°C for ca. 24 s. Such composite PA membranes can be used in spiral wound modules for reverse osmosis and nanofiltration.

5.5.2 Sensors

In an amperometric sensor for uric acid, PAAm is used to fix catalase, uricase and ferrocenecarboxylic acid on a working electrode (21). Uric acid is detected by the production of hydrogen peroxide through the reaction of the enzyme and the uric acid. The sensing window for the electrode is prepared by (21):

1. Mixing PAAm and a phosphate solution
2. Forming a PAAm solution by mixing the PAAm with ferrocenecarboxylic acid
3. Diluting uricase with the PAAm
4. Dripping a the diluted uricase obtained on the sensing window.

5.5.3 Flocculants

PAAms are widely used as flocculants to enhance the clarification of drinking waters. Methods and apparatus how to use flocculants have been described (22). Such treatment compositions do not require the use of filters.

A hybrid flocculant consisting of PAAm and aluminum hydroxide has been synthesized by the polymerization of AAm in colloidal solution containing $Al(OH)_3$ (23). Experiments on the settling rate of kaolin suspensions show that the hybrid flocculant is superior in comparison to pure PAAm. Starch grafted or microcrystalline cellulose grafted types are advantageously used (24).

In addition, ionic copolymers are in use (6). Cationic and anionic comonomers are shown in Table 5.1. When anionic and cationic aqueous solutions of polyelectrolytes are combined, a binding of the ions between the oppositely charged polyelectrolytes takes place that results in aqueous gels.

Flocculants are used also in hydrometallurgical processing (25). Flocculants used in hydrometallurgy are mostly copolymers of AAm and sodium acrylate. The activity of PAAm polymers with respect to flocculation processes may be influenced by aging processes (26).

The amide group may serve as a nitrogen source for microorganisms. Several bacterial strains have been isolated (27). Aerobic bacteria can utilize the nitrogen in PAAm (28). So PAAm stimulates methanogenesis reactions if it is the major source of nitrogen in a carbon rich medium.

5.5.4 *Hydrogels*

The hydrogels made from AAm based copolymers may be used for the treatment of incontinence and vesicouretal reflux (29) and as soft tissue filler endoprostheses (30).

In particular, a hydrogel has been described that is obtained using AAm and *N,N'*-methylenebisacrylamide as comonomers. These are polymerized by a free radical polymerization. The polymerization is carried out in aqueous solution with ammonium persulfate as free radical initiator. *N,N,N',N'*-Tetramethylene ethylenediamine is used as accelerator, or coinitiator, respectively, for the redox initiation reaction.

After polymerization, it is important to remove essentially all amounts of the monomers. The monomers are toxic to the patient as well as detrimental to the stability of the hydrogel. More details can be found elsewhere (29,30).

In the treatment of urinary incontinence, the hydrogel is typically injected into the urethra, specifically under the submucosal membrane of the urethra. Injection is via the external surface of the urethra and toward the submucosal membrane (29).

The method of administration for soft tissue filler endoprostheses and the clinical experience has been reported (30). The injection of the gel may be performed under local anaesthesia. However, for the correction of wrinkles and folds, local anaesthesia is not necessarily required. For lip augmentation, anaesthesia through the nerve block is recommended.

Approximately 900 patients underwent facial corrections with the gel described above. The overall cosmetic results have been claimed to be excellent. Only in a few cases, adverse events have been reported (30).

5.5.5 *Agriculture*

Soil erosion, i.e., the detachment of particles of soil and sediments, is a serious problem recognized worldwide. Essentially, the erosion occurs via the forces of wind and water that facilitate the movement of topsoil from one place to another (31). Of such forces, water erosion is generally considered more detrimental to soils both by the volume of soil removed, and the area of land influenced.

Several attempts have been made to prevent soil erosion. These include agricultural practices, such as contour farming and terracing, no-till cultivation, strip farming, and polyvarietal cultivation. Other approaches consist of adding organic matter to the soil. These materials produce polysaccharides that may cause the soil particles to stick together and resist potential erosion.

Compositions have been developed which are essentially an aqueous mixture of cellulose, mulch and PAAm or a copolymer of AAm (31). The latter is a hydrophilic polymer. Particularly useful are comonomers that can bear ionic groups, for instance, sodium acrylate.

In addition to providing superior soil erosion resistant properties, the compositions exhibit fire-retardant properties. For this reason, they are believed to be useful in reducing potential damage to vegetation by wildfires (31).

The hydration in plants, whether potted or in natural soils, may be enhanced by adding hygroscopic polymers. A formulation may include a substrate treated with a binder and securing a layer of hydrating particles (32).

Typical binders may include lignite, or other naturally occurring materials such as sugars, molasses, corn syrup, or gelatin. A by-product of wood, lignite has been found to be very effective. Various materials are serving as a hydrating agents. PAAm has been found to serve well as hydrating agent (32).

PAAm is used in hydro-seeding mixtures. A hydro-seeding composition is an aqueous mixture from seeds, mostly grass seeds and an anionic PAAm type (33). It is advantageous to use potassium or ammonium salts instead of sodium salts of PAAm. The use of potassium or ammonium based PAAm, dramatically improves the flocculation of the soil particles and further the seed germination.

5.5.6 *Remediation of Acid Spills*

The release of large amounts of a liquid acid creates an emergency situation. In general, the acid is neutralized to form a less corrosive material. However, neutralizing a large amount of acid, with e.g., sodium hydroxide, produces a highly exothermic reaction.

A method to respond to a release of liquid acids consists of the use of a PAAm powder (34). The amide groups in PAAm are claimed to neutralize the released acid in a controlled way that does not result in an instantaneous liberation of large quantities of thermal energy (34).

5.5.7 *Concrete Compositions*

The placement of concrete into forms is a critical phase in building projects. Typically, it is desired to place the concrete into forms as soon as possible after mixing while the concrete is in a plastic and workable state. In many projects it is also required that concrete placement be continuous.

The placement of the concrete is achieved by concrete pumps. Concrete pumping allows the direct placement of concrete without rehandling. This results in significantly reduced costs as well as improved concrete quality. In the priming a concrete pump line several steps are necessary (35):

- Providing a solid particulate mixture comprised of solvatable polymeric material and urea
- Mixing the solid particulate mixture with water to form a flowable composition
- Pumping the flowable composition through the concrete pump line.

The solvatable polymeric material includes a wide variety of polymers that either dissolve in water or at least form a colloidal dispersion in water. The most preferred polymer is a mixture of PAAm, a copolymer of AAm, and an acrylate monomer (35).

5.5.8 *Paper Additives*

Paper products made from untreated cellulose fibers lose their strength rapidly when they become wet, i.e., they have very little wet strength. The wet strength of ordinary paper is only about 5% of its dry strength. To overcome this disadvantage, various methods of treating paper products have been employed.

Additives are classified into the permanent type and into the temporary type, according to how long the paper retains its wet strength after immersion in water.

The retention of the wet strength is a desirable property in packaging materials. On the other hand, wet strength retention presents a disposal problem. A composition for enhancing the wet strength of paper has been described that consists of two basic components, a amidoamine polymer and a stabilized glyoxalated PAAm (36).

The modification with glyoxal is done at a pH of 7–10. Glyoxal (OHC–CHO) reacts with the pendant amide groups in different macromolecules in forming crosslinks. When the desired viscosity is reached, the glyoxalation reactions and the crosslinking reactions can be substantially terminated by acid quenching.

5.5.9 *Oil Field Applications*

In drilling fluids, thermally stable water-soluble polymers are required to impart viscosity into the fluids. These are advantageously binary and ternary copolymers from PAAm. The monomers used are summarized in Table 5.3.

Table 5.3 Monomers for Copolymers in Drilling Muds (37,38)

Vinyl monomer
Acrylamide
2-Acrylamido-2-methyl-1-propane sulfonic acid
2-Acrylamido-2-methyl-1-propane sulfonic acid sodium salt
N-Vinyl lactam
N-Vinyl-2-pyrrolidone
N-Acryloyl-*N*′-methyl piperazine
N-Acryloyl morpholine
N-Acryloyl-*N*′-(3-sulfopropyl)-*N*′-methyl piperazinium inner salt

Some of these copolymers are suitable as gelling compositions in oil field applications. Apart from drilling fluids, workover fluids and completion fluids, gelling compositions are used for permeability corrections, matrix acidizing, and fracture acidizing (37,39).

The methods for the preparation of numerous copolymers from the monomers listed in Table 5.3 as well as the preparation of some monomers have been described in detail (37).

The hydrolysis of AAm-based polymers proceeds rapidly under alkaline conditions. In alkaline-surfactant-polymer flooding this can be a major drawback. In the course of hydrolysis, carboxyl groups are formed. These may react by an intermolecular reaction with multivalent cations (40). Besides multivalent cations, residual oxygen is a critical factor for the degradation behavior.

However, in the absence of dissolved oxygen and multivalent cations, the polymer backbone can remain stable. The solutions can maintain at least half their original viscosity for more than 7 years at 100°C and around 2 years at 120°C. The stability is not dependent on whether oil is present (41).

Partially hydrolyzed PAAm and copolymers are frequently used in water-based drilling muds (42–44). They act as shale stabilizers. In addition, hydroxamated PAAm polymers are useful as chelating agents in the formation of iron complexes. In addition, hydroxamated polymers can be used as flocculants (8).

Partially hydrolyzed PAAm is pH sensitive. At low pH the polymer molecules are coiled tightly. This results in a very low solution viscosity. This feature allows the polymer solution to be injected into the reservoir more easily. After injection, the acid may be neutralized by the minerals of the formation. This effects an increase of the pH. Eventually, the polymer chains uncoil which leads to a large increase of the viscosity of the solution (45). Rheological studies revealed that both shear and oscillatory viscosities are considerably depending on the pH in a reversible manner.

In oil field applications there is either the demand of improved stability of the polymer or the desire for the polymer to degrade after time. The desired property depends on the field of specific task in hand.

5.5.9.1 *Improved Stability*

Aqueous drilling fluids are commonly employed for downhole operations (46). The lost circulation of such aqueous fluids is easier to control than that of oil-based drilling fluids. Further, the use of aqueous drilling fluids avoids the environmental issues associated with the clean-up and disposal of drill cuttings coated with oil.

Usually, aqueous drilling fluids contain clays or polymers, such as polysaccharides or PAAms to increase the viscosity. Polymer-based aqueous fluids are preferred over the clay-based fluids due to the problems associated with the clay-based fluids. For instance, the downhole equipment may be clogged by the clay that is present in the drilling fluid.

Unfortunately, the thermal stability of polymer-based aqueous drilling fluids is often less than desired. The temperature in a subterranean formation generally rises with a gradient of 30°C km^{-1}. Thus, the polymers may undergo severe degradation, which leads to an undesirable reduction in fluid viscosity.

Formulations with enhanced thermal stability have been developed. It has been found that a composition containing both xanthan gum and a relatively high concentration of PAAm retains its stability when exposed to temperatures of 120°C (46). This is much higher than for those compositions that are not containing both types of polymers.

5.5.9.2 *Controlled Degradation*

In the petroleum industry it is a common practice to use hydraulic fracturing. This method stimulates the production from a well by pumping water at high rates and consequently pressures into the well, thus creating a fracture in the productive formation. Practical and cost considerations for these treatments require the use of materials to reduce pumping pressure by reducing the frictional drag of the water against the well tubulars. PAAm is very widely used for this purpose (47).

A drawback of these polymers is their persistence. A large fraction of the PAAm used in the treatment, frequently stays within the formation. The polymer residue may reduce the permeability of the rock. It may also jeopardize the recovery of the water used in the treatment. PAAm provides a source of nitrogen which supports the growth of bacteria in the well. PAAm that stays in solution will make the disposal of the recovered water difficult.

Some of these problems can be overcome by the use of a vis breaker, i.e., a substance that degrades the polymer in the formation. PAAm can be degraded by the use of peroxides. For example, hydrogen peroxide is added at a concentration of about 0.004%, to a well treating fluid containing freshwater and ca. 0.01% PAAm (47).

The reduction of the viscosity with time with fluids with various concentrations of peroxide is shown in Table 5.4.

Table 5.4 Change of Viscosity at 32°C (47)

Poly(acrylamide) /[%]	0.01	0.01	0.01
Hydrogen Peroxide /[%]	0.007	0.0035	0.002
Time /[h]	Viscosity /[$mPa\,s$]		
0	1.01	1.00	1.01
1	0.95	0.94	0.96
2	0.93	0.90	0.90
4	0.90	0.87	0.90
6	0.87	0.84	0.87
18	0.80	0.81	0.83

The viscosities of the fluids return to the approximate viscosity of water, which is about 0.79 $mPa\,s$ at a temperature of 32°C, within about 18 h.

5.5.9.3 *Drag Reduction*

Polymeric nanoemulsions facilitate the flow as they reduce drag and friction in multiphase pipelines containing both oil and water (48). Conventional water-soluble emulsion polymers have viscosities greater than 1000 $mPa\,s$, with particle sizes larger than 1 μ.

These emulsions are only kinetically stable, i.e., they will separate with time. Nanoemulsions have particle sizes below 200 nm and are thermodynamically stable. Nanoemulsions are visually clear and have viscosities less than 200 $mPa\,s$.

The nanoemulsions are generally made by combining the component parts with agitation under high shear conditions. In general, more surfactant is used for nanoemulsions as compared to conventional emulsions. In Table 5.5, the viscosities of conventional and nanoemulsions made from PAAm polymers are compared.

Table 5.5 Comparison of Viscosities of Conventional and Nanoemulsions (48)

Shear rate /[s^{-1}]	Viscosity /[$mPa\ s$]			
	Micro	Micro	Nano	Nano
0.5	5900	9000	43	50
1	3800	6000	39	48
10	1200	1200	39	48
100	580	250	37	48

5.5.10 *Protein Analysis*

5.5.10.1 *Gel Electrophoresis*

PAAm gel electrophoresis has become one of the most frequently used techniques for the separation of biological macromolecules such as proteins, nucleic acids, and polysaccharides. There is a wide variety of equipment and methods for many types of high-resolution separation of these biological macromolecules for both analytical and preparative purposes. Two of the most widely used classes of separation methods involve (49):

- Separating protein molecules according to molecular weight using sodium dodecyl sulfate denaturation, often referred to as SDS-PAGE
- Separation of amphoteric molecules, such as proteins, using isoelectric focusing on stabilized pH gradients, where the molecules migrate to the isoelectric point.

These two techniques can be combined wherein a complex mixture of amphoteric molecules are initially focused in a one dimensional pH gradient means, such as a strip or tube of gel, to the isoelectric points of the components. The focused amphoteric molecules in the one dimensional pH gradient are then subsequently separated by molecular weight using sodium dodecyl sulfate denaturing electrophoresis in an orthogonal direction to the first dimension (49). This two-dimensional separation of complex mixtures using PAAm electrophoresis is a very powerful technique.

A significant improvement in the ease of handling has been described. Two-dimensional gel electrophoresis can be implemented as a vertical mini-gel. Details of the apparatus are given elsewhere (49).

5.5.10.2 *Microarrays*

Data from DNA microarrays have been used to elucidate the control mechanisms of cells and organisms based on gene expression profiles. Conventionally, protein analysis consists of isolating, purifying, and characterizing a single protein or a small group of related proteins.

However, many groups of researchers have expanded this analysis to include hundreds or thousands of proteins in a single experiment through the development of protein microarrays. The microarrays are generated on coated glass slides. The coatings include metals, organic and anorganic polymers. Microarrays are built up by a wide range of specific and nonspecific protein-surface interactions, including electrostatic attachment, affinity-based binding, and covalent bond formation.

Oligonucleotides and proteins can be immobilized as microarrays in PAAm gel pads. The PAAm gel acts by immobilizing the molecules into an activated PAAm gel matrix. Proteins can be also immobilized by a polymerization procedure. Numerous examples have been reported that detail this procedure (50,51).

The reaction between the peptide, protein, immobilized on the reaction surface can be detected by using fluorescent, chemiluminescent, colorimetric, antibody, or radioactive labels (50).

5.6 Suppliers and Commercial Grades

Some suppliers and commercial grades are shown in Table 5.6. There are both homopolymers, as well as cationically and anionically modified copolymers. For example, Bio-Gel® P gels are porous PAAm beads prepared by the copolymerization of AAm and *N,N′*-methylene bisacrylamide. They are used in column chromatography.

Table 5.6 Examples for Commercially Available Poly(acrylamide) Polymers

Tradename	Supplier
Alcomer	Ciba Specialty Chemicals Corp.
Bio-Gel® P	Biorad
Kayafloc® Series	Nippon Kayaku Co., Ltd.
Magnafloc®	Ciba Speciality Chemicals Water Treatments Ltd.
Polyflex	Cytec
Polyplus	MI Swaco
Reten® Series	Hercules Inc.
Sumikagel F®	Sumitomo Chemical Co.

5.7 Safety

PAAm is incompatible with strong oxidizing agents, aluminium, copper, iron, and iron salts. The toxicity is considered as low (52).

5.8 Environmental Impact and Recycling

PAAm and related polymers are of importance as oil field chemicals. Processes have been described in that solids and fatty acids are flocculated in completion and workover fluids. These processes free the solids from adhering oils. The oil free particulate solids as well as the produced waters are thereby rendered more environmentally acceptable for their discharge to the environment (53).

It has been shown that chitosan (CS) can be used in place of poly-(ethyleneimine) to crosslink AAm based polymers. Its usefulness as a crosslinking agent has been limited by its relative poor solubility in aqueous solutions (54).

While this is a step forward in the effort to provide more environmentally acceptable systems, the major component, or base polymer, of such a gel system is still a non-biodegradable polymer. The major component of such gel system is generally a homopolymer or a copolymer of acrylate-type monomers, such as acrylic acid, AAm, *N*-vinyl-2-pyrrolidone, etc.

The backbone of such polymers contains continuous carbon-carbon single bonds, which are of poor biodegradability. Since the

CS crosslinking agent is only a minor component of the gel composition, the total system is still predominantly non-biodegradable due to poor biodegradability of the synthetic base polymer. Therefore, instead of using PAAm, it has been proposed to use oxidized polysaccharide based polymers for well treatment fluids (54).

Tradenames appearing in the references are shown in Table 5.7.

Table 5.7 Tradenames in References

Tradename Description	Supplier
ALL-TEMP® Acrylate tetrapolymer (38)	Baker Hughes Drilling Fluids
Amres® Polyamide-epichlorohydrin wet strength resin (36)	Georgia-Pacific Resins, Inc.
BIO-LOSE™ Complexed polysaccharide, filtration control agent (42)	Baker Hughes
BIO-PAQ™ Water soluble polymer (42)	Baker Hughes INTEQ
Bore-Drill™ Anionic polymer (38)	Borden Chemicals
Chek-Loss® PLUS Ultra-fine lignin (38,42)	Baker Hughes
Chemtrol® X Blend of ground lignitic earth and synthetic maleic anhydride copolymers (38)	Baker Hughes
Clay Sync™ Shale stabilizer (46)	Baroid
ClaySeal® Shale stabilizer (46)	Baroid Fluid Services
DFE-129™ Acrylamide/AMPS copolymer (38)	Baker Hughes Drilling Fluids
DFE-243 Partially hydrolyzed polyacrylamide/trimethylaminoethyl acrylate (42)	Baker Hughes INTEQ
Driscal® D Water soluble polymer (38)	Drilling Specialties Comp.
Filter-Chek® Modified Cellulose (46)	Halliburton Energy Services, Inc.
Grabber® Flocculant (46)	Baroid

Table 5.7 (cont.)

Tradename Description	Supplier
Hydagen® HCMF Chitosan lactate (54)	Cognis
Hydro-Guard® Inhibitive Water-Based Fluid (46)	Halliburton Energy Services, Inc.
Hyperdrill™ CP-904L Acrylamide copolymer (42,44)	Hychem, Inc.
Hypermere® Polymeric surfactant (48)	Uniqema
Irgacure® 2959 1-[4-(2-Hydroxyethoxy)-phenyl]-2-hydroxy-2-methyl-1-propane-1-one, Photoinitiator (9)	Ciba
Isopar® (Series) Isoparaffinic solvent (20)	Exxon
KEM-SEAL® PLUS NaAMPS/*N*,*N*-dimethylacrylamide copolymer (38)	Baker Hughes Drilling Fluids
Kemseal® Fluid loss additive (38)	Baker Hughes Norge
Ligco® Lignite (38)	Baker Hughes
Ligcon® Causticized lignite (38)	Milchem Inc.
Max-Trol® Sulfonated resin (38)	Baker Hughes Drilling Fluids
Mil-Bar® Barite weighting agent (38)	Baker Hughes
Mil-Carb® Ground marble (38,42,44)	Baker Hughes Drilling Fluids
Mil-Gel-NT® Bentonite quartz mixture (38)	Baker Hughes
Mil-Gel™ Ground montmorillonite (38)	Baker Hughes
Mil-Temp® Maleic anhydride copolymer (38)	Baker Hughes
N-Dril™ HT Plus Filtration control agent (46)	Baroid
New-Drill® PLUS Partially hydrolyzed poly(acrylamide) (42)	Baker Hughes INTEQ
PAC™ -L Filtration control agent (46)	Baroid

Table 5.7 (cont.)

Tradename Description	Supplier
Parez® (Series)	Kemira Oyj Comp.
Cationic poly(acrylamide) copolymers (36)	
Polydrill®	Degussa AG
Anionic polymer (38)	
Protecto-Magic™	Baker Hughes
Ground asphalt (38)	
PYRO-TROL®	Baker Hughes
Acrylamide/AMPS copolymer (38)	
Rev Dust	Milwhite, Inc.
Artificial drill solids (38)	
Soltex®	Chevron Phillips Chemical Comp.
Sulfonated asphalt (38)	
Sulfa-Trol®	Baker Hughes Drilling Fluids
Sulfonated asphalt (38)	
Superfloc™	Cytec Industries, Inc.
Acrylamide copolymer (42, 44)	
Teflon®	DuPont
Tetrafluoro polymer (39)	
Xan-Plex™ D	Baker Hughes INTEQ
Polysaccharide viscosifying polymer (42)	
Xanvis™	Baker Hughes INTEQ
Polysaccharide viscosifying polymer (42, 44)	

References

1. H. Fikentscher and A. Steinhofer, Verfahren zur Herstellung von in kaltem Wasser löslichen hochmolekularen Polymerisationsprodukten, DE Patent 697 481, assigned to IG Farbenindustrie AG, October 15, 1940.
2. L.M. Minsk and W.O. Kenyon, Process for preparing water-soluble polyacrylamide, US Patent 2 469 696, assigned to Eastman Kodak Co., May 10, 1949.
3. Anonymous, Process for producing shrinkage effects in textiles, GB Patent 502 907, assigned to IG Farbenindustrie AG, March 28, 1939.
4. M. Martan, Preparation of acrylamide, US Patent 4 069 255, assigned to UOP Inc. (Des Plaines, IL), January 17, 1978.

5. Y. Kambara, I. Oonaka, K. Asao, and K. Fukushima, Preparation process of acrylamide, US Patent 5 015 766, assigned to Mitsui Toatsu Chemicals, Incorporated (Tokyo, JP), May 14, 1991.
6. J. Melzer and H.-U. Schenck, Retention aids and flocculants based on polyacrylamides, US Patent 4 104 226, assigned to BASF Aktiengesellschaft (Ludwigshafen, DE), August 1, 1978.
7. S.I.J. Reekmans and P. Cornet, Surfactant composition for inverse emulsion polymerization of polyacrylamide and process of using the same, US Patent 6 686 417, assigned to Imperial Chemical Industries PLC (London, GB), February 3, 2004.
8. M.E. Lewellyn and D.P. Spitzer, Preparation of modified acrylamide polymers, US Patent 4 902 751, assigned to American Cyanamid Company (Stamford, CT), February 20, 1990.
9. G.I. Naylor and A. Flisher, Method of treating polymers, US Patent 7 615 258, assigned to Ciba Specialty Chemicals Water Treatments Ltd. (West Yorkshire, Bradford, GB), November 10, 2009.
10. A. Moradi-Araghi, Gelation of acrylamide-containing polymers with hexamethylenetetramine and an aminobenzoic acid compound or phenol, US Patent 5 905 100, assigned to Phillips Petroleum Company (Bartlesville, OK), May 18, 1999.
11. B.A. Baker, R.L. Murff, and V.T. Milam, Tailoring the mechanical properties of polyacrylamide-based hydrogels, *Polymer*, 51(10):2207–2214, May 2010.
12. D. Calvet, J.Y. Wong, and S. Giasson, Rheological monitoring of polyacrylamide gelation: Importance of cross-link density and temperature, *Macromolecules*, 37(20):7762–7771, October 2004.
13. J.M. Maerker, Shear degradation of partially hydrolyzed polyacrylamide solutions, *SPE J.*, 15(4):311–322, August 1975.
14. A.F. Prokop, S. Vaezy, M.L. Noble, P.J. Kaczkowski, R.W. Martin, and L.A. Crum, Polyacrylamide gel as an acoustic coupling medium for focused ultrasound therapy, *Ultrasound in Medicine & Biology*, 29(9): 1351–1358, September 2003.
15. M.-H. Yang, The two-stages thermal degradation of polyacrylamide, *Polym. Test.*, 17(3):191–198, May 1998.
16. H.D. Burrows, H.A. Ellis, and S.I. Utah, Adsorbed metal ions as stabilizers for the thermal degradation of polyacrylamide, *Polymer*, 22(12): 1740–1744, December 1981.
17. S.P. Vijayalakshmi and G. Madras, Thermal degradation of water soluble polymers and their binary blends, *J. Appl. Polym. Sci.*, 101(1):233–240, 2006.
18. M.J. Caulfield, X. Hao, G.G. Qiao, and D.H. Solomon, Degradation on polyacrylamides. Part II. Polyacrylamide gels, *Polymer*, 44(14):3817–3826, June 2003.

19. D. Simon, D. Lazzari, S.M. Andrews, and H. Herbst, Process for preparing a stabilized polyester, US Patent 7 514 526, assigned to Ciba Specialty Chemicals Corp. (Tarrytown, NY), April 7, 2009.
20. W.E. Mickols and C. Zhang, Polyamide membrane with coating of polyalkylene oxide and polyacrylamide compounds, US Patent 7 815 987, assigned to Dow Global Technologies Inc. (Midland, MI), October 19, 2010.
21. S.-K. Hsiung, J.-C. Chou, T.-P. Sun, and M.-L. Cheng, Amperometric sensor for uric acid and method for the same, US Patent 7 794 778, assigned to Chung Yuan Christian University (Taoyuan County, TW), September 14, 2010.
22. W.Y. Young, Water treatment mixture and methods and system for use, US Patent 7 591 952, September 22, 2009.
23. W. Yang, J. Qian, and Z. Shen, A novel flocculant of $Al(OH)_3$-polyacrylamide ionic hybrid, *J. Colloid Interface Sci.*, 273(2): 400–405, May 2004.
24. Z. Cai, H. Yan, and Y. Tao, Method for manufacturing grafted polyacrylamide flocculant of cationic/ampholytic ions, US Patent 5 990 216, assigned to Guangzhou Institute of Environmental Protection Sciences (Guangzhou, CN), November 23, 1999.
25. G. Moody, The use of polyacrylamides in mineral processing, *Miner. Eng.*, 5(3-5):479–492, 1992.
26. A.T. Owen, P.D. Fawell, and J.D. Swift, The preparation and ageing of acrylamide/acrylate copolymer flocculant solutions, *Int. J. Miner. Process.*, 84(1-4):3–14, October 2007.
27. K. Nakamiya and S. Kinoshita, Isolation of polyacrylamide-degrading bacteria, *J. Ferment. Bioeng.*, 80(4):418–420, 1995.
28. M.E. Haveroen, M.D. MacKinnon, and P.M. Fedorak, Polyacrylamide added as a nitrogen source stimulates methanogenesis in consortia from various wastewaters, *Water Res.*, 39(14):3333–3341, September 2005.
29. J. Petersen, Polyacrylamide hydrogel for the treatment of incontinence and vesicouretal reflux, US Patent 7 780 958, assigned to Contura SA (Soborg, DK), August 24, 2010.
30. J. Petersen, Polyacrylamide hydrogel as a soft tissue filler endoprosthesis, US Patent 7 790 194, assigned to Contura A/S (Soborg, DK), September 7, 2010.
31. S. Harrison, Compositions and methods for resisting soil erosion and fire retardation, US Patent 7 666 923, February 23, 2010.
32. T.K. Thrash, Hydration maintenance apparatus and method, US Patent 7 726 070, June 1, 2010.
33. C. Collin, Hydro-seeding mixture, US Patent 7 504 445, assigned to SNF S.A.S. (Saint Etienne, FR), March 17, 2009.

34. D. Weinhold, Apparatus, composition, and methods to remediate an acid and/or liquid spill, US Patent 7 662 081, assigned to Phase III, Inc. (Chandler, AZ), February 16, 2010.
35. P. Inglese and D.R. Hurst, Concrete pump primer, US Patent 7 704 314, assigned to Pat Inglese (Smyrna, GA), April 27, 2010.
36. C. Hagiopol, Y. Luo, D.F. Townsend, J.W. Johnston, C.E. Ringold, and K.D. Favors, Blends of glyoxalated polyacrylamides and paper strengthening agents, US Patent 7 488 403, February 10, 2009.
37. I. Ahmed, A. Moradi-Araghi, A.-A. Hamouda, and O.I. Eriksen, Compositions and processes for treating subterranean formations, US Patent 6 051 670, assigned to Phillips Petroleum Company (Bartlesville, OK), April 18, 2000.
38. M. Jarrett and D. Clapper, High temperature filtration control using water based drilling fluid systems comprising water soluble polymers, US Patent 7 651 980, assigned to Baker Hughes Incorporated (Houston, TX), January 26, 2010.
39. B.R. Reddy, L.S. Eoff, J. Chatterji, S.T. Tran, and E.D. Dalrymple, Preventing flow through subterranean zones, US Patent 6 176 315, assigned to Halliburton Energy Services, Inc. (Duncan, OK), January 23, 2001.
40. D. Levitt, G.A. Pope, and S. Jouenne, Chemical degradation of polyacrylamide polymers under alkaline conditions, in *SPE Improved Oil Recovery Symposium*, Tulsa, OK, April 2010. Society of Petroleum Engineers.
41. R.S. Seright, A. Campbell, and P. Mozley, Stability of partially hydrolyzed polyacrylamides at elevated temperatures in the absence of divalent cations, in *SPE International Symposium on Oilfield Chemistry*, The Woodlands, TX, April 2009. Society of Petroleum Engineers.
42. T. Xiang, Drilling fluid systems for reducing circulation losses, US Patent 7 226 895, assigned to Baker Hughes Incorporated (Houston, TX), June 5, 2007.
43. J.R. Hayes, High performance water-based mud system, US Patent 7 351 680, April 1, 2008.
44. T. Xiang, Methods for reducing circulation loss during drilling operations, US Patent 7 507 692, assigned to Baker Hughes Incorporated (Houston, TX), March 24, 2009.
45. C. Huh, S.L. Bryant, M.M. Sharma, and S.K. Choi, ph sensitive polymers for novel conformance control and polymerflood applications, in *SPE International Symposium on Oilfield Chemistry*, The Woodlands, TX, April 2009. Society of Petroleum Engineers.
46. J.L. Maresh, Wellbore treatment fluids having improved thermal stability, US Patent 7 541 316, assigned to Halliburton Energy Services, Inc. (Duncan, OK), June 2, 2009.

47. O.L. Valeriano and R.J. Dyer, Viscosity breaker for polyacrylamide friction reducers, US Patent 7 621 335, assigned to Chemplex, Ltd. (Snyder, TX), November 24, 2009.
48. J. Yang and S.J. Weghorn, Polymeric nanoemulsion as drag reducer for multiphase flow, US Patent 7 468 402, assigned to Baker Hughes Incorporated (Houston, TX), December 23, 2008.
49. J.T. Champagne, Facile method and apparatus for the analysis of biological macromolecules in two dimensions using common and familiar electrophoresis formats, US Patent 7 517 442, assigned to Life Technologies Corporation (Carlsbad, CA), April 14, 2009.
50. S.B. Brueggemeier, S.J. Kron, and S.P. Palecek, Glass-immobilized, protein-acrylamide copolymer and method of making thereof, US Patent 7 560 258, assigned to Wisconsin Alumni Research Foundation (Madison, WI) and The University of Chicago (Chicago, IL), July 14, 2009.
51. S.B. Brueggemeier, S.J. Kron, S.P. Palecek, L. Parker, and S.B.H. Kent, Hydrogels for biomolecule analysis and corresponding method to analyze biomolecules, US Patent 7 588 906, assigned to Wisconsin Alumni Research Foundation (Madison, WI) and The University of Chicago (Chicago, IL), September 15, 2009.
52. The physical and theoretical chemistry laboratory. Chemical and other safety information, Safety web pages, Oxford University, Oxford, 2010. [electronic:] http://ptcl.chem.ox.ac.uk/MSDS/.
53. S.R. Luxemburg, Process for cleaning fluids and particulate solids, US Patent 6 267 893, July 31, 2001.
54. B.R. Reddy, L.S. Eoff, and E.D. Dalrymple, Well treatment fluid and methods with oxidized polysaccharide-based polymers, US Patent 7 007 752, assigned to Halliburton Energy Services, Inc. (Duncan, OK), March 7, 2006.

6

Poly(vinylamine)

Poly(*N*-vinylamine) (PVAm) polymers are mostly used as precursor polymers for poly(vinyl acetal)s and poly(vinyl aminal)s. The latter are dealt within a separate chapter. However, there are also applications for PVAm polymers as such.

6.1 Monomers

Monomers for *N*-vinylamine (VAm) polymers are summarized in Table 6.1. Some important monomers are shown in Figure 6.1.

Table 6.1 Monomers for *N*-Vinylamine polymers

Vinyl Monomers	Remarks
N-Vinyl formamide (NVF)	Common
N-Vinyl acetamide	Common
N-Vinyl-2-pyrrolidone (NVP)	
N,N'-Divinylethyleneurea (= 1,3-Divinylimidazolid-2-one)	Crosslinking agent
N,N'-Methylenebisacrylamide	Crosslinking agent
N,N'-Methylene bismethacrylamide	Crosslinking agent
Ethylene glycol dimethacrylate	Crosslinking agent
Trimethylolpropane triacrylate	Crosslinking agent
Glutaraldehyde	Crosslinking agent
Ethylene glycol diglycidyl ether	Crosslinking agent
Methyl chloroformate	Functionalization agent

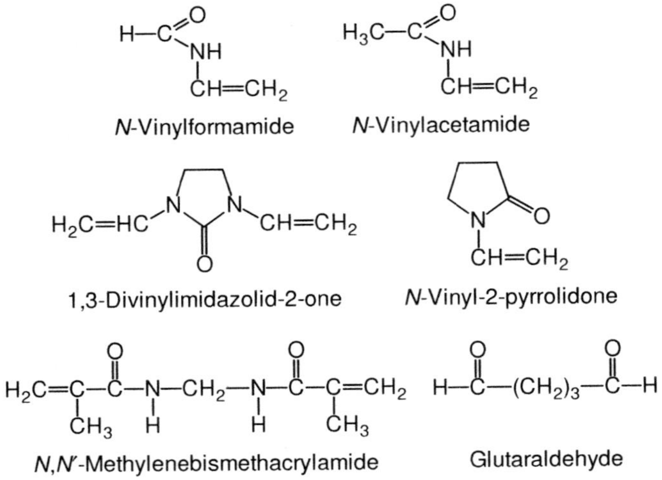

Figure 6.1 Monomers for *N*-Vinylamine Polymers

6.2 Polymerization and Fabrication

6.2.1 *Poly(N-vinylamine)*

A basic procedure for the synthesis of PVAm is the precipitation polymerization of NVF in a mixture of isopropanol/acetone. The water-soluble monomer is dissolved in the isopropanol/acetone solvent and polymerization is initiated.

The polymer formed in the course of polymerization precipitates from the solution. The solid is collected and dried. The powder is then hydrolyzed to PVAm using a 2-fold excess of NaOH and refluxing the aqueous solution for 4 *h*. The pH of the solution of PVAm is then adjusted to pH 7 (1).

A water-soluble poly(amine) could be obtained by the treatment of poly(vinyl chloride) (PVC) with $NaNO_2$ and subsequent catalytic reduction with hydrazine hydrate (2). However, the modification of PVC with sodium nitrite gives in the first step a potentially explosive modified polymer.

6.2.1.1 *Copolymers*

Copolymers that contain VAm and vinyl alcohol units can be prepared from vinyl acetate and NVF as comonomers by radical copolymerization and subsequent hydrolysis. In the hydrolysis reaction, the amide groups are converted to amine groups and the ester groups are converted to hydroxyl groups.

6.2.1.2 *Hydrolysis*

A PVAm polymer is produced by hydrolysis of poly(*N*-vinylformamide), under either acid or basic conditions. PVAm also can be produced from other poly(*N*-vinyl amide)s, like poly(*N*-vinylacetamide), poly(*N*-vinylpropionamide), and poly(*N*-vinylsuccinamide). It is desirable that the hydrolysis of the poly(*N*-vinyl amide) is substantial to essentially complete. If residual monomer or other impurities, such as aldehydes, are present in the poly(*N*-vinyl amide), hydrolysis conditions can lead to a crosslinking, which increases the molecular weight of the PVAm and can result in undesirable and unpredictable gel formation.

Therefore, the methods of synthesizing PVAm require either a rigorous purification of the poly(*N*-vinylformamide), or an extremely long reaction time and a relatively high reaction temperature to ensure that all the residual NVF monomer is consumed during the polymerization (3).

The problem of residual monomer content and the presence of other impurities, can be overcome by the use of scavenging agents to remove the residual monomer and other impurities from the poly(*N*-vinyl amide).

The scavenging agent is basically a compound capable of reacting with *N*-vinyl amides, like NVF and other aldehyde impurities, such as formaldehyde or acetaldehyde, under hydrolysis conditions. It has been found that sodium borohydride is a highly active scavenging agent.

The use of scavenging agents has the advantage of greatly reducing the process time and costs, currently invested to ensure that all the *N*-vinyl amide monomer and other impurities are consumed prior to hydrolysis.

In general, the hydrolysis is achieved in alkaline medium. Hydrolysis is conducted above room temperature in the range of 45–65°C for 4–8 *h* in the presence of the scavenger.

The degree of hydrolysis is controlled by the amount of acid or base, the reaction temperature, and the reaction time. In general, greater amounts of acid or base higher reaction temperatures and longer reaction times result in higher degrees of hydrolysis.

6.2.2 Popcorn Polymers

The name *popcorn polymer* stands for foam-like, crusty polymer particles having a cauliflower-like structure. Owing to their strong crosslinking, popcorn polymers are insoluble and scarcely swellable (4). They can be used as adsorbents, ion exchangers, carrier materials, and filter assistants.

NVP, NVF with small amounts of a crosslinking agent, such as *N,N′*-divinylethyleneurea can be polymerized into popcorn polymers (5, 6). In the course of polymerization, the crosslinking agent is not effective to withstand the swelling and chain scission of the polymer in solution. The chain scission reaction effects the generation of a huge number of reactive radicals that are active sites for polymerization. Due to concurrent chain scission, transfer reactions, and recombination, a highly branched and brittle polymer is formed at a high rate of polymerization that is only hardly swellable in the solvent. In the absence of bases the appearance of a popcorn polymer takes several hours.

Popcorn polymers have the property that on contact with the monomers of which they are composed or with other monomers, they can convert these into popcorn polymers. They act as a seed for the polymerization. Trapped radicals are formed on the popcorn matrix. The activity is lost if they come into contact with air.

After acidic hydrolysis, the amine groups in the popcorn polymers are generally present as salts. The corresponding anions of the liberated carboxylic acids, for example formate, are the counter ions. In order to obtain polymers with free amino groups, the polymers are deprotonated in aqueous suspension by adding bases. The salts formed in the neutralization remain in the aqueous solution.

6.2.3 *Carbamates*

VAm polymers can be functionalized with methyl chloroformate and other chloroformates to convert the pendent amino group into carbamate groups (7). The reaction runs in aqueous solution at a pH of 9.5 at 30°C. The basic reaction is shown in Figure 6.2.

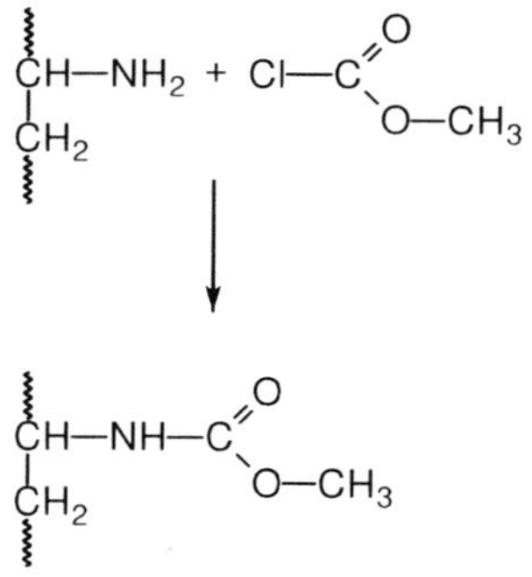

Figure 6.2 Formation of Pendent Carbamate Groups

The resulting carbamate-functionalized VAm polymers can be used as retention, drainage, and flocculation aids and as fixatives in papermaking, as protective colloids for the preparation of aqueous alkyldiketene dispersions and as dispersants for the preparation of aqueous filler slurries.

6.2.4 *Phosphonomethylated Poly(N-vinylamine)s*

The phosphonomethylation of amines is carried out in similar fashion to a Mannich reaction by reacting primary or secondary amines with phosphorous acid and formaldehyde at an acidic pH (8). The basic reaction is shown in Figure 6.3. The phosphonomethylation is followed the hydrolysis of a NVF homopolymer with concentrated hydrochloric acid.

The hydrolyzate thus obtained is diluted with water, followed by phosphorous acid, the amount depending on the degree of the desired phosphonomethylation. Under reflux, an aqueous 30% formaldehyde solution is slowly added. On completion of the addition, the mixture is refluxed for additional 15 *h*. The volatiles and excess

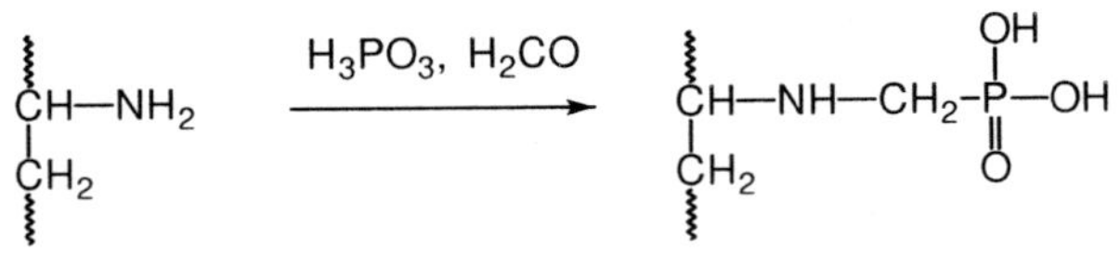

Figure 6.3 Phosphonomethylation of Amines (8)

reagents are removed by a steam treatment for 2 *h*. The solution is concentrated to about a third of its volume.

To purify the phosphonomethylated polymers, the polymers are precipitated in the form of the sodium salt by adding the still acid concentrated solution dropwise to ten times the volume of methanolic sodium hydroxide solution. The amount of NaOH should correspond to the acid number equivalent of the concentrated reaction mixture. Such products can be used as complexing agents for polyvalent metal ions.

6.3 Applications

Applications of PVAms cover a wide range of fields. They are used in or as

1. Flocculants and demulsifiers
2. Anti-scaling agents
3. Water absorbent materials
4. Membrane modifiers (9)
5. Medical materials
6. Paper industry
7. Analytical column materials
8. Well treatment fluids (10).

6.3.1 *Flocculants and Demulsifiers*

Flocculants work by gathering the particles in a net, bridging from one surface to the other and binding the individual particles into large agglomerates. Flocculation not only increases the size of the particles, but it also affects the physical nature of the agglomerates, so that the slurry will loose water at a faster rate because of the reduction of the gelatinous structure of the agglomerate.

The aggregation kinetics of negatively charged poly(styrene) (PS) latex particles as model compounds in the presence of a PVAm has been measured by time-resolved simultaneous static and dynamic light scattering. The effects of polymer dose, ionic strength, and the pH dependence have been studied (11).

In the same way like phase diagrams stability maps can be set up. The aggregation rate depends strongly on the ionic strength. In contrast, the electrophoretic mobility shows a much weaker effect. For this reason, the electrophoretic mobility is a poor indicator of the stability of a suspension (11).

6.3.1.1 *Coal Slurries*

Coal slurries can be dried by a two component system consisting of an anionic flocculant, such as copolymers of acrylamide and acrylic acid (AA), followed by a coagulant, which is a copolymer from VAm and vinyl alcohol (12).

6.3.1.2 *Food Processing Wastewater*

Food processing wastewater containing suspended solids can be cleared by the addition of PVAm. The suspended solids are coagulated, flocculated, and separated from the wastewater (1).

6.3.1.3 *Paper Wastewaters*

Wastewaters and circulation waters in the papermaking industries can be treated by adsorption of water-soluble anionic substances by finely divided adsorbents which consist of popcorn polymers that contain VAm units.

The insoluble popcorn polymers can be used for removing anionic compounds, such as water-soluble polyanionic substances and colored substances, as well as insoluble impurities and resins from circulation waters and wastewaters in the papermaking process. The insolubility combines the advantages of an adsorbent with the advantages of a cationic polymer (6). The popcorn polymers are the finely divided to a mean particle diameter from 0.5 to 5 *mm*.

In acidic to weakly alkaline waters, as encountered in the paper industry, the amino groups on the large surface of the polymers are for the mostly protonated and carry a positive charge. Due to the high charge density, all anionic substances which come into contact with the cationic surface, are immediately adsorbed regardless of their chemical composition. This is true also for anionic water-insoluble dispersed particles such as abietic acid or phenol resins.

6.3.1.4 Oil-Containing Wastewaters

Wastewater that is produced from steel and aluminum mills emerging from hot rolling mills contains lubricating and hydraulic pressure hydrocarbons. Wastewater from cold rolling mills contains oil that lubricates the sheets and reduces rust formation. For example, in cold rolling mills, oil-in-water emulsions are sprayed on the metal during rolling to act as coolants.

The emulsified oil in the wastewater is typically present in the range of several hundred to tens of thousands of *ppm*. It is critical to remove this oil from an environmental standpoint. Historically, dry polymers, solution polymers, inverse emulsion latexes, and metal ions have been used to treat the produced water. Each material has its own advantages and disadvantages.

While dry polymers have the benefit of being extremely concentrated, thereby reducing shipping costs, the equipment to dissolve the polymers is expensive and is not available to all end-users on site. Latex polymer preparations include 30–35% solids dispersed in oil. The latex polymer must be also inverted prior to use.

Numerous problems associated with this feeding method have caused many customers to avoid latex polymers. In addition, the latexes generally have a very narrow treating range often resulting in over-treatment at higher dosages. Furthermore, latex polymers add even more oil to the stream to be treated. Of course, adding more oil is something most customers would not want to do when treating their wastewater streams.

Although solution polymers require no prior make up, the percent solids and molecular weight are severely limited due to the nature of the material. These materials are often used to break oil-in-water emulsions but they are unable to flocculate the dispersed oil, thus requiring another chemical to accomplish this.

Metal ions, such as, Fe^{3+}, Zn^{2+}, Al^{3+}, etc., have long been used to break oil-in-water emulsions but recent government regulations have restricted their levels in discharged streams. Although effective at breaking oil-in-water emulsions, they too, require another chemical to flocculate the oil.

The use VAm polymers offers many solutions to these problems. The polymers are water soluble and, unlike latex polymers, there is no oil solvent. This is important, because (13):

- The polymers do not present a fire hazard
- Oil is not added to the water to be treated, which is more environmentally friendly
- Dissolution of the polymer requires only the addition of water; no special activators are needed
- The ability for these materials to dissolve or to invert is superior to that of oil dispersion latexes, and
- The polymers may be diluted to virtually any concentration by using water.

6.3.2 *Anti-scaling Agents*

Low-molecular weight phosphonates are often used as anti-scaling agents. Examples are summarized in Table 6.2.

Table 6.2 Phoshonic acid Based Anti-scaling Agents

Compound
Nitrilotrismethylene phosphonic acid
Ethylenediaminetetramethylene phosphonic acid
Hexamethylenediaminetetramethylene phosphonic acid
Diethylene triaminepentamethylene phosphonic acid
Hydroxyethane diphosphonic acid

The main disadvantage of low-molecular weight compounds is that they precipitate in the form of their calcium salts. In practice high calcium concentrations are encountered sometimes.

Phosphonomethylated PVAms can be used as anti-scaling agents (8). These polymeric phosphono compounds are more compatible with calcium ions owing to their higher molecular weights and more effectively way they are absorbed on the sewage sludge in water treatment plants. They are capable of suppressing the scale formation processes so that shutdown times for cleaning the plant, for example by boiling out, can be significantly reduced.

A further application of phosphonomethylated PVAms is their use in detergent compositions. There they act as builders, complexing agents, bleach stabilizers, and incrustation inhibitors.

6.3.3 Water Absorbent Materials

Water-absorbing resins are used in a wide field as summarized in Table 6.3.

Table 6.3 Fields of Use of Water-absorbing Resins

Application
Sanitary and hygienic goods
Wiping cloths
Water-retaining agents
Dehydrating agents
Sludge coagulants
Disposable towels and bath mats
Disposable door mats
Thickening agents
Disposable litter mats for pets
Condensation-preventing agents
Release control agents for various chemicals

Water-absorbing resins are available in a variety of chemical forms, for example, hydrolysis products of starch acrylonitrile graft polymers, carboxymethyl cellulose, crosslinked poly(acrylate)s, sulfonated PSs, hydrolyzed poly(acrylamide)s, poly(vinyl alcohol)s, poly(ethylene oxide)s, poly(vinylpyrrolidine)s, and poly(acrylonitrile).

Such a water-absorbing resin is also termed as a superabsorbent polymer (SAP), SAPs are typically slightly crosslinked hydrophilic polymers. SAPs are capable of absorbing and retaining large amounts of aqueous fluids equivalent to many times their own weight, even under moderate pressure. The ability to absorb aqueous fluids under a confining pressure is an important requirement for a SAP used in a hygienic article, like a diaper. The dramatic swelling and absorbent properties of SAPs are attributed to (3)

1. Electrostatic repulsion between the charges along the chains of the polymer
2. Osmotic pressure of the counter ions.

The absorption properties are drastically reduced in solutions containing electrolytes, such as saline, urine, and blood. The salt poisoning effect has been explained as follows. Water-absorption and water retention characteristics of SAPs are attributed to the presence of ionizable functional groups in the polymer structure. The ionizable groups typically are carboxyl groups, a high proportion of these groups are in the salt form when the polymer is dry.

The groups undergo dissociation and solvation upon contact with water. In the dissociated state, the polymer chains contain a plurality of functional groups having the same electric charge and thus repel one another. This electronic repulsion leads to expansion of the polymer structure which, in turn, permits further absorption of water molecules. Polymer expansion, however, is limited by the crosslinks in the polymer structure, which are present in a sufficient number to prevent solubilization of the polymer.

The presence of a significant concentration of electrolytes interferes with the dissociation of the ionizable functional groups and leads to the salt poisoning effect. Dissolved ions, such as sodium and chloride ions, therefore, have two effects on SAP gels. The ions screen the polymer charges and the ions eliminate the osmotic imbalance due to the presence of counter ions inside and outside of the gel. The dissolved ions, therefore, effectively convert an ionic gel into a non-ionic gel and the swelling properties are lost.

PVAm-based superabsorbent gels can be used in conjunction with an acidic water-absorbing resin, like poly(acrylic acid) (PAA), to help overcome the salt poisoning effect. A salt of a PVAm polymer can also be used alone as a SAP.

PVAm does not function as a SAP in its neutral form because there is no ionic charge on the polymer. Thus, the driving force for water absorption and retention is lacking. PVAm can be converted to the hydrochloride salt by treatment with hydrochloric acid. Another possibility is the use together with an acidic water-absorbing resin, like a PAA, then PVAm behaves like a SAP. Such two component SAP materials are considered as an improved class of SAPs. Typically, one component is a water-absorbing resin with the second component acting as an ion exchanger to remove electrolytes from an aqueous media.

In a modification of this principle, two uncharged slightly crosslinked polymers, each of which is capable of swelling and absorbing aqueous media can be used. When contact is made with water the two uncharged polymers neutralize each other to form a superabsorbent material.

Neither the polymer in its uncharged form behaves as a SAP by itself when it comes in contact with water. Thus, such a two component superabsorbent material contains two resins, one acidic and one basic, which are capable of acting as an absorbent material in their polyelectrolyte form. While PAA makes up the acidic resin, PVAm is the basic resin.

6.3.3.1 *Preparation of a Poly(N-vinylamine) SAP*

Preferred comonomers for NVF for crosslinking are *N,N'*-methylenebisacrylamide, *N,N'*-methylene bismethacrylamide, ethylene glycol dimethacrylate, and trimethylolpropane triacrylate (3). In this case, the crosslinks are formed during radical polymerization.

Water-soluble initiators can be used to start the radical polymerization, such as 2,2'-azobis-(2-amidinopropane)dihydrochloride. The reaction temperature is increased from 45°C in steps up to 65°C for the completion of the polymerization reaction. The hydrolysis is conducted in presence of sodium borohydride solution which acts as a scavenger for potential impurities that could effect gelling in the course of hydrolysis.

PVAm as such, either linear or slightly crosslinked, can be further crosslinked by multifunctional aldehydes such as glutaraldehyde or by ethylene glycol diglycidyl ether (EGDGE). A special technique to achieve this type of crosslinking is referred to as surface crosslinking. Surface crosslinking is achieved by spraying PVAm particles with an isopropanol solution of a surface crosslinking agent to wet only the outer surfaces of the PVAm particles.

Surface crosslinking and drying of the polymer is then performed, preferably by heating the wetted surfaces of the PVAm particles. Often a slightly crosslinked PVAm is subjected to a surface crosslinking. It has been found that surface crosslinking of a PVAm enhances the ability of the polymer to absorb and retain aqueous media under load.

The two component SAP is prepared by mixing powdered PVAm with a particle size of 180–710 μm with a lightly crosslinked PAA with a particle size of 210–710 μm in a weight ratio of 25% PVAm to 75% PAA.

6.3.4 *Papermaking*

VAm based polymers are used in several fields of papermaking:

1. Retention and drainage aids
2. Creping adhesives
3. Dry strength enhancers.

6.3.4.1 *Retention and Drainage Aids*

The manufacture of paper or paperboard involves the processing of an aqueous pulp fiber suspension often referred to as the *furnish* to produce a uniform dry paper sheet (14). Numerous additives which affect the final sheet properties of the finished paper are used to treat the furnish. For example, pigments, sizing agents, fillers are commonly added to the furnish to improve brightness, opacity, color, and ink receptivity. Other common additives include starches, polymers, china clay, titanium dioxide, optical brighteners, etc.

Retention is a term used in papermaking to denote the extent to which the pulp fibers, contaminants, and papermaking additives, which are added to the furnish are retained in the finished paper. The retention of pulp fibers, sizing agents, fillers, anionic trash, other anionic contaminants from recycled paper or other additives in the paper sheet during its formation in a papermaking machine is an important requirement to paper makers.

A retention aid acts by increasing the flocculating tendency of the pulp fibers, contaminants, and additives to inhibit their loss during drainage through the paper machine wires or screens. A high degree of retention is advantageous. However, too strong flocculation may be disadvantageous as it may result in a poor paper appearance.

Water-soluble VAm copolymers in combination with microparticles are suitable drainage and retention aids in the paper manufacturing process. The combination of a high molecular weight polymeric flocculating agent together with microparticles is often referred to as a *microparticle system*.

A copolymer from NVF and *tert*-butyl acrylamide is obtained with 2,2′-azobis-(*N*,*N*′-dimethyleneisobutyramidine) dihydrochloride as a radical initiator in aqueous solution. The polymerization is carried out from 55–60°C in nitrogen atmosphere. The resulting polymer is then hydrolyzed with hydrochloric acid (14). The polymer is combined with bentonite or organic crosslinked ionic microbeads (15) to form a microparticle system.

6.3.4.2 *Creping Adhesives*

Wet strength resins are often added to paper and paperboard at the time of manufacture. In the absence of wet strength resins, paper normally retains only 3–5% of its strength after being wetted with water. However, paper made with wet strength resin generally retains at least 10%–50% of its strength when wet. Wet strength is useful in a wide variety of paper applications, some examples of which are towelling, milk and juice cartons, paper bags, and liner board for corrugated containers.

The creping process typically involves adhering a cellulose web to a rotating creping cylinder, known as a Yankee dryer, and then dislodging the adhered web with a doctor blade. The impact of the web against the doctor blade ruptures some of the fiber-to-fiber bonds within the web and causes the web to wrinkle or pucker.

A mixture of a poly(amino amide) and a VAm copolymer can be reacted with epichlorohydrin as crosslinking agent in order to serve as creping adhesive (16). The poly(amino amide) is prepared from adipic acid and diethylenetriamine. The VAm copolymer is composed from 6 *mol*-% VAm and 94 *mol*-% vinyl alcohol.

Fibrous webs can be creped using the composition by (16).

1. Applying the composition described above to a drying surface for the web or to the web

2. Pressing the fibrous web against the drying surface to effect adhesion of the web to the drying surface
3. Dislodging the web from the drying surfaces with a creping device such as a doctor blade to crepe the fibrous web.

Poly(vinyl alcohol/VAm) copolymers can be crosslinked with ammonium zirconium carbonate (17). The crosslinking density can be adjusted by the ratio of the comonomers, the molecular weight, and the amount of crosslinking agent. Increased crosslinking generally will increase the glass transition temperature, the brittleness, and the hardness. Further, increased crosslinking provides a different response to mechanical stresses than uncrosslinked polymers.

6.3.4.3 Dry Strength Enhancers

The reaction products of starch with cationic polymers are suitable dry strength enhancers for paper. Examples of suitable polymeric cationizers are PVAms with high molecular weight.

A drainage aid such as an ethyleneimine-grafted water-soluble poly(amidoamine) is incorporated into the formulation. The poly-(amidoamine) is formed from adipic acid and a triamine and crosslinked with a bischlorohydrin ether. A modified anionic maize starch is produced by heating oxidized maize starch having carboxyl groups as anionic groups together with the poly(amidoamine) and the PVAm cationizer in aqueous solution (18).

The modified starches obtained are used as paper, paperboard, and cardboard dry strength enhancers. They are added to the paper stock in amounts of from 1–6%, based on dry paper stock.

6.3.5 *Tanning Materials*

In the making of leather the hide is customarily tanned with mineral tanning materials, such as basic chromium, aluminum, and zirconium salts either singularly or combined with synthetic tanning materials. A subsequent retanning with natural or synthetic tanning materials has the purpose of improving leather properties such as hand, softness, grain constitution, and fullness. Copolymers of NVF with maleic anhydride or AA are suitable retanning materials (19). The polymers are hydrolyzed to set free the amine group.

6.3.6 Delayed Drug Release

The use of the VAm moiety along with other monomers has been described in a wide variety of medical applications in hydrogels (20). Oral drug forms with a delayed release of the active ingredient have the advantage owing to a reduced frequency of intake, more uniform levels of the medicinal substance in the blood, and the avoidance of local irritation (21).

Coated slow-release forms are formulations of cores containing the medicinal substance. They are coated with a film which is either insoluble in water but semipermeable, or contains pores through which the medicinal substance diffuses. Besides coated slow-release forms, it is also possible to achieve control and prolongation of release by embedding the medicinal substance in a matrix. Embedding the medicinal substance in a matrix offers the advantage of simple and low cost production and high drug safety because dose-dumping effects cannot occur.

The ancillary substances which are usually employed for this, such as hydroxypropyl cellulose, hydroxypropyl methyl cellulose, alginic acid, alginates, and xanthan, exhibit some disadvantages with use. These derive, on the one hand, from the release of medicinal substance being dependent on the pH or ionic strength and, on the other hand, from the unsatisfactory direct tabletability. This is because the small binding effect and the poor flow properties of these polymers, in a low hardness or are inhomogeneous.

Moreover, since some of these ancillary substances are of natural origin, variations in batch conformity may occur and thus an unfavorable effect on the performance of the pharmaceutical preparation (21).

It has been proposed to use NVP or NVF together with stearyl acrylate as a matrix for controlled drug release formulations. Thus, both hydrophilic and hydrophobic components are included into the polymer. The rate of release of the active ingredients can be altered as required as a function of amount, composition, and of the swelling properties of the polymers used.

6.3.7 *Biomaterial Surfaces*

Synthetic materials used in blood-contacting medical devices can promote thrombotic phenomena that are induced by the surface properties. A thrombus can be formed that impairs the function of an implanted device. Therefore, the surface properties of blood contacting synthetic biomaterials are to be modified to become less thrombogenic.

The surface of natural biological cells termed as glycocalyx is not thrombogenic. From this consideration, polymers that mimetic the glycocalyx have been developed (22). The polymers are of the PVAm type. The PVAm is modified by the reaction with *N*-(hexanoyloxy)succinimide and dextran lactone. Thus, pendant hexanoyl and dextran aldonamide groups are obtained.

6.3.8 *Biocides*

It has been demonstrated that VAm polymers and copolymers modified with γ-butyrolactone of ϵ-caprolactone show biocidal activity (23). The minimum effective concentration value is the lowest polymer concentration that can prevent the bacterial growth over a period of 100 *h*. The results for PVAm are shown in Table 6.4.

The biocidal action is determined by measurements in a Malthus apparatus (23). The bacteria are allowed to grow in a tryptone soya broth with different concentrations of biocidal polymer. The growing cells produce CO_2 which is trapped in a KOH solution and thereby changing the electric conductivity of the KOH solution. The alteration of the conductivity is measured by the Malthus apparatus. The detection time for the start of growth is recorded and evaluated.

Modified PVAms are prepared as follows: To an aqueous solution of PVAm, ϵ-caprolactone is added dropwise within 20 *min* under vigorous stirring. Subsequently, the reaction mixture is heated to 80°C for 5 *h* (24). The ϵ-caprolactone reacts with some of the pendent amine groups to functionalize the polymer as composed from *N*-vinyl-6-hydroxycapramide units. In the same way, γ-butyrolactone has been used to functionalize the amine groups. These polymers show a similar biocidal activity than pure PVAm.

It has been found that in a laundry detergent composition, the combined action of an oxidoreductase enzyme and a PVAm, results

Table 6.4 Antibacterial Activity of Poly(*N*-vinylamine) (23)

Bacterium	MIC[a]/[*ppm*]
Gram-positive test microorganisms	
Bacillus subtilis	500
Listeria monocytogenes	1000
Staphylococcus aureus	500
Streptococcus mutans	500
Gram-negative test microorganisms	
Escherichia coli	4000
Pseudomonas aeruginosa	4000
Pseudomonas fluorescens	2000
Shewanella putrefaciens	1000
Vibrio parahaemolyticus	1000

[a] Minimum effective concentration (MIC) is the lowest polymer concentration in each case, which prevents the growth over a period of 100 *h*

in synergistic antimicrobial effect (25). In addition, the biocidal action of PVAm modified with epoxides, has been described. In this case, the polymers contain pendent β-hydroxyalkyl-VAm units (26). These polymers serve also as emulsifiers for aqueous filler suspensions and as corrosion inhibitors for metals (27).

6.3.9 *Chromatographic Supports*

Crosslinked PVAm is used as a coating for capillaries used in capillary electrophoresis (CE) for polyanionic acids (28). In the analysis of ions with high mobility by CE with an uncoated capillary, a serious problem emerges as these substances can migrate against the electroosmotic flow (EOF) because of their electrophoretic mobility. When the electrophoretic velocity is in the same magnitude of the flow, these substance would never leave the column.

The EOF and electrophoretic mobility change in the same way with pH. Therefore, changing the pH would not cause elution of the respective substances. However, in capillaries with a reverse EOF, both electroosmotic and electrophoretic forces drive the substances towards the anode. This results in shorter elution times. The EOF in the capillary can be reversed by the addition of cationic surfactants to the electrophoretic buffer. In contrast, no additives are needed when the wall of the capillary is modified with a coating that effects a strong anodic EOF.

For the analytical application, a PVAm with a special grade of purity is synthesized by polymerization and acidic hydrolysis of *N*-vinyl-*tert*-butylcarbamate. The polymer is dialyzed to remove the low-molecular products of hydrolysis. The capillary is modified by allowing to absorb the polymer on the wall followed by crosslinking with *N,N'*-methylenebisacrylamide and endcapping of the primary amino groups with an acrylamido derivative bearing quaternary amino functions, such as *N,N,N*-trimethylaminoethylacrylamide.

Crosslinking is necessary to improve the service time of the capillary. Endcapping of the primary amino groups is necessary to avoid the adsorption of the organic acids that are analyzed.

6.4 Suppliers and Commercial Grades

Suppliers and commercial grades are shown in Table 6.5.

Table 6.5 Examples for Commercially Available Vinylamine Polymers

Tradename	Producer	Remarks
Celvam®	Celanese International Corp.	Vinylamine polymers for papermaking, et. al.
Diafix®	Mitsubishi Chemical Corp.	Vinyl alcohol/vinylamine copolymers for hair care
Airvam®	Air Products and Chemicals, Inc.	Vinylamine polymers for papermaking

6.5 Safety

N-Vinylacetamide is an irritant to skin, eyes, and respiratory system. It can be absorbed through the skin. Similarly, NVF is an irritant, with additional serious risk to the eyes. About the corresponding polymer essentially no safety data are found in the literature.

Tradenames appearing in the references are shown in Table 6.6.

Table 6.6 Tradenames in References

Tradename Description	Supplier
Airvol-107 Poly(vinyl alcohol) (MW 40,000) (17)	Air Products and Chemicals, Inc.
Airvol-200 Poly(vinyl alcohol) (MW 40,000) (17)	Air Products and Chemicals, Inc.
Airvol-350 Poly(vinyl alcohol) (MW 155,000) (17)	Air Products and Chemicals, Inc.
Airvol-540 Poly(vinyl alcohol) (MW 155,000) (17)	Air Products and Chemicals, Inc.
Amrezs Poly(aminoamide)-epichlorohydrin (17)	Georgia-Pacific Corp.
Bacote® 20 Ammonium zirconium carbonate solution, crosslinking agent (17)	Magnesium Elektron, Inc.
Cascanid Poly(aminoamide)-epichlorohydrin (17)	Borden Chemicals
Crepetrol® (Series) Poly(aminoamide)/epichlorohydrin, creping adhesive (16)	Hercules, Inc.
Kollidon® CL Poly(vinylpyrrolidone), crosslinked, super-disintegrant in tablets (21)	BASF AG
Kymene® 450 Adipic acid, polymer with 1-chloro-2,3-epoxypropane and diethylenetriamine, wet strength resin (16)	Hercules, Inc.
Kymene® 557 LX Poly(aminamide)-epichlorohydrin condensation adduct cationic polymer (16)	Hercules, Inc.
Kymene® 557H Adipic acid-diethylenetriamine copolymer, reaction product with epichlorohydrin, Wet strength resin (16)	Hercules, Inc.
Kymene® 736 Amine polymer-epichlorohydrin adduct, wet strength resin (16)	Hercules, Inc.
Nalco® 7134 Low molecular weight poly(amine), wastewater refinery (13)	Nalco Chemical Comp.
Nalco® 8105 Diallyldimethylammonium chloride/acrylamide copolymer, polymeric flocculant (13)	Nalco Chemical Comp.

Table 6.6 (cont.)

Tradename Description	Supplier
Nalco® 9806	Nalco Chemical Comp.
Polymeric flocculant (12)	
Nalco® 9810	Nalco Chemical Comp.
Water-in-oil emulsion of a sodium acrylate-acrylamide copolymer, partially hydrolyzed poly(acrylamide), flocculant (12)	
Polymin® SK	BASF AG
Ethyleneimine-grafted water-soluble poly(amidoamine) formed from adipic acid and a triamine and crosslinked with a bischlorohydrin ether (18)	
Primid® XL-552	Ems Chemie AG
β-Hydroxyalkylamide (3)	
Quasoft® 202-JR	Quaker Chemical Corp.
Mixture of linear amine amides and imidazolines, softener (17)	
Quasoft® 209-JR	Quaker Chemical Corp.
Mixture of linear amine amides and imidazolines, softener (17)	
Rezosol ® 8290	Houghton Co.
Poly(aminoamide)-epichlorohydrin, adhesive (17)	

References

1. A.G. Sommese and D.K. Chung, Use of polymers containing vinylamine/vinylformamide for the treatment of food processing wastewater, US Patent 6 610 209, assigned to Ondeo Nalco Company (Naperville, IL), August 26, 2003.
2. N. Bicak, D.C. Sherrington, and H. Bulbul, Vinylamine polymer via chemical modification of PVC, *Eur. Polym. J.*, 37(4):801–805, April 2001.
3. M.A. Mitchell, T.W. Beihoffer, and R.S. Sultana, Poly(vinylamine) - base superabsorbent gels and method of manufacturing the same, US Patent 6 603 055, assigned to BASF Aktiengesellschaft (Ludwigshafen, DE), August 5, 2003.
4. A. Ernst, H. Meffert, A. Sanner, S. Stein, and F. Ruchatz, Styrene-containing popcorn polymers, method for producing same and utilisation, US Patent 6 525 156, assigned to BASF Aktiengesellschaft (Ludwigshafen, DE), February 25, 2003.
5. H. Hartmann, W. Denzinger, M. Kroener, and C. Nilz, Insoluble, only slightly swellable polymers containing amino groups, their prepara-

tion and their use, US Patent 5 599 898, assigned to BASF Aktiengesellschaft (Ludwigshafen, DE), February 4, 1997.

6. F. Linhart, M. Niessner, M. Rubenacker, and C. Nilz, Process for cleaning wastewater and recirculating water in paper production, de-inking and bleaching of pulp, US Patent 6 190 503, assigned to BASF Aktiengesellschaft (Ludwigshafen, DE), February 20, 2001.
7. J. Utecht, M. Niessner, S. Weiguny, P. Lorencak, and A. Stamm, Method for producing polymers containing carbamate units and the use thereof, US Patent 6 184 310, assigned to BASF Aktiengesellschaft (Ludwigshafen, DE), February 6, 2001.
8. J. Mohr, K. Oppenlaender, W. Denzinger, H. Hartmann, R. Baur, C. Gousetis, and A. Kud, Phosphonomethylated polyvinylamines and preparation and use thereof, US Patent 6 264 839, assigned to BASF Aktiengesellschaft (Ludwigshafen, DE), July 24, 2001.
9. C. Yi, Z. Wang, M. Li, J. Wang, and S. Wang, Facilitated transport of CO_2 through polyvinylamine/polyethlene glycol blend membranes, *Desalination*, 193(1-3):90–96, May 2006.
10. L.S. Eoff and M.J. Szymanski, Well treatment fluid and methods for blocking permeability of a subterranean zone, US Patent 7 128 148, assigned to Halliburton Energy Services, Inc. (Duncan, OK), October 31, 2006.
11. W.L. Yu, F. Bouyer, and M. Borkovec, Polystyrene sulfate latex particles in the presence of poly(vinylamine): Absolute aggregation rate constants and charging behavior, *J. Colloid Interface Sci.*, 241(2):392–399, September 2001.
12. A.G. Sommese and K.J. Pillai, Vinylamine copolymer coagulants for use in coal refuse dewatering, US Patent 5 622 533, assigned to Nalco Chemical Company (Naperville, IL), April 22, 1997.
13. A.G. Sommese and A. Sivakumar, Polymers containing vinylamine/vinylformamide as demulsifiers in oily wastewaters, US Patent 5 702 613, assigned to Nalco Chemical Company (Naperville, IL), December 30, 1997.
14. L.L. Kuo, R.Y. Leung, S.R. Prescott, and T. Hassler, Production of paper and paperboard, US Patent 6 273 998, assigned to Betzdearborn Inc. (Trevose, PA), August 14, 2001.
15. D.S. Honig and E. Harris, Charged organic polymer microbeads in paper-making process, US Patent 5 274 055, assigned to American Cyanamid Company (Stamford, CT), December 28, 1993.
16. W.W. Maslanka, Resin compositions for making wet and dry strength paper and their use as creping adhesives, US Patent 5 994 449, assigned to Hercules Incorporated (Wilmington, DE), November 30, 1999.

17. P.V. Luu, C.M. Neculescu, and D.M. Mews, Crosslinkable creping adhesive formulations, US Patent 6 815 497, assigned to Fort James Corporation (Atlanta, GA), November 9, 2004.
18. P. Lorencak, A. Stange, K. Diehl, and N. Mahr, Modifying starch with cationic polymers and use of the modified starches as dry-strength agent, US Patent 6 746 542, assigned to BASF Aktiengesellschaft (Ludwigshafen, DE), June 8, 2004.
19. A. Negele, G. Wolf, A. Kistenmacher, and G. Igl, *N*-vinyl-containing polymeric tanning materials, US Patent 6 652 597, assigned to BASF Aktiengesellschaft (Ludwigshafen, DE), November 25, 2003.
20. A.S. Hoffman, Hydrogels for biomedical applications, *Adv. Drug Deliv. Rev.*, 54(1):3–12, January 2002.
21. H. Meffert and F. Ruchatz, Use of *N*-vinyllactam or *N*-vinylamine containing copolymers as matrix for producing solid pharmaceutical and cosmetic presentations, US Patent 6 436 440, assigned to BASF Aktiengesellschaft (Ludwigshafen, DE), August 20, 2002.
22. A. Sen Gupta, S. Wang, E. Link, E. Anderson, C. Hofmann, J. Lewandowski, K. Kottke-Marchant, and R. Marchant, Glycocalyx-mimetic dextran-modified poly(vinyl amine) surfactant coating reduces platelet adhesion on medical-grade polycarbonate surface, *Biomaterials*, 27(16):3084–3095, June 2006.
23. N. Gebhardt, D. Zeller, C. Nilz, U. Steuerle, and C. Johansen, Use of polymers as biocides, US Patent 6 261 581, assigned to BASF Aktiengesellschaft (Ludwigshafen, DE), July 17, 2001.
24. J. Tropsch, D. Zeller, A. Negele, N. Mahr, and J. Decker, Use of water-soluble polymers as biocides, US Patent 6 458 348, assigned to BASF Aktiengesellschaft (Ludwigshafen, DE), October 1, 2002.
25. C. Johansen, Methods for laundry using polycations and enzymes, US Patent 6 287 585, assigned to Novozymes A/S (Bagsvaerd, DK), September 11, 2001.
26. J. Tropsch, D. Zeller, A. Negele, N. Mahr, and J. Decker, Use of polymers containing β-hydroxyalkylvinylamine units as biocides, US Patent 6 214 885, assigned to BASF Aktiengesellschaft (Ludwigshafen, DE), April 10, 2001.
27. J. Utecht, M. Niessner, D. Monch, and M. Rubenacker, Polymers containing β-hydroxyalkylvinylamine units, preparation and use thereof, US Patent 6 057 404, assigned to BASF Aktiengesellschaft (Ludwigshafen, DE), May 2, 2000.
28. M. Chiari, L. Ceriotti, G. Crini, and M. Morcellet, Poly(vinylamine)-coated capillaries with reversed electroosmotic flow for the separation of organic anions, *J. Chromatogr. A*, 836(1):81–91, March 1999.

7

Poly(vinylpyridine)

The first patent on the polymerization of vinylpyridine appeared in the 1950th and the polymer was used for coating of photographic material (1).

7.1 Monomers

Monomers and comonomers are summarized in Table 7.1 and Figure 7.1.

Table 7.1 Monomers

Monomers	Remarks
2-Vinylpyridine	Basic monomer
4-Vinylpyridine (4-VPy)	Most common
2,6-Divinylpyridine	Crosslinking agent
Comonomers	**Remarks**
Butadiene	Latex compositions
Styrene	Latex compositions
Acrylic acid	Adhesion promoter
Acrylonitrile	Grafting aid
Ethylene glycol dimethacrylate	Crosslinking agent
Divinylbenzene	Crosslinking agent

2-Vinylpyridine (2-VPy) is an important monomer used in synthesizing various polymers. Butadiene and styrene monomers are used with 2-VPy to form a latex terpolymer that is used for bonding fabric cords to the rubber matrix of tires. Combining methanol and 2-VPy, 2-(2-methoxy-ethyl)pyridine is a veterinary anthelmintic.

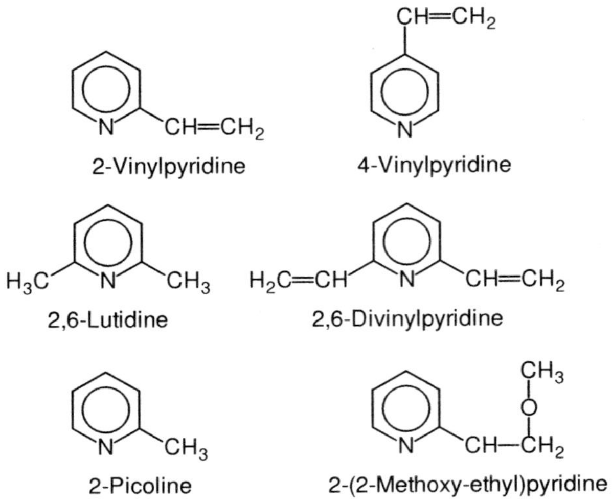

Figure 7.1 Vinylpyridine Monomers and Precursors

Initially pyridines (Py)s were isolated from coal tars from coking operations. The synthesis of 2-VPy can be achieved by a two-step procedure which involves a base catalyzed addition of 2-picoline to formaldehyde to give 2-(2-hydroxyethyl)pyridine followed by the dehydration to 2-VPy monomer. Modified zeolites may serve as a catalyst. 3-Picoline is an intermediate in the production of niacin, one of the B vitamins. Commercial 2-VPy is prepared from acetylene and acrylonitrile using a cobaltocene catalyst or by oxidative dehydrogenation of 2-ethylpyridine on a Cr/Nb catalyst (2).

2,6-Divinylpyridine (2,6-DVPy) is used in the preparation of ion exchange resins containing divinyl substituted heterocyclic comonomers as crosslinking agents. 2,6-DVPy and 2-methyl-6-vinylpyridine were synthesized by the condensation of 2,6-lutidine and formaldehyde using potassium salts as catalysts. Zeolites are also suitable catalysts (3). A wide variety of Pys are produced by Koei Chemical Comp.

7.2 Polymerization and Fabrication

7.2.1 *Suspension Polymerization*

4-VPy can be radically polymerized in aqueous suspension. A suitable suspension agent can be poly(*N*-vinyl-2-pyrrolidone) having a K value of around 30 or a poly(vinyl alcohol) (PVA), or other inert water-soluble polymers (4). Two radical initiators are added. In the initial stage, *tert*-butyl peroxypivalate (Lupersol® 11) is added at 85°C. After 1 *h* polymerization time, 2,5-dimethyl-2,5-di-(*tert*-butylperoxy)hexane (Lupersol® 101) is added and the temperature is raised to 110.°C. The reaction is complete after 8 *h* (5).

7.2.2 *Quaternization*

The nitrogen in poly(4-vinylpyridine) (P4-VPy) can undergo quaternization. Quaternized water-soluble vinylpyridine carboxylate polymers can be obtained by the reaction with aqueous of sodium chloroacetate (5). The reaction is complete at 85°C after 2 *h*.

The quaternization reaction can be also performed by the Michael Addition with α,β-unsaturated acids followed by adding water (6). The reaction is exemplified with crotonic acid in Figure 7.2. A betaine structure emerges.

7.2.3 *Solution Polymerization*

As an alternative to the synthesis by suspension polymerization, the polymerization and quaternization steps can be reversed where the vinylpyridine monomer can be initially quaternized and the quaternized monomeric product is then polymerized in the presence of a free radical initiator. Since the quaternized product is converted to a water-soluble state, no suspension agent is needed in the polymerization (7).

7.2.4 *Spontaneous Polymerization*

As styrene and methacrylate esters, 4-VPy undergoes spontaneous polymerization. However, the mechanism of spontaneous polymerization in presence of alkylating agents is of ionic type. In the

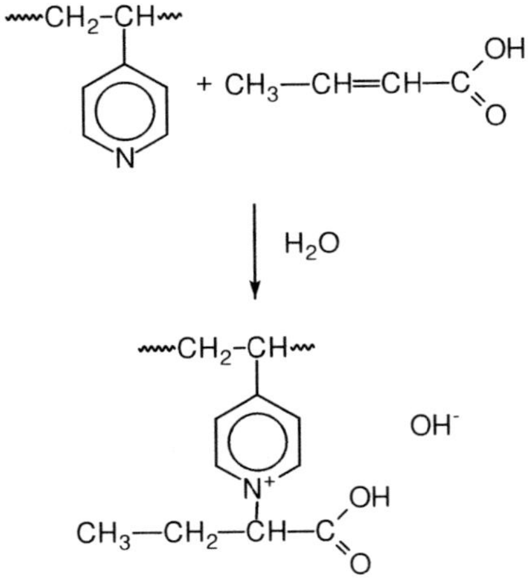

Figure 7.2 Michael Addition with Crotonic acid

presence of a strong nucleophilic counter ion, the initiation occurs by addition of the counter ion to the β-position of the double bound of the 4-vinylpyridinium salt to result in a resonance-stabilized zwitterion (8). The propagation occurs by a ionic mechanism.

The technique of spontaneous polymerization was used to prepare side chain liquid crystalline polymeric salts by quaternization with mesogenic groups (9,10). In contrast to the polymers obtained from 4-VPy, which always gives an atactic structure, the polymers obtained from 4-VPy in presence of quaternization agents have an isotactic character (11).

When acrylic acid (AA) and concentrated 4-VPy are mixed a considerable amount of heat is generated by mixing. The temperature rise effects a spontaneous copolymerization that results in carbonized products (12). When the monomers are diluted in a solvent, such as a mixture of ethanol and water, an alternating copolymer with a relatively low number average molecular weight (M_n) of 3.6 *k Dalton* is obtained.

7.2.5 *Dispersion Polymerization*

Manufacture of micron-sized monodisperse microspheres is difficult by conventional emulsion or suspension polymerization because these methods are unable to control the particle size and uniformity simultaneously. In contrast, dispersion polymerization is a valuable tool to prepare micrometer-scale monodisperse polymer particles in a single process.

In dispersion polymerization, the reaction mixture initially consists of monomers that are soluble in the dispersion medium (13). When the monomers react polymers are formed that are insoluble in the dispersion medium. Thus, dispersion polymerization is a kind of precipitation polymerization.

7.2.5.1 *Core-shell Polymer Particles*

Crosslinked core-shell polymer particles were synthesized by free radical dispersion copolymerization of 4-VPy and ethylene glycol dimethacrylate (EGDM) as a crosslinking reagent with methacryloyl-terminated poly(styrene) macromonomers (PS-M)s in cyclohexane and 1,4-dioxane (14). The PS-Ms are prepared by coupling living poly(styrene) (PS) endcapped with ethylene oxide, initiated by *n*-butyllithium, with methacryloyl chloride in a benzene tetrahydrofuran mixed solvent.

The PS-Ms act not exclusively as comonomer but function also as an emulsifier. The particle diameters can be controlled in the range 40–1600 *nm*. A possible application is the construction of polymeric superstructures that can be used as tunable wavelength modulators (15).

7.2.6 *Atom Transfer Radical Polymerization*

P4-VPy and a block copolymer from 4-VPy and styrene have been prepared by atom transfer radical polymerization (ATRP). As initiator, 1-phenylethyl chloride is used and as catalyst CuCl is used. 5,5,7,12,12,14-Hexamethyl-1,4,8,11-tetraazamacrocyclotetradecane (Me6[14]aneN4) is a suitable ligand (16). The preparation of the ligand is shown in Figure 7.3.

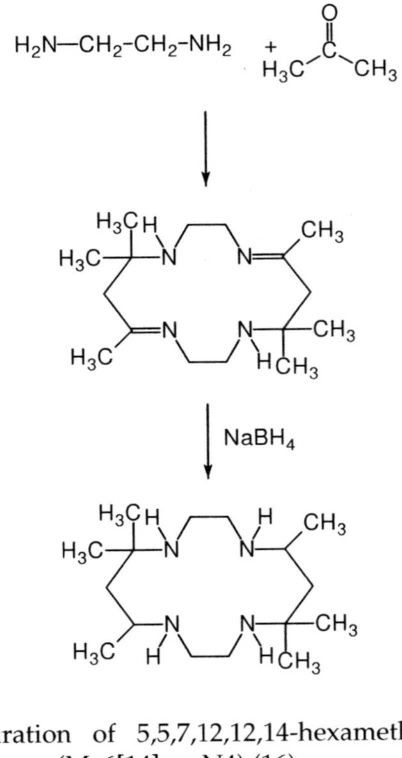

Figure 7.3 Preparation of 5,5,7,12,12,14-hexamethyl-1,4,8,11-tetraaza-macrocyclotetradecane (Me6[14]aneN4) (16)

The homopolymerization of 4-VPy is performed in 2-propanol at 40°C. The molecular weight of P4-VPy increases linearly with the conversion. A block copolymer with a low polydispersity index of $M_w/M_n = 1.2$ can be obtained by ATRP of styrene in *N,N*-dimethylformamide (DMF) at 110°C. In this case, a chloride quaternized P4-VPy is used as macroinitiator. CuCl with Me6[14]aneN4 is used as catalyst.

Functional polymers of this type and their block copolymers with controlled molecular weight and narrow molecular weight distribution, have attracted much interest, due to their potential applications as coordination reagents for transition metals.

7.2.7 *RAFT Polymerization*

Block copolymers containing styrene and 4-VPy have been prepared by a reversible addition-fragmentation chain transfer (RAFT) polymerization. In a first step, a polymeric chain transfer agent (CTA) suitable for RAFT polymerization is prepared by ordinary radical polymerization of styrene with 2,2′-azobisisobutyronitrile (AIBN) in the presence of 4-toluic acid dithiobenzoate as RAFT CTA (17). In the second step, 4-VPy is added and inserted into the active sites to form a block copolymer. The procedure is illustrated in Figure 7.4.

The block copolymer can be used as a stabilizer agent in the dispersion polymerization of methyl methacrylate (MMA) to get monodisperse poly(methyl methacrylate) (PMMA) particles. Methanol was used as a solvent for the monomer and AIBN was used as a radical initiator (17). Poly(styrene-*b*-4-VPy) is used as a stabilizer in dispersion polymerization.

7.2.8 *Electropolymerization*

The application of anodic or cathodic polarization to a highly concentrated solution of 4-VPy leads to the formation of a polymer film grafted on the metallic electrode. Acetonitrile or DMF is used as a solvent and the supporting electrolyte is tetraethylammonium perchlorate (18).

Divinylbenzene (DVB) can be used as a crosslinking agent. The crosslinking agent is added to get a network between the grafted

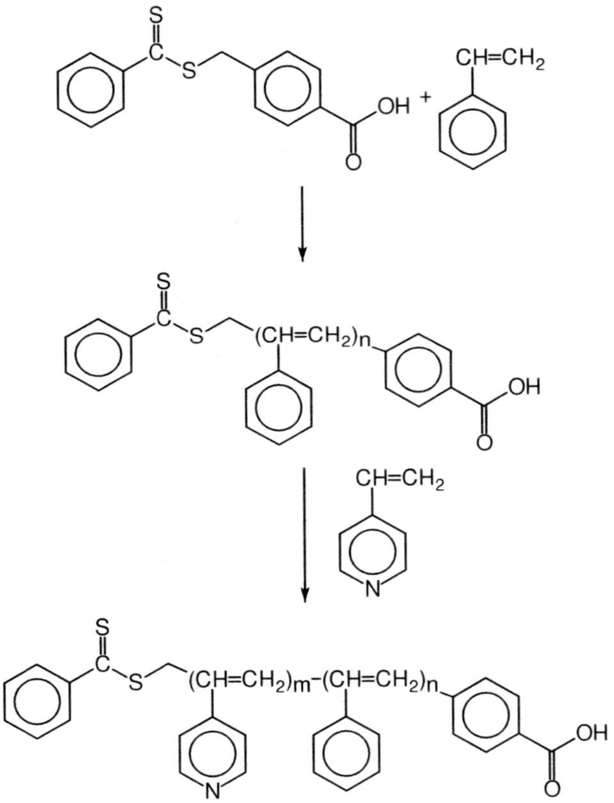

Figure 7.4 Block Copolymers from Styrene and 4-VPy by RAFT Polymerization (17)

polymeric chains, which leads to controlled and reproducible film thicknesses (19).

The mechanism of electropolymerization of poly(2-vinylpyridine) (P2-VPy) in aqueous solution has been elucidated. By surface-enhanced Raman scattering spectroscopy and cyclic voltametry (20), the 2-VPy monomer becomes protonated in acidic aqueous solutions and is selectively adsorbed on cathodic surfaces. The adsorbed ions then undergo electrochemical reduction to free radicals. These radicals initiate the polymerization reaction with neutral 2-VPy molecules that are also present in the solution.

Metallic substrates coated with P4-VPy films have a great potential to remove pollutants from aqueous solutions. The polymer easily complexes organic or mineral compounds.

Infrared reflection-absorption spectroscopy (IRRAS) is a well established technique for studying substances adsorbed on metal surfaces. When the samples are immersed in an aqueous solution of synthetic sulfate, the polymer swells and the Py complexes the copper ions. New Py vibrations at 1425 and 1617 cm^{-1} in the IRRAS spectra are observed. It is possible to extract copper ions from the polymer films by using concentrated aqueous ammonia solutions (19).

7.2.9 *Graft Polymerization*

Grafting of 4-VPy on to various polymers is performed in order to change the surface properties of these polymers. In particular, hydrophobic polymers can be modified to get a hydrophilic surface.

7.2.9.1 *Radiation Graft Polymerization*

Radiation-induced graft copolymerization of 4-VPy and its binary mixture with AN on to isotactic poly(propylene) (PP) has been performed in aqueous medium (21). The grafting in solution effects a considerable crosslinking (22). In contrast, solid phase grafting initiated by dibenzoyl peroxide minimizes crosslinking.

7.2.9.2 Graft Polymerization on Activated Sites

Poly(vinylidene fluoride) (PVDF) membranes are frequently used in microfiltration (MF) and ultrafiltration due to their good processability and well-controlled porosity. An improved membrane can be obtained by combining the superior bulk properties of the hydrophobic PVDF with hydrophilic materials. Such materials are obtained by grafting hydrophilic monomers on the surface of the hydrophobic bulk material.

Active sites on PVDF can be established by ozone treatment of PVDF. For this purpose, PVDF powders are dissolved in *N*-methyl-2-pyrrolidone (NMP) and a continuous stream of a mixture of ozone and oxygen was bubbled through the solution at 25°C. Afterwards, 4-VPy is grafted on to the active sites by a thermal graft copolymerization process in argon atmosphere (23). 2-VPy exhibits a higher graft copolymerization efficiency than 4-VPy (24).

From the polymer, a MF membrane is prepared by phase inversion in aqueous media. The polymer is dissolved in NMP and the solution is cast on to a glass plate, which is then immersed in an aqueous coagulation bath. Aqueous solution has a defined pH.

The surface graft concentration of the vinylpyridine moieties of the MF membrane increases with increasing pH value of the coagulation bath. The situation is reverse, when AA is used for grafting. In this way, MF membranes with different surface composition and different mean pore size can be obtained from the same copolymer by adjusting the pH of the casting bath. The permeability of the MF membranes is dependent on the pH of the filtration medium.

7.2.9.3 Plasma-induced Graft Polymerization

The surface of poly(ethylene terephthalate) (PET) track membranes (TM) can be modified with a plasma-induced graft polymerization technique. TM are polymer films with micrometer-sized pores of around 0.1–0.2 μ in diameter which are produced by heavy-ion bombardment.

2-Methyl-5-vinylpyridine in aqueous solution was grafted on to PET that was earlier exposed to a RF-discharge of 13.56 *MHz* at a gas pressure 0.13 *Pa* of air (25). To suppress homopolymerization, Cu^{2+} was added to the solution.

Depending on the time of plasma treatment, the pore sizes are reduced due to the grafted layer from initially 0.18 μ down to 0.05 μ after 10 *min* of plasma treatment. In addition, the thickness of the membrane increases. The permeability of water of the grafted polymeric membranes can be controlled by changing the pH of the filtrate. The nitrogen of the Py moiety can be present in the neutral form and in the ionized form. An increase of the pH favors the neutral nitrogen atom and modifies thus the properties of the membrane.

7.2.10 *Poly(vinylpyridine N-oxide)*

Copolymers of 4-VPy *N*-oxide can be obtained by copolymerizing 4-VPy with *N*-vinyl-2-pyrrolidone (NVP), *N*-vinyloxazolidone and other vinyl monomers and subsequently oxidizing the vinylpyridine units to vinylpyridine *N*-oxide units (26).

The oxidation of poly-4-VPy is performed in acetic acid by treatment with peracetic acid. Peracetic acid is obtained from acetic acid by mixing with hydrogen peroxide with sulfuric acid as catalyst (27). The mixture is kept at first for 30 *min* at ambient temperature and then heated to 80–85°C for 3 *h*. After completion of the reaction the polymer solution is mixed with acetone and a yellow brown viscous syrup is precipitated. This precipitate is washed with acetone to yield a pale crystalline solid.

7.3 Properties

7.3.1 *Miscibility*

Poly(vinyl phenyl ketone) where the keto groups are partly reduced to hydroxy groups and poly(styrene-*co*-4-VPy) are miscible in a wide range of composition (28). This can be detected by differential scanning calorimetry (DSC) and thermo mechanical analysis. The compositions show only a single glass transition temperature. This behavior results from the fact that the Py groups are strong proton acceptors.

The miscibility of PS with poly(butyl acrylate) is poor but can be improved by the introduction of 4-VPy monomer into the PS by copolymerization, as well as by the introduction of acrylate groups into poly(butyl acrylate). PSs with quaternary ammonium salt of 4-VPy and poly(butyl acrylate)s with potassium acrylate groups become miscible. The miscibility is raised by increasing ion content. When the ion content exceeds 11 *mol-%*, the polymers tend to become completely miscible (29).

Blends of P4-VPy with some proton donor polymers, such as hydroxylated poly(methacrylate)s (30), poly(vinyl acetate-*co*-vinyl alcohol) (31), poly(vinylphenol) (PVPh) (32) and poly(ethylene-*co*-vinyl alcohol) (33) are miscible.

Often, in miscible blends, hydrogen bonds between Py and hydroxyl groups are detected. In contrast, with immiscible blends, self-association of hydroxyl groups diminishes contacts between Py and hydroxyl groups, which cause the miscibility. In fact, the system P4-VPy and PVA results in immiscible blends (34).

The intermolecular interaction through hydrogen bonding can be characterized by fourier transform infrared spectroscopy (FTIR). Namely, specific interactions will affect the local electron density and the corresponding bands shift in frequency.

In blends of P2-VPy with aliphatic dicarboxylic acids, e.g., succinic acid, suberic acid, and dodecanedioic acid shifts of several infrared bands have been observed. The Py skeleton vibration band at 1588 cm^{-1} shifts to a higher frequency due to the formation of hydrogen bonds (35). New bands emerge at higher frequencies and their intensities increase with increasing acid content.

The carbonyl band of the diacids is found around 1700 cm^{-1}. This is at a lower frequency than an ordinary ester carbonyl band and is caused due to the dimerization of the carboxylic groups. When in the blend with poly(vinylpyridine) a hydrogen bonding interaction takes place and the carbonyl band shifts to a somewhat higher frequency. This shift is explained due to the disruption of the dimerization of the diacids.

FTIR studies on the hydrogen bonding interaction between the hydroxyl group of a phenolic resin and the Py ring of poly(vinylpyridine)s at various compositions and temperatures indicate a greater inter-association for P4-VPy than for P2-VPy (36). The absorption band under discussion is located in the range of 4000–2700 cm^{-1}. This difference arises from the steric hindrance effect in P2-VPy.

In addition to FTIR, X-ray photoelectron spectroscopy and solid state ^{13}C nuclear magnetic-resonance spectroscopy (NMR) are suitable experimental techniques to elucidate inter-polymer interactions in polymer blends and complexes. Using high-resolution solid state ^{13}C NMR indicate changes in chemical shift or line shape of the peaks.

Poly(methacrylic acid) (PMAA) forms complexes with P4-VPy. The carboxyl carbon peak of PMAA at 181 *ppm* exhibits a high-field shift in these complexes (37). Similarly, in blends of P4-VPy and PVPh studies of the miscibility with ^{13}C NMR have been reported (38).

7.3.2 *Thermal Properties*

7.3.2.1 *Thermogravimetric Studies*

Thermogravimetry (TG) of P4-VPy in nitrogen atmosphere shows an onset of degradation at around 380°C (39). Only one single degradation step is observed. The activation energy of degradation is 73 *kJ*. The main degradation product of P4-VPy is the monomer, with traces of Py and 4-methylpyridine (40).

The thermal stability of a series of crosslinked copolymers made from 2-VPy, styrene and DVB was determined in nitrogen atmosphere from TG (41). All these copolymers degrade in two stages. A pronounced weight loss rate is observed around 400°C. The weight loss at heating up to 750°C is around 70–80%. When the nitrogen in vinylpyridine is oxidized, the thermal stability is lowered. On the other hand, the weight loss is somewhat less.

In very thin samples of 100–400 *nm*, the glass transition temperature of P2-VPy and PMMA using a high sensitivity DSC method (42), is shifted toward higher temperatures by 10–20 *K*. The energy of activation enthalpy of the glass transition is shifted to lower values by a factor of 2–4. However, the experiments are done at high heating rates as 16–18 $K\,ms^{-1}$.

7.3.2.2 *Flame Retardancy*

In copolymers of styrene and 4-VPy, the limiting oxygen index (LOI) increases linearly from around 18% for pure PS up to 24% for pure

P4-VPy. In contrast, the modification of the copolymer with vanadyl dichloride has a dramatic effect on the LOI. LOI values near 50% are reached (43).

7.3.3 *Pharmaceutical Properties*

Some vinylpyridines are biological active. 2,6-Di-[2-(thien-2-yl)vinyl]pyridine has be shown to exhibit in vitro anti-tumor activity (44).

Quaternary ammonium salts exhibit bactericidal activity. In particular, cationic polymers with quaternary ammonium groups exhibit higher antimicrobial activities than the corresponding low-molecular weight model compounds. Both water-soluble polymers with quaternary pyridinium salts and quaternary ammonium salt complexes immobilized on ion exchange resins are active as bactericides (45).

Various pyridinium-type functional polymers, including both soluble and crosslinked (46, 47), have been synthesized and extensively characterized with respect to their antibacterial activity (48, 49). Hexadecylpyridinium chloride immobilized on Amberlite™ IRC-50 is considered as the best one with respect to bactericidal activity. Amberlite™ IRC-50, a copolymer from methacrylic acid (MA) and DVB, is a weak acidic resin with carboxylic acid functionality.

Various poly(*N*-benzyl-4-vinylpyridinium) salts have been prepared by homopolymerization of 4-VPy, followed by quaternization of the polymers with benzyl chloride, benzyl bromide, butyl bromide, and others (50). The resulting macromolecules showed strong antibacterial activity against gram-positive bacteria, but a minor activity against gram-negative bacteria.

A composite microporous membrane made of poly(*N*-benzyl-4-vinylpyridinium chloride) can be used as a filter material for removing airborne bacteria (51). Macromolecules of this type have been found to produce coagulation and sediment and are useful for the removal of microorganisms from water.

Copolymers from styrene and 4-VPy, quaternized with *n*-octyl iodide exhibit antibacterial properties. Block copolymers are more active than random copolymers (52).

7.4 Applications

Vinylpyridine containing polymers find many applications in the adhesive field as adhesion promoters, and in the field of laundry detergents. Further, they are used in special applications as catalyst supports.

7.4.1 *Adhesion Promoters*

7.4.1.1 *Automotive Tires*

Reinforcing fibers to be used to increase the strength or durability of rubber tires or rubber belts are coated with film to increase the adhesion between the fiber and the rubber base material. Such adhesion promoters may consist of a vinylpyridine/styrene/butadiene terpolymer (53). PET is widely used as fiber material.

A vinylpyridine/styrene/butadiene terpolymer can be used in combination with other adhesion promoters, such as resorcinol/formaldehyde condensation products, epoxides, and blocked poly-(isocyanate)s (54).

During vulcanization, a portion of the styrene/butadiene/vinylpyridine copolymer is coupled to the rubber via sulfur crosslinking, thus increasing the adhesion.

7.4.1.2 *Circuit Boards*

Poly(imide) (PI)-based materials are widely used in the microelectronics packaging industry, especially in printed circuit boards (PCB)s, because of their superior mechanical properties, high-temperature resistance, solvent resistance, and low dielectric resistance. Copper is preferably used in PCBs due to its excellent conductivity and low electromigration property.

Electromigration is denoted as the transport of material due to the momentum transfer between conducting electrons and metal atoms that can move by diffusion in the solid state. This effect is important in small-scale electronic devices where large current densities appear.

For flexible PCBs, copper is coated on to PIs by physical vapor deposition techniques, such as vacuum evaporation, metal sputtering, ion sputtering, or laminated over the copper. PI-copper laminates have a poor adhesion strength.

The adhesion strength can be improved by a surface modification of the respective PIs. 4-VPy can be attached to a PI film, such as poly(*N*,*N*′-(oxydiphenylene) pyromellitimide), with thermal graft copolymerization. The PI is subjected to an argon plasma pretreatment. The copolymerization occurs with simultaneous lamination to the copper foils (55).

The single lap shear strength of the copper to PI bond with P4-VPy as the adhesive was increased from 4.67 to 8.50 $N\,mm^{-2}$ using the optimized condition of argon plasma pretreatment time of 30 s, a curing temperature of 140°C and curing time of 4 h.

7.4.2 *Dye Transfer Inhibitors*

Dye complexing polymers have been used in laundry detergent compositions and fabric softener compositions to stabilize and minimize the leaching of dyes in colored fabrics. During washing of colored and white fabrics, some of the dyes can bleed out of a colored fabric and a portion of the leached dye may deposit on white or lighter colored fabrics. The degree of bleeding is influenced by the character of the dye, the type of cloth and the pH, temperature, and the mechanical efficiency of the agitation process.

Although in some cases the bled dye in the wash liquor can be washed off without altering the color of lighter colored fabrics, the dyed fabric looses a degree of brilliance resulting in a somewhat faded appearance.

Often it is found that the fugitive dye deposits either on to the same fabric or on to another fabric leading to patches and streaks in the washed material. The deposition of the bled dye can be inhibited by the use of a dye transfer inhibitor (DTI) compound, which can complex the dye and minimize leaching or prevent the redeposition on the fabrics. Quaternized vinylpyridine carboxylate salt polymers are particularly useful in a laundry detergent to inhibit the migration of dye (5). Also, poly(vinylpyridine *N*-oxide) has been described as color transfer inhibitor (27).

The effect of various DTIs is shown in Table 7.2. Instead of sodium 2-chloropropionate, crotonic acid can be used for the quaternization

Table 7.2 Effect of Various Dye Transfer Inhibitors (56)

Samples	ΔE[a]
White cloth (no test solution, as standard)	0
No polymer	33
Poly(4-vinylpyridine) (P4-VPy), 100% quaternized[b]	6.6
Poly(4-vinylpyridine), 75% quaternized[b]	7.7
Poly(4-vinylpyridine), 50% quaternized[b]	10.4
Copolymer of P4-VPy + NVP, (50:50), 100% quaternized[b]	10.9
Copolymer of 4-VPy + NVP, (25:75), 100% quaternized[b]	14.3
Poly(*N*-vinyl-2-pyrrolidone)	23.7
Poly(vinylpyridine *N*-oxide) (PVPNO)	11.9
poly(1-vinylimidazole) (PVI)	10.1
PVP + poly(1-vinylimidazole) (60:40)	8.2

[a] Reflectance readings
[b] Quaternized with sodium chloroacetate
Poly(4-vinylpyridine) sodium carboxymethyl betaine chloride

to result in useful DTIs (6). This reaction runs via a Michael Addition, as shown in Figure 7.2.

7.4.2.1 *Test Procedure of the Dye Transfer Inhibitors*

The action of the DTIs was tested by the following procedure (56): A test solution was prepared by mixing 10 *ppm* of the polymer, 10 *ppm* of a dye (Direct Red 80) and 1 $g l^{-1}$ of a laundry detergent, which contained a mixture of both an anionic and a non-ionic surfactant. The solution was diluted with water to 1000 *ml*.

Three white cotton cloth swatches #400 (bleached and desized) were immersed in the test solution at 37°C (100°F) and the solutions were agitated for 10 *min* in a terg-o-tometer (Instrument Marketing Services Co.). A terg-o-tometer is a laboratory instrument used for wash tests. It is used for modelling the wash conditions encountered in a top-loading washing machine.

The cloths were then removed, excess solution squeezed out, the cloths washed again in clean water for 3 *min*, squeezed again and dried. Reflectance measurements were taken on this test material on a colorimeter. The reflectance readings were recorded as ΔE, which is a composite of the degree of whiteness, redness, and blueness indices in the dyed cloth. These readings were taken as a direct measure of the degree of dye deposition under the test washing conditions.

7.4.2.2 *Laundry Compositions*

Laundry compositions are subdivided in detergent compositions and softener compositions which are subsequently described in detail.

Detergent Compositions. Clearly in practice laundry compositions contain more components beyond a DTI (57, 58). Common ingredients are summarized in Table 7.3. A wide variety of non-ionic

Table 7.3 Components in Liquid Laundry Detergent Compositions (57)

Component Type	%
Non-ionic surfactant detergent	30 to 80
Acid terminated surfactant	0 to 20
Builders	10 to 60
Hexylene glycol	5 to 30
Propylene carbonate	0 to 5
Anti-settling agent	0 to 1.5
Rheology additive	0 to 1.5
Phosphoric acid alkanol ester	0 to 1
Anti-incrustation agent	0 to 10
Anti-redeposition agent	0 to 4
Alkali metal perborate bleaching agent	5 to 15
Bleach activator	1 to 8
Sequestering agent for bleach	0 to 3
Optical brightener	0.05 to 0.75
Enzymes	0.75 to 1.25
Perfume	0.1 to 1.0
Anti-foam agent	
Thickening agent	

surfactants are available, for example, ethoxylated fatty alcohols. Acid terminated non-ionic surfactants are the reaction products of a non-ionic surfactant and succinic anhydride. Detergent builders, such as poly(trisodium phosphate) are substances added to increase the cleaning efficiency of the surfactant by inactivating water hardness minerals. Builders may be classified in to:

1. Sequestering Builders. They inactivate water hardness mineral ions and hold them tightly in solution. Examples are complexing agents, such as phosphates and citrates.
2. Precipitating Builders. A precipitating builder removes water hardness ions by forming an insoluble substance from the ions that contribute to hardness. Examples are sodium carbonate or sodium silicate that precipitate calcium ions.
3. Ion Exchange Builders. A class of ion exchange builders are aluminosilicate compounds, also known as zeolites.

Hexylene glycol and propylene carbonate substantially reduces the apparent viscosity of the composition and improves the dispersibility of the composition (57). These ingredients are typical for liquid compositions.

Organic derivatives of magnesium aluminum silicate function as an anti-settling agent. As a rheology additive to improve the physical stability of the product, distearyl dimethyl ammonium chloride may be used. Phosphoric acid alkanol ester acts as a stabilizing agent. Anti-incrustation agents are copolymers of acrylate and maleic anhydride in the form of alkali metal salts.

Anti-redeposition agents act in preventing loosened soil from redepositing on to the already cleaned fabrics. Examples of anti-redeposition agents are carboxymethyl cellulose, or poly(ethylene glycol) and poly(acrylate) salts. Corrosion inhibitors such as sodium silicate, protect the engine washer from corrosion.

Bleaching agents provide active oxygen. Most widely used is sodium perborate tetrahydrate. The bleach activator lowers the decomposition temperature of the peroxide bleaching agent. Preferred activators are tetracetyl ethylenediamine or pentaacetyl glucose. The bleach activator interacts with the peroxy compound to form a peroxy acid bleaching agent in the wash water.

In the presence of metal ions in the wash water, a sequestering agent of high complexing power may be added to inhibit undesired reactions between the peroxy acid and hydrogen peroxide in the wash solution. Suitable sequestering agents include the sodium salts of nitrilotriacetic acid, ethylenediamine tetraacetic acid, diethylene triamine pentaacetic acid, diethylene triamine pentamethylene phosphonic acid, and ethylenediaminetetramethylene phosphonic acid (57).

Another widely used bleaching agent is based on sodium hypochlorite. Sodium hypochlorite and oxygen bleaches should not be used together.

Optical brighteners adhere to fabrics in the same manner as dyes. Ultraviolet energy is absorbed, converted, and emitted as visible blue light by a fluorescence reaction. This enhances the brightness of the fabrics. Suitable optical brighteners include stilbene, triazole, and benzidine sulfone compositions.

Enzymes degrade organic adherents as grass and blood. These degraded soils can be more easily removed by the other detergent ingredients in the formulation. Examples for proteolytic enzymes are bromelin, papain, trypsin, and pepsin. Other enzymes belong to the class of amylases and lipases. Most widely used are proteolytic enzymes.

Anti-foam agents consist of poly(siloxane)s. Only small amounts are needed. Perfumes mask the chemical odor of the detergent and the odor of soils in the washing solution.

Softener Compositions. Fabric softening compositions are well known for depositing fabric softening actives on fabrics during the laundry operation and thereby imparting a softened feel of the laundered fabrics. Fabric softening compositions for the use in the washing machine are typically formulated in bulk liquid formulations that are either dispensed directly into the rinse water at the beginning of the rinse cycle, or placed in a dispensing device at the beginning of the wash cycle for delayed dispensing of the composition (59).

Polymers and copolymers of quaternized vinylpyridine are used in softener compositions for fabrics. They act in softening by enhancing the electrostatic repulsion of the materials. Very effective are poly(vinylpyridine) or copolymers of vinylpyridine and acrylamide (AAm) (60:40), with a molecular weight about 40 *k Dalton* with about 60% of the available Py nitrogens quaternized (60).

7.4.3 *Catalysts*

7.4.3.1 *Alkylation Catalysts*

Anhydrous hydrogen fluoride is widely used as a catalyst in the petrochemical industry. It is particularly effective as alkylation catalyst such as in the production of high-octane gasoline via isoalkane-olefin alkylation. Similarly, detergent alkylates are produced by alkylating aromatic compounds, such as benzene.

These technologies have achieved significant application in industry. However, because of the high volatility of hydrogen fluoride with boiling point of about 20°C, the environmental and health risks are increasingly unacceptable.

Therefore, the industry has reverted either to the use of sulfuric acid, which is a less suitable but otherwise less volatile alkylation catalyst, or has used e.g., liquid onium poly(hydrogen fluoride) complexes where the hydrogen fluoride is less volatile. In general, solid fluoride complexes are still more advantageous than liquid catalysts.

This can be achieved by using a polymeric onium poly(hydrogen fluoride) complex. A solid polymeric poly(hydrogen fluoride) catalyst can be prepared by cooling slightly crosslinked P4-VPy to about -78°C, and adding anhydrous hydrogen fluoride in small portions. Thus, a poly(4-vinylpyridinium) poly(hydrogen fluoride) (P4-VPyHF) catalyst is formed (61).

The alkylation of isobutylene has been demonstrated with the P4-VPyHF catalyst. The analysis of the alkylate product showed approximately a 70% yield with a research octane number of about 90. The alkylate product was found to contain about 63% of octane isomers, with 36% of 2,2,4-trimethylpentane.

Catalysts of these types are also suitable for complete isomerization of pivalaldehyde to methyl isopropyl ketone (62). The mechanism of the carbocationic rearrangement under superacidic activation in solution is believed to be similar to the *McLafferty* rearrangement reported in the gaseous phase.

7.4.3.2 Transition Metal Complexes

The first synthesis of esters by the reaction of an olefinic substrate with carbon monoxide and alcohols in the presence of transition metal complexes goes back to *Reppe* and *Kröper* (63). The reaction involves the treatment of an olefin with an alcohol in order to get an ester, as shown in Eq. 7.1.

$$R{-}CH{=}CH_2 + CO + ROH \rightarrow R{-}CH_2{-}CH_2{-}CO{-}OR \qquad (7.1)$$

Today important technical carbonylation processes are the carbonylation of methanol to make acetic acid, and processes for the carbonylation of methyl acetate to make acetic anhydride. The performance and process applications of Pd *Reppe*-carbonylation catalysts have been recently reviewed (64).

Metal Catalyst Sequestering. One of the processes of carbonylating methanol uses a homogeneous iridium catalyst in the reactor. One drawback of the iridium catalyzed homogeneous system is the tendency of the catalyst to form volatile species which leads to a loss of the catalyst.

Vinylpyridine and NVP resins can be used for sequestering of the volatile catalyst species. These resins are crosslinked to a degree of crosslinking of about 20%. They are insoluble in the reaction medium (65). The catalyst may be recovered by digesting the polymer, e.g., by combustion.

Metal Catalyst Support. Likewise, carbonylation catalysts can be supported on crosslinked vinylpyridine resins. A supported rhodium catalyst can be prepared as follows. In an autoclave, a 4-VPy/DVB resin is swelled in methanol. Acetic acid, methyl iodide, and rhodium acetate are then added. After deaeration with nitrogen, the autoclave is heated to 190°C and carbon monoxide is charged up to a total final pressure of 50 *at*. After 30 *min*, the autoclave is cooled and purged with nitrogen gas. The solids are washed several times with methanol to obtain a rhodium loaded resin catalyst (66).

In the course of the reaction, the Py rings of the resin are quaternized with the methyl iodide to form a pyridinium salt. To this salt structure, the rhodium carbonyl complex is attached. Other polymeric supports are built from 2-VPy and methyl acrylate (67).

Rhodium(I) complexes with 1,5-cyclooctadiene, Py or other amines and PF_6 are immobilized on P4-VPy (68). They are catalysts for the hydroformylation of 1-hexene, the water-gas shift reaction (69) and the reduction of nitrobenzene (70,71). These catalysts are preferred due to their ease of preparation and good stability. The immobilization of transition metal complexes into polymers combines the good activity, selectivity, and reproducibility.

Poly(4-vinylpyridine-*co*-acrylic acid) (P4-VPy A) or poly(4-vinylpyridine-*co*-NVP), serve as complexing agents for palladium. The copolymers are treated with palladium dichloride in the presence of sodium borohydride (72,73). P4-VPy, is an alcohol soluble synthetic polymer. Moreover, the polymer can form a homogeneous system in ethanol with the palladium complex. These complexes are active catalysts in the homogeneous hydrogenation of aromatic nitro compounds. The activity of the catalyst is enhanced in strong alkaline medium, such as potassium hydroxide in ethanol.

Rare earth metal triflates, i.e., trifluoromethanesulfonates including lanthanide (III) triflates ($Ln(OTf)_3$) have been employed as mild Lewis acid in many organic transformations, such as Diels-Alder reactions, aldol condensations, Michael Additions, etc. The catalysts can be used as homogeneous catalysts in organic solvents. Since these catalysts are more soluble in water than in common organic solvents, they can be recovered by aqueous extraction for reuse. Alternatively, these catalysts can be immobilized on solid supports of crosslinked 4-VPy/styrene copolymers (74).

Certain alcohols can be selectively oxidized to their corresponding aldehydes or ketones at room temperature using a P4-VPy supported sodium ruthenate catalyst (75). As cooxidants, iodosyl benzene or tetrabutylammonium periodate are used.

The oxidation of 2,6-dimethylphenol to diphenoquinone (4,4′-biphenyldione) with H_2O_2 is catalyzed by certain transition metal-salphen complexes that are supported on poly(vinylpyridinium salts) (76). Salphen complexes are *N,N′-o*-phenylene di(salicylaldimine) moieties, as shown in Figure 7.5.

Most reactive is a poly(4-vinylpyridinium) methyl iodide supported Mn-salphen catalyst, and a poly(4-vinylpyridinium) *n*-butyl

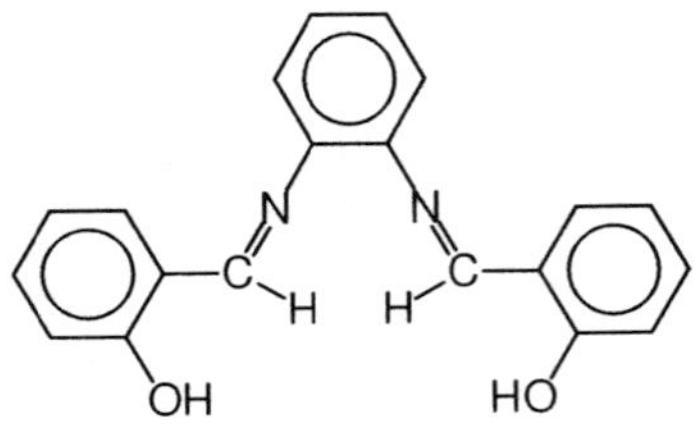

Figure 7.5 *N,N'-o*-Phenylene di(salicylaldimine), Salphen, Salophene (76)

bromide supported Cu-salphen is most selective as catalyst. No reaction occurs in the absence of these polymers.

7.4.4 *Toner Resins*

Charge control agents act to impart a positive charge to the toner particles. Quaternary ammonium compounds and diazo-type compounds can be used as charge control agents. However, some charge control additives are incompatible with the thermoplastic toner resin, or they adversely affect the electrical properties of the resin. Low-molecular weight charge control agents may leach out some of the toner composition and contaminate the carrier surface.

Improved toner compositions have been described that use partially quaternized vinylpyridine polymers as charge control agents (77). P2-VPy was partially quaternized with isopropyl bromide and melt-blended with a styrene-*n*-butyl methacrylate (65:35) copolymer resin as matrix resin together with carbon black. In addition, P4-VPy and a copolymer from 4-VPy and styrene that is partially quaternized with *n*-butyl bromide is a suitable charge control agent.

7.4.5 *Photolithography*

A copolymer from *N*-Dodecylacrylamide and *N*-ω-acryloylundecyl-4-vinylpyridinium salt as crosslinkable group is photo crosslinkable. *Langmuir-Blodgett* films can be prepared on quartz substrates and silicon wafers. These films can be crosslinked by UV irradiation. The polymer becomes insoluble at small conversions. The crosslinking reaction consists of a dimerization of the acryloyl groups in the film (78).

The copolymer is prepared by the radical copolymerization of *N*-dodecylacrylamide with 4-VPy. Then the resulting copolymer is quaternized with 11-bromo-1-undecanol. In the third step, the crosslinkable copolymer is synthesized by the reaction of acryloyl chloride with the hydroxyl group in the quaternized copolymer (79). The structure of the copolymer is shown in Figure 7.6.

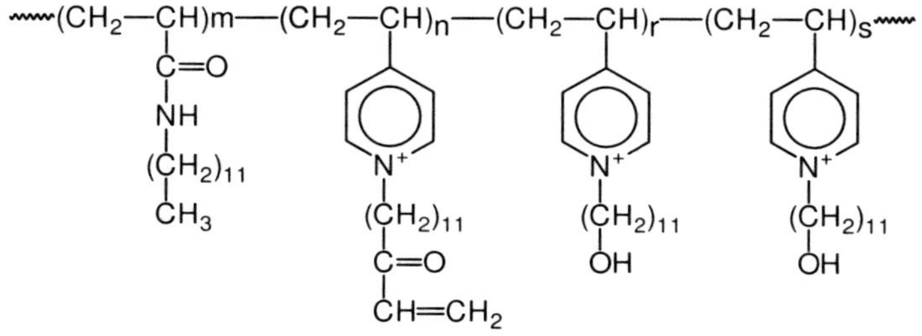

Figure 7.6 Crosslinkable Copolymer from *N*-Dodecylacrylamide and 4-Vinylpyridine (79)

Molecules in *Langmuir-Blodgett* films can be aligned with a highly ordered orientation. Therefore, a nanoscale resolution is theoretically possible. Molecular patterning with a polymer film due to formation of a two-dimensional network polymer is also possible. A molecular pattern from these polymers is shown in Figure 7.7.

7.4.6 *Optoelectronic Devices*

Polymers containing π-conjugated structures are well known for their electroluminescent properties. In contrast to non-heterocyclic π-conjugated polymers, in polymers that contain nitrogen heteroatoms either in the main chains or in the side chains, quaternization reactions and protonation of the nitrogen sites are possible. For this reason the emission spectra can be readily tuned by adding a charged species to a nitrogen site. In this way the electronic structure is changed (80).

Most common are Py moieties that are built directly in the backbone of the polymer as 2,5-pyridine units or 3,5-pyridine units (81). However, the manipulations of the Py unit are not restricted to these

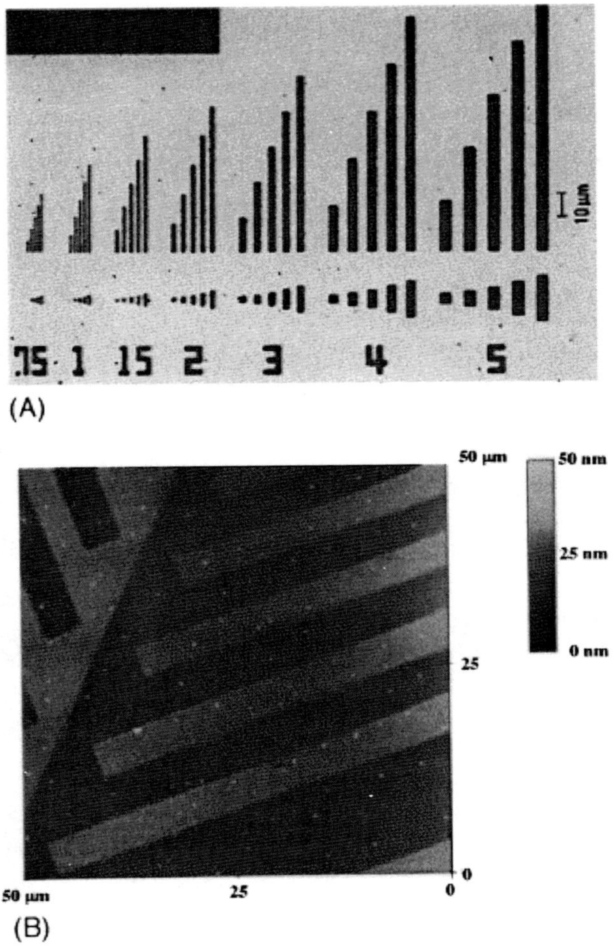

Figure 7.7 Patterns with Two-dimensional Polymer *Langmuir-Blodgett* Films. (A) The Optical Micrograph for the Film with 40 layers. (B) The AFM image for the film with 10 layers. Reprinted from (79) with permission from Elsevier

types of polymers. Polymers from vinylpyridine have been demonstrated to exhibit electroluminescent properties.

A solution of P4-VPy in pure Py forms a gel. The photoexcitation with 385, 464, and 557 *nm* produces blue emissions at 470 *nm*, green emissions at 527 *nm* and red emissions at 598 *nm* respectively (82). The gel formation occurs by physical crosslinking of the hydrogen bonds. The self-protonation of the polymer in Py solution and a quaternization reaction effect the formation of emitting centers.

Unfortunately a typical gel film is too thick and too inhomogeneous to result in an electroluminescent device. However, a special film deposition technique, i.e., electrostatic self-assembly deposition, allows the formation of such devices.

P4-VPy is dissolved in methanol and hydrochloric acid is added to achieve a certain degree of protonation. Multilayer structures are prepared by successive dipping of a charged substrate in to a solution of poly(styrene-4-sulfonate) and in to a solution containing poly(4-VPy-*co*-vinylpyridinium chloride). Film thicknesses of 20–40 *nm*, are prepared on indium tin oxide coated glass followed by the deposition of aluminum as counter-electrode.

Optionally additional electron transport layers, such as 2-(4-biphenylyl)-5-(4-*tert*-butylphenyl)-1,3,4-oxadiazole (PBD), and layers for hole transport, such as poly(*N*-vinylcarbazole) (PVK) can be used (83,84).

Advantageously, PVK is deposited on the anode to improve hole injection, PBD is deposited next to the cathode in order to improve electron injection. The electroluminescence spectra of the self-assemble film show a maximum emission around 500 *nm* at a voltage of 8 *V*.

7.4.7 *Chromatographic Resins*

7.4.7.1 *Porosity Control*

Macroporous copolymers are widely employed as column packing materials for gel permeation chromatography and as the supports for ion exchange resins. Their performance is better in comparison to conventional gel-type materials.

The formation of the porous structure can be controlled by the addition of porogens that are essentially solvent/non-solvent systems.

In copolymers based on 2-VPy and DVB, the effect of *n*-heptane, diethyl phthalate, and toluene and mixtures of these substances as porogens have been studied (85,86).

The resins synthesized in the presence of pure toluene with a low heptane proportion are transparent and gel like materials, although toluene is a non-solvating diluent for P2-VPy. As the heptane content in the diluent mixture is increased, at a fixed DVB content, the pore volume increases and the apparent density decreases.

In comparison with copolymers based on styrene and DVB, in copolymers based on styrene and 2-VPy, the formation of the porous structure is influenced more by the nature and composition of the diluent. This fact can be explained by the increased difference of the polarity of the monomers styrene and 2-VPy.

7.4.7.2 *Separation of Acids*

It is often necessary to separate carboxylic acids from other compounds by treating mediums containing their admixture. For example, citric and lactic acid are manufactured by fermentation in large scale. Such fermentations provide broths containing sugars and other compounds from which the desired acids must be separated.

Ampholytic base polymers can be used to treat mediums to chromatographically separate acids from other compounds. Such polymers are prepared by modifying a P2-VPy or P4-VPy resin crosslinked with DVB. The polymers can be quaternized with monochloroacetic acid, 3-chloro-2,4-pentanedione, diethylbromo malonate, epichlorohydrin, etc., (87).

Due to the high crosslinking density, the polymers do not swell to any significant extent during the separation of the acids from the other compounds.

7.4.7.3 *Solid Phase Extraction*

Solid phase extraction (SPE) is a chromatographic technique in the preparation of samples for quantitative analysis, for example, via high performance liquid chromatography (HPLC) or gas chromatography (GC). SPE can be used to separate a component of interest in a complex solution from potentially interfering matrix elements and to concentrate the analyte to levels amenable to detection and measurement.

For example, SPE is used in the analysis of environmental samples or in the analysis of pharmaceutical agents in blood plasma, where various other components may interfere with the analysis of trace organic materials. SPE of an aqueous solution is typically performed by passing the solution through a single-use cartridge containing a chromatographic adsorbent.

Most widely used adsorbents consist of porous silica particles that are functionalized on their surface with hydrophobic octyl and octadecyl functional groups. Prior to use, the adsorbents must be wetted with a water-miscible polar organic solvent to solvate the alkyl chains present in the adsorbent.

Adsorbents, which are not pre-wetted or have dried out, display a poor solute retention and inadequate separation of the components. The requirement that the adsorbent must remain wetted during the extraction procedure complicates SPEs and substantially slows sample analysis.

The wetting problem can be solved by using another class of adsorbents. This may be a copolymer prepared from DVB as the hydrophobic monomer and NVP or vinylpyridine as the hydrophilic monomer (88).

The polymer beads are obtained by suspension polymerization. The organic phase with the monomers 4-VPy and DVB and toluene as diluent is polymerized with AIBN as initiator (89). The sieved copolymer particles with a size of 32–50 μ are suitable as stationary phase.

In solid phase microextraction, a polymer-coated tube or a monolithic capillary tube can be directly coupled as a pre-column with GC and HPLC. A poly(acrylamide/vinylpyridine/*N*,*N*′-methylenebisacrylamide) monolithic capillary has been developed as extraction medium (90).

Before polymerization the capillary was derivatized with 3-(triethoxysilyl) propyl methacrylate. The monomers together with AIBN and dimethyl sulfoxide (DMSO) as solvent (to provide porosity) were then filled directly into the tube. The tube was sealed and heated to 60°C to initiate the polymerization reaction.

The total length of the polymer monolith in the capillary was

15 *cm*. Satisfactory extraction efficiency and adequate permeability was obtained. The device has been tested with acidic drugs, environmental pollutant phenols, and other analytes. A remarkable improvement in the limits of detection for acidic and neutral analytes was obtained. This was explained by the greater sorbent loading amount in comparison to open-tubular capillaries.

7.4.7.4 *Metal Sorption*

Divinyl sulfide copolymerizates with 4-VPy, 1-methyl-4-vinylpyridinium methyl sulfate, and 1,2-dimethyl-5-vinylpyridinium methyl sulfate according to a radical mechanism. Poly(divinyl sulfide-*co*-4-vinypyridine) with a content of 12.67% sulfur exhibits a high sorption capacity with respect to metal ions (91). The sorption capacities are shown in Table 7.4.

Table 7.4 Sorption Capacity of Poly(divinyl sulfide-*co*-4-vinypyridine) for Metal Ions (91)

Metal Ion	Condition	Amount/[$mg\, g^{-1}$]
Gold	1 M HCl and H_2SO_4 solutions	740–1200
Silver	1 M HNO_3 solution	340
Platinum	1 M HCl and H_2SO_4 solutions	672
Palladium	1 M HCl and H_2SO_4 solutions	414

7.4.7.5 *Imprinted Polymers*

Molecular Imprinting. The specific action of enzymes or deoxyribonucleic acid (DNA) is caused by the special topological structure of the molecule. Some sites are shaped in a fashion so that some special molecules (and only these molecules) fit in these sites and are absorbed. The selective absorption causes the particular reaction of the enzyme. Molecular imprinting tries to synthesize molecules with a special topology under laboratory conditions.

The molecular imprinting method consists of the polymerization of functional monomer with the crosslinking agent in the presence of a target template molecule. The functional monomer forms a complex with the template molecule prior to polymerization. After the completion of polymerization the template is extracted from the polymer using a suitable solvent. As a result of this procedure the polymer contains specific recognition sites for the template. Molecularly imprinted polymers (MIP) have gained much attention in biological and separation technology, such as drug assay, SPE, and biosensors (92).

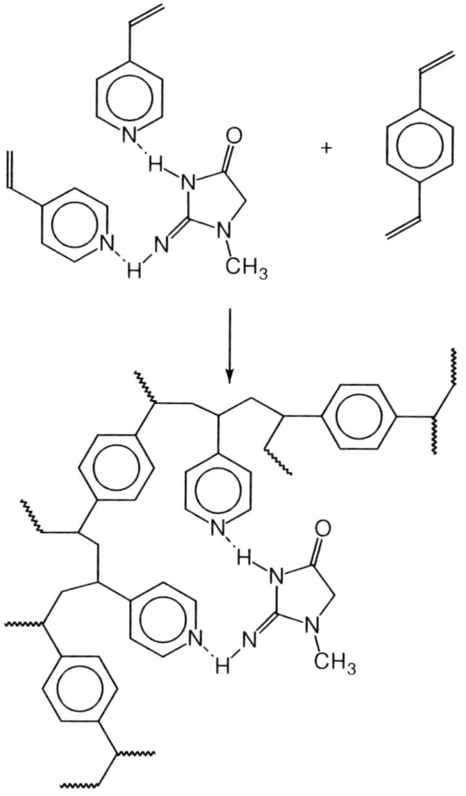

Figure 7.8 Principle of Molecular Imprinting

The basic principle of molecular imprinting is illustrated in Figure 7.8. Vinylpyridine is mixed with creatinine in a solvent such as DMSO in a proper ratio. The nitrogen atoms in vinylpyridine form a complex with the creatinine molecules. This results in a loose orientation of the vinylpyridine molecules. In the next step, the vinylpyridine molecules are fixed in this particular orientation by the copolymerization with a crosslinking agent, such as DVB. Finally, the creatine molecules are washed out from the polymer. The polymer formed in this way bears now active sites for the selective absorption of creatinine. The polymer can be used as a selective stationary phase.

The adsorption of creatinine on MIP and not imprinted polymers has been measured. The adsorption follows a *Langmuir* isotherm, c.f. Eq. 7.2.

$$A = A_{max}\frac{CK}{1 + CK} \tag{7.2}$$

A/A_{max}	Adsorbed amount of creatinine relative to saturation
C	Concentration of creatinine in solution
K	Equilibrium constant

The adsorption isotherm of creatinine, both for not imprinted polymer and imprinted polymer is shown in Figure 7.9. Inspection of Eq. 7.2 reveals that the reciprocal will result in a linear plot.

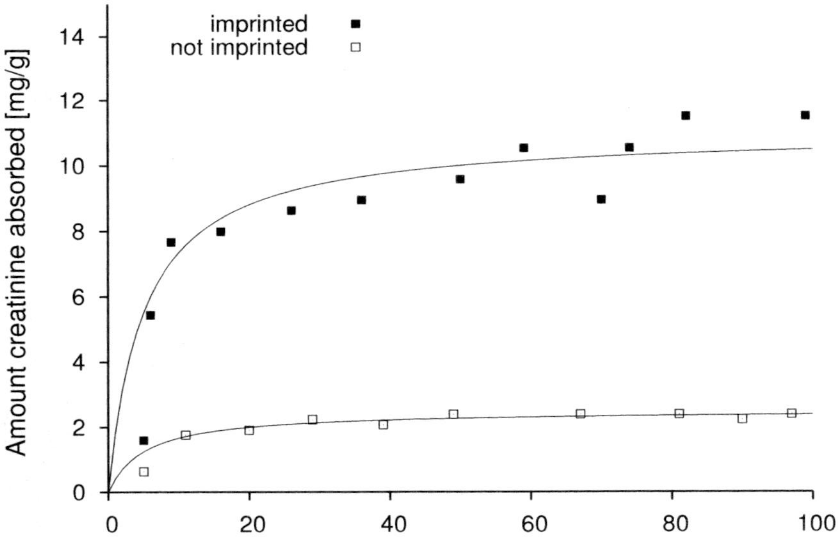

Figure 7.9 Adsorption Isotherm of Creatinine, for Not Imprinted Polymer and Imprinted Polymer (92)

$$\frac{1}{A} = \frac{1}{A_{max}CK} + \frac{1}{A_{max}}. \tag{7.3}$$

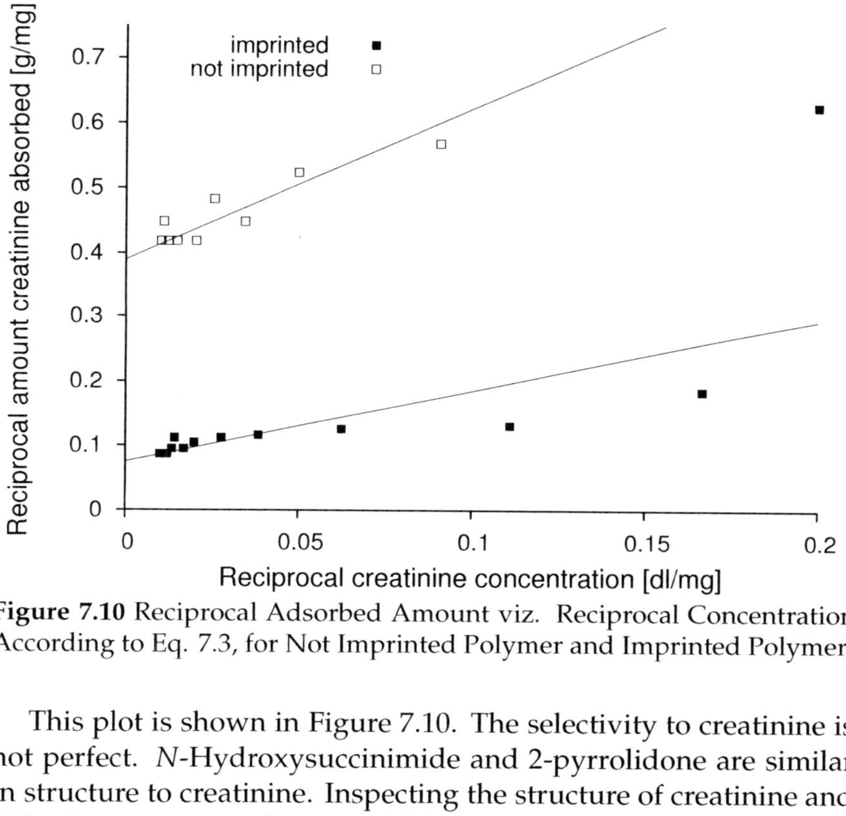

Figure 7.10 Reciprocal Adsorbed Amount viz. Reciprocal Concentration According to Eq. 7.3, for Not Imprinted Polymer and Imprinted Polymer

This plot is shown in Figure 7.10. The selectivity to creatinine is not perfect. *N*-Hydroxysuccinimide and 2-pyrrolidone are similar in structure to creatinine. Inspecting the structure of creatinine and *N*-hydroxysuccinimide, it could be expected the N–H groups of creatinine may form stronger hydrogen bonds than the hydroxyl groups of *N*-hydroxysuccinimide. In fact, performing experiments with solution mixtures of creatinine, *N*-hydroxysuccinimide, and 2-pyrrolidone, selective absorption of creatinine is still observed (92). Molecularly imprinted enantioselective polymer membranes which are prepared by photo-copolymerization of PP membranes with the functional monomer 4-VPy have been described (93).

Monolithic MIP have been tested for the separation of *Z*-phenylalanine by liquid chromatography (94). *Z*-*L*-Phenylalanine was used as template in the polymerization of a series of experiments.

The major monomer was trimethylolpropane trimethacrylate. Co-monomers were MA, 4-VPy, and AAm. 4-VPy exhibited the best performance.

Metal Ion Imprinting. In the same way as molecular imprinting, the metal ion imprinted polymer (MIIP) technique works. Hg^{2+} imprinted and non-imprinted copolymers have been prepared by copolymerizing mercury chloride, diazoaminobenzene, and 4-VPy using EGDM as crosslinking agent in the presence of AIBN as initiator (95).

In order to fix the vinylpyridine units in the proper position, diazoaminobenzene is used as auxiliary agent. The complex is shown in Figure 7.11. In the second stage, 4-VPy is copolymerized with EGDM.

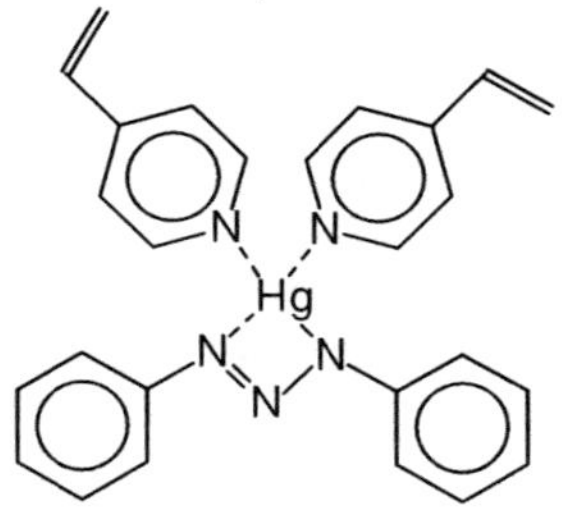

Figure 7.11 Mercury complex with Diazoaminobenzene and 4-VPy (95)

The separation characteristics of the copolymers for Hg^{2+} have been investigated by batch and column procedures. The selectivity for Hg^{2+} in imprinted copolymers in an electrolyte solution composed from sodium chloride, potassium bromide, sodium nitrate, sodium phosphate, sodium sulfate, calcium nitrate, and magnesium nitrate is greatly improved with the tolerance limits 10–100 fold of the non-imprinted copolymers. Thus, Hg^{2+} imprinted copolymers can be safely used as an adsorbent for Hg^{2+} in high concentrated electrolytes.

Batch experiments indicate that an equilibration time of about 60 *min* is required for 95% sorption. The amount of Hg^{2+} adsorbed increases with increasing pH. At a pH below 3, sorption quantity is very low, due to the protonation of the nitrogen groups.

The adsorption properties of the imprinted polymers were tested with respect to several other parameters, such as effect of temperature, effect of flow rate, selectivity of copolymers for matrix ions (95). The results cannot be reproduced here in detail.

It has been demonstrated that the quaternization of P4-VPy beads with 2-chloroacetamide results in resins that extract selectively Hg^{2+} with respect to a series of bivalent ions without using the MIIP technique (96).

Selenium is an important element from the ecotoxicological view due to the narrow concentration range between its essential and toxic effects. Selective extraction and pre concentration of Se^{4+} from aqueous solution can be achieved by a polymer synthesized from 2-VPy and EGDM. Selenium is complexed with *o*-phenylenediamine (97).

7.4.8 Ion Exchange Membranes

Ion exchange membrane technologies are used not only for separation and purification but also in energy conversion devices, storage batteries and sensors (98). The methods of preparation of ion exchange membranes have been reviewed (99).

7.4.8.1 Pore Modification

The pores of PP MF membranes have been modified with polymers from 4-VPy and DVB. The polymers are deposited in the pores by the *in situ* photoinitiated copolymerization of 4-VPy and DVB. In this way, a polyelectrolyte is anchored in the pores of the microporous PP support (100). The transport properties and the cation selectivity of the membranes can be enhanced by quaternization.

The properties of the membranes can be tuned for a variety of applications, for example, for an ultra-low-pressure water softening technique (101, 102). Moreover, these membranes could be used in diffusion dialysis.

Diffusion dialysis relies on the difference in the chemical potential of species on either side of a membrane. The membrane allows the selective transport of anions across the membrane while ideally remaining impermeable to cations other than protons. External power is merely required to circulate the solutions.

The membranes exhibit a so-called *pH-valve effect*, which can exceed three orders of magnitude in permeability. The valves are open at pH > 5. Here, the incorporated polyelectrolyte is largely in its unionized form. In contrast, at low pH, the anchored P4-VPy is protonated. Therefore, at low pH, the membranes can separate inorganic salts from aqueous solutions (103).

7.4.8.2 Bipolar Membranes

Bipolar membranes are generally composed of one cation exchange and one anion exchange layer joined together (104). Likewise, they consist of a single material where the faces are modified with ion exchanging groups.

This particular design results in a high separation efficiency for ions. For example, when used in electrodialysis processes, the bipolar membrane provides a low energy tool for the production of acids and bases from their corresponding salts (105). Possible applications are the wastewater treatment arising in paper industries.

Bipolar membranes have been described that consist of PVA as polymer matrix (106, 107). Poly(sodium styrene sulfonate) is functional anionic polyelectrolyte and poly(*N*-ethyl-4-vinylpyridinium bromide) (PE4-VPyBr) is chosen as cationic polyelectrolyte. The composite can be crosslinked with 1,2-dibromoethane. The crosslinking reaction needs temperatures as high as 140°C.

Similar anion exchange membranes consisting of PVA, β-cyclodextrin and PE4-VPyBr have been crosslinked with formaldehyde at 60°C. When the crosslinking density of the membranes is increased with formaldehyde, the membranes will become more tight.

Due to the introduction of methylene groups into the membranes, the membranes become more hydrophobic (108). With increasing content of β-cyclodextrin in the membranes, the electrical resistance of the membrane decreases, whereas their water content and thickness increases.

7.4.8.3 Hollow Fibers

Multilayer composite membranes from poly(sulfone), P4-VPy and silicon rubber have been developed for the separation of carbon dioxide and methane (109). The hollow fibers are manufactured using a wet spinning technique.

Water is used as external coagulant to yield an outer skin layer. A mixture of an organic solvent in water is employed as the internal coagulant in order to create an open porous inner surface of the hollow fibers. A cross section of the hollow fibers is shown in Figure 7.12.

Figure 7.12 SEM Images of a Poly(sulfone) Hollow fiber Substrate. Cross Section of the Hollow Fibers. Reprinted from (109) with permission from Elsevier

The selectivity of the composite membrane in the binary carbon dioxide methane system increases with decreasing pore size of the substrate. A high performance multilayer composite membrane with a CO_2/CH_4 selectivity of 29 was demonstrated (109).

In multilayer composite hollow fiber membranes composed from silicon rubber, P4-VPy, and poly(ether imide), selectivities of H_2/N_2 of 117, CO_2/CH_4 of 62 and for O_2/N_2 of 5.8 are reported (110).

Gel-filled hollow fiber membranes have been proposed for water softening (111). The pores of microporous PP hollow fibers are filled with P4-VPy. The gels are fixed in the pores of the membrane by crosslinking the P4-VPy using 2,5-dichloro-1,4-xylene. Finally, a quaternization is performed using benzyl bromide.

7.4.9 Sensor Techniques

7.4.9.1 Humidity Sensors

Polymer electrolytes have been widely used in the preparation of resistive-type humidity sensors. The basic construction of a resistive-type humidity sensor is shown in Figure 7.13.

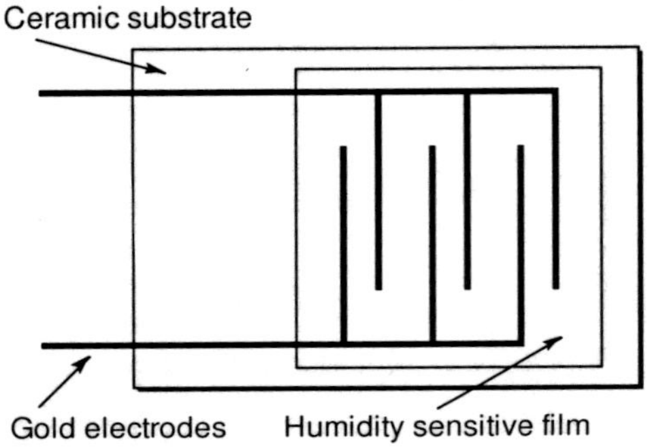

Figure 7.13 Basic Construction of a Resistive-type Humidity Sensor (112)

Several useful methods for improving the characteristics of polymer-based resistive-type humidity sensors have been proposed. For example, crosslinking of hydrophilic polymers or the built up of interpenetrated polymer networks with a hydrophobic polymer makes the hydrophilic polymers durable at high humidities. Water durable humidity sensors can be fabricated by graft polymerization (113).

P4-VPy is a hydrophilic polymer. It can be simultaneously crosslinked and quaternized (114). To improve the water resistance, P4-VPy was grafted on poly(tetrafluoroethylene) (115). Humidity sensors based on terpolymers of vinylpyridine/butyl acrylate/styrene copolymers have been described (116).

The hysteresis and response time of the sensor decreases with increasing content of BA and decreasing content of styrene. In the same way, with increasing quaternization, hysteresis and response time decrease. The reproducibility under various long-term test conditions is around 2–3% relative humidity.

When butyl methacrylate as a flexible monomer is copolymerized with 4-VPy the film formation ability and water resistance is improved (117). The stability of this copolymer in a humid environment can be further improved by crosslinking with dibromobutane (112). The impedance of such a sensor changes by four orders of magnitude in the range from 33 to 95% RH.

Humidity sensors that operate at even lower humidity consist of poly(*N*-vinyl-2-pyrrolidone) grafted on to carbon black. The grafting proceeds by trapping radicals on the carbon black (118). The polymer is further quaternized and crosslinked with alkyl halides (119).

7.4.9.2 DNA Sensitive Electrode

An osmium complex, $Os(5,6-dimethyl-1,10-phenanthroline)_2Cl_2$ that is bounded to P4-VPy is a highly selective and sensitive redox indicator for DNA. The polymeric indicator is much more sensitive to DNA than the corresponding monomeric indicator (120).

7.4.9.3 Glucose Sensors

The most widely used glucose biosensors work with amperometric enzymatic electrodes based on glucose oxidase. The reaction is based on the formation of hydrogen peroxide in presence of glucose. The amperometric detection of hydrogen peroxide requires high over-potentials and may cause interferences in biological fluids due to large amounts of other substances that are prone to oxidation.

The selectivity can be improved by introducing a membrane layer consisting of poly(4-VPy-*co*-styrene) as a filter for undesired interfering species (121). Due to the ionic properties of the Py unit, anionic species are completely repelled and neutral substances are partially stopped by the membrane.

7.4.9.4 Multi responsive Sensors

The term *chemical sensor* as used in the original paper alludes to chemical analysis (122). Actually, the sensor measures the mechanical response of a hydrogel that swells or shrinks when in contact with the medium to be characterized. The hydrogel consists of a block copolymer composed from 2-VPy or *N,N*-dimethylacrylamide and crosslinkers. The polymer is fixed on a pressure-sensitive device that transduces the changes of dimensions of the hydrogel into an electrical signal. Using this principle of measurement concentrations of salts and the pH can be monitored.

7.4.10 *Oilfield Applications*

7.4.10.1 *Paraffin Inhibitors*

Crude oils may contain large fractions of paraffins. At the temperature of the wells, the paraffins are liquids and dissolved in the crude oil. As the oil rises to the surface its temperature becomes lower and the paraffins may crystallize. This results in a loss of fluidity, which makes production, transportation, storage, and even the treatment of these oils difficult. In the worst case, the plugging of pipelines and treatment tools may occur.

For this reason, various paraffin inhibitors have been developed. Copolymers containing alkyl acrylate and vinylpyridine moieties are suitable paraffin inhibitors. The presence of 1–10% of 2-VPy or 4-VPy units in the copolymer chains improves the inhibiting efficacy significantly (123).

7.4.10.2 *Gas Hydrate Inhibitors*

The formation of gas hydrates in a gas pipeline, where an aqueous phase is inherently present, is a serious problem, especially in areas with low temperatures in the winter season or in the sea. Anti-freeze compounds, such as ethylene glycol (EG) or methanol, may be added during transport of such liquids and gases to minimize gas hydrate formation.

A copolymer of *N*-vinyl caprolactam and 4-vinylpyridine, and optionally with NVP, is an inhibitor for gas hydrate formation (124). The polymerization of the monomers may be carried out in solvent such as 2-butoxyethanol using di-*tert*-butyl peroxide as initiator. To the obtained polymer in solution, a carrier solvent is added, such as EG. This acts as an anti-freeze agent as well. The same formulation is also active as a corrosion inhibitor in pipelines (125).

7.4.10.3 Fluid Loss Control

Hydraulic cements in a well bore must be placed within or close to porous media, such as earthen strata. In this case, water tends to filter out of the slurry and into the strata during the setting of the cement. Problems arise with regard to an uncontrolled fluid loss, such as uncontrolled setting rate, improper placement of the slurry, impaired strength properties, and contamination of the surrounding strata.

Additives to reduce the water loss from cement slurries are copolymers prepared from 4-VPy, sodium acrylate solution, and AAm (126). The polymers are prepared in aqueous solution using sodium persulfate as initiator.

7.4.11 Lubricating Additives

1-Vinylimidazole, or 4-VPy can be grafted on to poly(olefin)s. In lubricating oils these grafted poly(olefin)s act as a viscosity index improver (127).

Such grafted poly(olefin)s require only minimal amounts of additives which increase the low-temperature viscosity of the lubricating oil blends. Therefore, lubricating oil blends can contain a higher viscosity base stock. The use of a higher viscosity base stock provides better lubrication at high operating temperatures and reduces the proportion of volatile species.

The grafting reaction may be carried out on the solid poly(olefin) in an extrusion reactor, in the molten poly(olefin), or in a solvent. A radical initiator, such as di-*tert*-butyl peroxide is used.

7.4.12 Corrosion Inhibition

Organic compounds with electronegative functional groups and π-electrons are often good corrosion inhibitors. In particular, Py derivatives are known as effective corrosion inhibitors. poly(4-vinylpyridine)/poly(oxyethylene) (128) and poly(4-vinylpyridine isopentyl bromide) (P4VPyIPBr) (129) are effective corrosion inhibitors for pure iron in sulfuric acid.

The corrosion rate is measured as the weight loss rate at a certain period of immersion of the metal in the aggressive medium. The inhibition efficiency E is calculated as

$$E = 100\left(1 - \frac{r_c}{r_0}\right). \tag{7.4}$$

E Inhibition efficiency
r_c Rate of corrosion with a certain amount of inhibitor added
r_0 Rate of corrosion with no inhibitor added

P4VPyIPBr with different ratios of quaternization have been tested with respect to their corrosion efficiency. The results are shown in Table 7.5. Inspection reveals that the efficiency of inhibition is independent of the degree quaternization of P4VPIPBr.

Poly(4-vinylpyridine-3-oxide ethylene tosylate) has been checked in a similar way with respect to its corrosion efficiency as P4VPyIPBr (130). This polymer is still more effective than the latter. P4VPyIPBr acts only as a cathodic inhibitor.

However, poly(4-vinylpyridine-3-oxide ethylene tosylate), due to the presence of tosylate group, induces an anodic action and a power protection. It is believed that the presence of the tosylate anion with its negatively charged oxygen atom enhances the adsorption on the iron surface and induces so the anodic action.

7.5 Suppliers and Commercial Grades

P4-VPy is available from Aldrich, Reilly. A vinylpyridine/styrene/butadiene latex type is Pliocord® originally by Goodyear, but the trademark is owned by Eliokem, Inc. Corp.

7.6 Safety

The effects of transient skin exposure amongst laboratory workers to 2-VPy and 4-VPy include an initial intense burning pain and a reddish-brown skin discoloration which abates over the course of about 4 weeks. A delay in onset of pain has been observed in one case.

Py derivatives substituted at the 4 position appear to be more toxic than those substituted at the 2 position, irrespective of their

Table 7.5 Corrosion Efficiency of Poly(vinylpyridine) Derivates (129)

Polymer[a] 1 M H_2SO_4	c/[10^{-8} M][b]	Rate[c] [$mg\,cm^{-2}h^{-1}$]	E[d]
No polymer	0	2.488	0
P4VPyIPBr 6%	1000	0.000	100
P4VPyIPBr 6%	500	0.000	100
P4VPyIPBr 6%	250	0.049	98
P4VPyIPBr 6%	100	0.149	94
P4VPyIPBr 6%	50	1.169	53
P4VPyIPBr 6%	10	1.692	32
P4VPyIPBr 6%	5	2.040	18
P4VPyIPBr 6%	1	2.214	11
P4VPyIPBr 18%	1000	0.000	100
P4VPyIPBr 18%	500	0.025	99
P4VPyIPBr 18%	250	0.124	95
P4VPyIPBr 18%	100	0.249	90
P4VPyIPBr 18%	50	1.742	30
P4VPyIPBr 18%	10	2.139	14
P4VPyIPBr 18%	5	2.239	10
P4VPyIPBr 18%	1	0.746	7
P4VPyIPBr 79%	1000	0.000	100
P4VPyIPBr 79%	500	0.099	96
P4VPyIPBr 79%	250	0.199	92
P4VPyIPBr 79%	100	0.348	86
P4VPyIPBr 79%	50	1.542	38
P4VPyIPBr 79%	10	1.991	20
P4VPyIPBr 79%	5	2.164	13
P4VPyIPBr 79%	1	2.239	10

[a] Poly(4-vinylpyridine isopentyl bromide) with % quaternization
[b] Concentration $mol\,l^{-1}$
P4VPIPBr 6%, $M = 4.2 \times 10^5$ *Dalton*
P4VPIPBr 18%, $M = 4.7 \times 10^5$ *Dalton*
P4VPIPBr 79%, $M = 7.2 \times 10^5$ *Dalton*
[c] Rate of corrosive ablation of iron at 298 K in 1 M H_2SO_4
[d] Corrosion efficiency Eq. 7.4

route of absorption. Isomers of vinylpyridine are water-soluble, colorless, volatile liquids at room temperature, with a pungent odor, which is detectable by humans at levels of 3 *ppm* (131). The toxicology of the polymer seems to be widely unexplored.

7.7 Environmental Impact and Recycling

7.7.1 *Biodegradable Poly(styrene)*

It has been demonstrated that the introduction of small amounts of *N*-benzyl-4-vinylpyridinium chloride units into a backbone of PS results in biodegradable polymers. These types of polymers are prepared by radical copolymerization of styrene and 4-VPy. The polymer is then quaternized by the reaction with benzyl chloride in toluene at 80°C for 4 *h*. Equimolar amounts of benzyl chloride with respect to 4-VPy in the copolymer are used (132).

Biodegradation of the PS with a pyridinium group in the main chain was carried out by placing films of the polymer in an aeration tank up to 270 days. After treatment the intrinsic viscosity of these polymers decreases. The reduction depends on the content the pyridinium group and reaches up to 50% at 4% pyridinium content.

It is believed that the degradation of PS with a pyridinium group is achieved by microorganisms contained in an activated sludge. However, the biodegradation must be catalyzed by extracellular enzymes, which are produced by microorganisms, because the polymer is insoluble in water and cannot penetrate through the cell wall of microorganisms. The incorporation of the pyridinium group into a hydrophobic synthetic polymer enhances the hydrophilicity of the polymer. This allows the enzymes to permeate into the polymer films and promote the biodegradation.

7.7.2 *Bacterial Coagulants*

The wilt of tomato caused by bacteria has been suppressed by coagulation of bacterial cells without disinfection. As polymeric coagulant for the bacterial cells, a copolymer of MMA with *N*-benzyl-4-vinylpyridinium chloride in a molar ratio of 3:1 is used.

When 10 *ppm* of polymer was added to the soil before transplanting of tomato seedlings and 2 *ppm* was again added one week after transplanting, a reduction of appearance and a reduction of the symptoms by around 50% was observed (133).

Still more effective is the addition of sawdust coated with the polymeric coagulant prior to transplantation (134). The polymer does not exhibit a bactericidal activity against the bacterium. Further, the polymeric coagulant is highly biodegradable. Its half-life is 5.1 days when treated with activated sludge in soil (133). Thus, a green chemical method for controlling soil-borne plant diseases without disinfection is developed in this way.

Tradenames appearing in the references are shown in Table 7.6.

Table 7.6 Tradenames in References

Tradename Description	Supplier
Adogen® 3690 1-Methyl-1-oleylamidoethyl-2-oleylimidazolinium methylsulfate (59)	Goldschmidt Chemical Corp.
Adogen® 415 Soyatrimethylammonium chloride (59,60)	Goldschmidt Chemical Corp.
Adogen® 472 Dioleyldimethylammonium chloride (59,60)	Goldschmidt Chemical Corp.
Ambrettolide Tradename for perfume a ingredient (a large list of perfume chemicals disclosed in this reference) (59)	—
Amidox® C5 Ethoxylated alkyl amide surfactant (59,60)	Stepan
Armeen® APA-10 Alkyl amido propylamine (59,60)	Akzo Nobel
Berol® 303 Ethoxylated amine surfactant (59,60)	Rhone-Poulenc, Inc.
Berol® 397 Ethoxylated amine surfactant (59,60)	Rhone-Poulenc, Inc.
BHA Butylated hydroxyanisole (60)	Eastman Chemical Products, Inc.
Brij® (Series) Ethoxylated fatty alcohols (59,60)	ICI Surfactants
Brij® 76 Poly(oxyethylene) (10) stearyl ether (59,60)	ICI Surfactants

Table 7.6 (cont.)

Tradename Description	Supplier
Bronopol®	Inolex Chemicals
2-Bromo-2-nitro-propane-1,3-diol, bactericide (59,60)	
CSM450	Sumitomo Seika Chemicals Co.
Chlorosulfonated poly(ethylene) (53)	
Decalin®	DuPont
Decahydronaphthalene (127)	
Denacol EX-611	Nagase Kaseikogyo K.K.
Sorbitol poly(glycidyl ether) (54)	
Dequest® 2010	Monsanto Co.
1-Hydroxyethylidene-1,1-diphosphonic acid (etidronic acid) (60)	
Dequest® 2046	Monsanto Co.
Pentasodium(ethylenediamine)tetramethylenephosphonate (26)	
Dequest® 2066	Monsanto Co.
Diethylene triamine pentamethylene phosphonic acid (DTPMP) (57)	
DM-6011	Meisei Kagakukogyo K.K.
Blocked poly(isocyanate) (54)	
Dobanol® 25-7	Shell
Ethoxylated C_{12}–C_{15} fatty alcohol (57)	
Dobanol® 91-5	Shell
Ethoxylated C_9–C_{11} fatty alcohol (57)	
Dowfax® 9N5	Dow
Poly(ethylene glycol) mono(nonylphenyl)ether, surfactant (59,60)	
DTPA®	Aldrich
Diethylenetriaminepentaacetic acid (60)	
Emersol® 223LL	Henkel
Reaction products of oleic acids with diethylenetriamine (59,60)	
Emersol® 7021	Henkel
Reaction products of oleic acids with diethylenetriamine (59,60)	
Emulphogene® BC-720	GAF
Poly(ethylene gylcol) monotridecyl ether (59,60)	
Emulphogene® BC-840	GAF
Poly(ethylene gylcol) monotridecyl ether (59)	
Ethodumeens® (Series)	Akzo Nobel
Ethoxylated amine surfactant (59,60)	
Ethomid® HT/60	Akzo Nobel
Ethoxylated alkyl amide surfactant (59,60)	
Ethomid® O/17	Akzo Nobel
Ethoxylated alkyl amide surfactant (59,60)	

Table 7.6 (cont.)

Tradename Description	Supplier
Ethoquad® (Series) Quaternary ammonium surfactant (59, 60)	Akzo Nobel
Fujicarbon® 203 Carbon black (54)	Fuji Shikiso K.K.
Hypalon 40 Chlorosulfonated poly(ethylene) (53)	DuPont Dow Elastomers LLC
Igepal® CO-620 Poly(ethylene glycol)mono(nonylphenyl)ether (59, 60)	
Igepal® CO-710 Alkyl-aryl alkoxylated surfactant (59)	Rhone-Poulenc, Inc.
Igepal® Alkylphenoxypoly(ethylenoxy)ethanol (60)	Rhone-Poulenc, Inc.
Irganox® 1010 Pentaerythritol tetrakis(3-(3,5-di-*tert*-butyl-4-hydroxyphenyl)propionate), phenolic antioxidant (60)	Ciba Geigy
Irganox® 1035 Thiodiethylene glycol bis[3-(3,5-di-*tert*-butyl-4-hydroxyphenyl)propionate] (60)	Ciba Geigy
Irganox® 1425 Calcium 3,5-di-*tert*-butyl-4-hydroxybenzyl monoethyl phosphonate (60)	Ciba Geigy
Irganox® 3114 Tris(3,5-di-*tert*-butyl-4-hydroxybenzyl)isocyanurate (60)	Ciba Geigy
Irganox® 3125 Tris[3-(3,5-di-*tert*-butyl-4-hydroxyphenyl)propionyloxyethyl] isocyanurate (60)	Ciba Geigy
Irganox® B 1171 *N*,*N*′-1,6-Hexanediylbis[3,5-bis(1,1-dimethylethyl)-4-hydroxybenzenepropanamide, mixed with tris[2,4-bis(1,1-dimethylethyl)phenyl]phosphite (60)	Ciba Geigy
Joppa H Class H cement (126)	General Portland
JSR 0652 Vinyl pyridine/styrene/butadiene rubber latex (54)	Japan Synthetic Rubber Co.
Kathon® Mixture or 5-chloro-2-methyl-4-isothiazoline-3-one and 2-methyl-4-isothiazoline-3-one, bactericide (59, 60)	Rohm & Haas

Table 7.6 (cont.)

Tradename Description	Supplier
KEX® (Series)	Koei Chemical Co.
Poly(vinylpyridine) (87)	
Lomar® D	Geo Specialty Chemicals, Inc. (Henkel)
Sodium salt of the formaldehyde condensation product of naphthalene sulfonic acid (126)	
Lucidol® CH 50	Akzo Nobel
Mixture of dibenzoyl peroxide and dicyclohexyl phthalate (123)	
Luperox® P	Arkema, Inc.
tert-Butyl peroxybenzoate (123)	
Lupersol® 11	Arkema, Inc.
tert-Butyl peroxypivalate (6,56)	
Lutensol® AP 14	BASF AG
Poly(ethylene glycol) mono(nonylphenyl) ether (59)	
Lutensol® AP 9	BASF AG
Poly(ethylene glycol) mono(nonylphenyl) ether (59,60)	
Milease® T	ICI
Ethylene/poly(oxyethylene) terephthalate copolyester (59,60)	
MR-200	Nippon Polyurethane K.K.
Poly(isocyanate) (53)	
Neocol SW	Daiichi Kogyoseiyaku K.K.
Dialkylsulfosuccinic ester sodium salt (54)	
Neodol® (Series)	Shell
Alkyl alkoxylated surfactants (59,60)	
Nipol-2518FS	Zeon Corp.
Vinylpyridine/styrene/butadiene terpolymer latex (53)	
Norsocryl® (Series)	Arkema, Inc.
Acrylates of C_22 alcohols (123)	
Plurafac® B-26	BASF AG
Ethoxylated straight chain alcohol nonionic surfactant (59,60)	
Plurafac® C-17	BASF AG
Alkyl alkoxylated surfactant, biodegradable (59)	
Plurafac®	BASF AG
Reaction product of a higher linear alcohol and a mixture of ethylene and propylene oxides (57)	
Pluronic® (Series)	BASF AG
Ethylene oxide/propylene oxide block copolymer, defoamers (59,60)	
Porapak® R	Waters Corp.
Divinylbenzene/*N*-vinylpyrrolidone copolymer (88)	

Table 7.6 (cont.)

Tradename Description	Supplier
Pyratex™	Nippon A&L Inc.
Vinylpyridine/styrene/butadiene terpolymer latex (53)	
Reillex® (Series)	Reilly Industries, Inc.
Poly(4-vinylpyridine), crosslinked with divinylbenzene (87)	
Rewopal® C6	Witco Corp.
Ethoxylated alkyl amide surfactant (59,60)	
Shell Swimm 11T	Shell
Paraffin inhibitor (123)	
Shell Swimm 5X	Shell
Paraffin inhibitor (123)	
Silwet®	O Si Specialities, Inc.
Organo silicone surfactants (60)	
Sokalan® CP (Series)	BASF AG
Water-soluble homo- and copolymers of maleic acid and acrylic acid (57)	
Sokalan® HP (Series)	BASF AG
Water-soluble homo- and copolymers of vinylpyrrolidone, vinylimidazole and nonionic monomers (26)	
Styragel®	Waters Corp.
Poly(divinylbenzene) (88)	
Sumikanol® 700S	Sumitomo Kagaku K.K.
Resorcinol/formaldehyde resin (54)	
Sustane® BHT	UOP
2,6-Di-*tert*-butyl-4-hydroxytoluene (60)	
Tenox® D TBHQ	Eastman Chemical Products, Inc.
tert-Butylhydroquinone (60)	
Tenox® PG	Eastman Chemical Products, Inc.
Gallic acid *n*-propyl ester, antioxidant (60)	
Tenox® S-1	Eastman Chemical Products, Inc.
Mixture of citric acid and gallic acid *n*-propyl ester, antioxidant (60)	
Tenox® -6	Eastman Chemical Products, Inc.
Mixture of BHT (butylated hydroxytoluene), BHA (butylated hydroxyanisole), propyl gallate, and citric acid (60)	
Tergitol® 15-S (Series)	Union Carbide Corp.
Ethoxylated C11-15-secondary alcohols, surfactant (59,60)	
Tetronic® (Series)	BASF-Wyandotte Corp.
Propoxylated ethylenediamine-poly(ethylene glycol) adduct, surfactant (59,60)	

Table 7.6 (cont.)

Tradename Description	Supplier
Tiron®	Kodak
4,5-Dihydroxy-*m*-benzene-sulfonic acid sodium salt (60)	
Trigonox® C	Akzo Nobel
tert-Butyl peroxybenzoate (123)	
Triton® N-111	Union Carbide Corp.
Alkyl-aryl alkoxylated surfactant (59, 60)	
Triton® N-150	Union Carbide Corp.
Alkyl-aryl alkoxylated surfactant (59)	
Tween® (Series)	Uniqema
Ethoxylated fatty acid ester surfactants (60)	
Variquat® -66	Goldschmidt Chemical Corp.
Tallow alkyl bis(polyoxyethyl)ammonium ethyl sulfate (59, 60)	
Varisoft® 222LT	Witco Corp. (Goldschmidt Chemical Corp.)
Fatty amidoamine based softener (59, 60)	
Varisoft® 3690	Witco Corp. (Degussa)
1-Methyl-2-noroleyl-3-oleyl amidoethyl imidazolinium methosulfate (59, 60)	
Varisoft® 417	Witco Corp.
Monotallow trimethyl ammonium chloride (59, 60)	
Varisoft® 471	Witco Corp.
Monolauryl trimethyl ammonium chloride (59, 60)	
Vazo® (Series)	DuPont
Azonitriles, radical initiators (56)	
Wickenol® (Series)	Wickhen Products, Inc.
Higher alcohol fatty esters (59, 60)	
Zelcon® 4780	DuPont
Ethylene/poly(oxyethylene) terephthalate copolyester (59, 60)	
Zetpol 2000	Zeon Corp.
Middle hydrogenated nitrile rubber (53)	

References

1. Improvements in the coating of photographic film, GB Patent 647 093, assigned to Eastman Kodak Co., Thomas Gentry Veal, and George Palmer, December 06, 1950.

2. S.J. Kulkarni, G. Madhavi, V. Viswanathan, and K.V. Raghavan, Process for the preparation of vinylpyridine from picoline over modified zeolites, US Patent 6 727 365, April 27, 2004.
3. S.J. Kulkarni, M. Gangapuram, and V.R. Kondapuram, Process for preparing 2,6-divinylpyridine and 2-methyl-6-vinylpyridine from 2,6-lutidine over modified zeolites, US Patent 6 897 320, assigned to Council of Scientific and Industrial Research (New Delhi, IN), May 24, 2005.
4. J.S. Shih, E.T. Yap, J.C. Hornby, and B. Srinivas, Process for producing a suspension of a vinylpyridine polymer in high polymer purity, US Patent 6 011 096, assigned to ISP Investments Inc. (Wilmington, DE), January 4, 2000.
5. J.S. Shih, E.T. Yap, J.C. Hornby, and B. Srinivas, Product and process for making quaternized, water soluble vinylpyridine carboxylate polymers, US Patent 6 156 829, assigned to ISP Investments Inc. (Wilmington, DE), December 5, 2000.
6. B. Srinivas and J.S. Shih, Laundry detergent compositions containing water soluble dye complexing polymers, US Patent 6 482 790, assigned to ISP Investments Inc. (Wilmington, DE), November 19, 2002.
7. J.S. Shih, E.T. Yap, J.C. Hornby, and B. Srinivas, Process for making quaternized vinylpyridine carboxylate polymers using suspension polymerization, US Patent 6 201 046, assigned to ISP Investments Inc. (Wilmington, DE), March 13, 2001.
8. W.L.F. J. C. Salamone, B. Snider, Polymerization of 4-vinylpyridinium salts. I. The counterion initiation mechanism, *J. Polym. Sci., Part B: Polym. Lett.*, 9(1):13–17, 1971.
9. D. Navarro-Rodriguez, D. Guillon, A. Skoulios, Y. Frère, and P. Gramain, Structural behaviour of thermotropic liquid crystals from poly-(4-vinylpyridine) fully and partially quaternized with a mesogenic group, *Makromol. Chem.*, 193(12), 3117-3128 1992.
10. C. Chovino, Y. Frere, D. Guillon, and P. Gramain, Effect of smectogenicity of the ionic groups on the thermotropic liquid crystalline behavior of pyridinium and poly(4-vinylpyridinium) salts quaternized with mesogenic groups, *J. Polym. Sci., Part A: Polym. Chem.*, 35 (13):2569–2577, September 1997.
11. C. Chovino and P. Gramain, Stereoregularity of poly(4-vinyl-*n*-alkyl-pyridinium) salts prepared by spontaneous polymerization, *Polymer*, 40(17):4805–4810, August 1999.
12. S. Masuda, K. Minagawa, M. Tsuda, and M. Tanaka, Spontaneous copolymerization of acrylic acid with 4-vinylpyridine and microscopic acid dissociation of the alternating copolymer, *Eur. Polym. J.*, 37(4): 705–710, April 2001.

13. G. Bosma, C. Pathmamanoharan, E.H.A. de Hoog, W.K. Kegel, A. van Blaaderen, and H.N.W. Lekkerkerker, Preparation of monodisperse, fluorescent PMMA-latex colloids by dispersion polymerization, *J. Colloid Interface Sci.*, 245:292–300, 2002.
14. K. Ishizu, M. Yasuda, and T. Tamura, Synthesis of cross-linked core-shell polymer particles by free-radical dispersion copolymerization of 4-vinylpyridine with polystyrene macromonomers in nonaqueous media, *J. Colloid Interface Sci.*, 267(2):320–325, November 2003.
15. K. Ishizu, Architecture of polymeric superstructures constructed by mesoscopically ordered cubic lattices, *Macromol. Rapid Commun.*, 24 (4):291–302, March 2003.
16. R. Yang, Y. Wang, X. Wang, W. He, and C. Pan, Synthesis of poly(4-vinylpyridine) and block copoly(4-vinylpyridine-*b*-styrene) by atom transfer radical polymerization using 5,5,7,12,12,14-hexamethyl-1,4,8,11-tetraazamacrocyclotetradecane as ligand, *Eur. Polym. J.*, 39(10):2029–2033, October 2003.
17. J.M. Lee, B.H. Lee, and S. Choe, The effect of polystyrene-block-poly(4-vinylpyridine) prepared by a RAFT method in the dispersion polymerization of MMA, *Polymer*, 47(11):3838–3844, May 2006.
18. C. Lebrun, G. Deniau, P. Viel, and G. Lécayon, Electrosynthesis of poly(4-vinylpyridine) films on metallic surfaces under anodic and cathodic polarizations: Structure and properties of the organic coatings, *Surf. Coat. Technol.*, 100-101:474–479, March 1998.
19. P. Viel, S. Palacin, F. Descours, C. Bureau, F. Le Derf, J. Lyskawa, and M. Salle, Electropolymerized poly-4-vinylpyridine for removal of copper from wastewater, *Appl. Surf. Sci.*, 212-213:792–796, May 2003.
20. X. Ling, M.D. Pritzker, C.M. Burns, and J.J. Byerley, A mechanism for electropolymerization of 2-vinylpyridine coatings on metal surfaces, *Macromolecules*, 31(26):9134–9140, December 1998.
21. I. Kaur and R. Barsola, Grafting onto polypropylene. V. Graft copolymerization of 4-vinyl pyridine and binary mixture of 4-vinyl pyridine with acrylonitrile by preirradiation method, *J. Appl. Polym. Sci.*, 41(9-10):2067–2076, 1990.
22. R.B. Brahmbhatt, A.C. Patel, R.C. Jain, and S. Devi, Solid phase grafting of 4-vinylpyridine onto isotactic polypropylene, *Eur. Polym. J.*, 35 (9):1695–1701, September 1999.
23. L. Ying, G. Zhai, A.Y. Winata, E.T. Kang, and K.G. Neoh, pH effect of coagulation bath on the characteristics of poly(acrylic acid)-grafted and poly(4-vinylpyridine)-grafted poly(vinylidene fluoride) microfiltration membranes, *J. Colloid Interface Sci.*, 265(2):396–403, September 2003.

24. G. Zhai, E.T. Kang, and K.G. Neoh, Poly(2-vinylpyridine)- and poly-(4-vinylpyridine)-graft-poly(vinylidene fluoride) copolymers and their pH-sensitive microfiltration membranes, *J. Membr. Sci.*, 217(1-2): 243–259, June 2003.
25. S.N. Dmitriev, L.I. Kravets, V.V. Sleptsov, and V.M. Elinson, Plasma-induced graft polymerisation of 2-methyl-5-vinylpyridine on the surface of poly(ethylene terephthalate) track membranes, *Polym. Degrad. Stabil.*, 90(2):374–378, November 2005.
26. D. Boeckh, S. Stein, A. Funhoff, J.A. Lux, and H.-U. Jäger, Use of water-soluble copolymers comprising *N*-vinylimidazole units as color transfer inhibitors in detergents, US Patent 6 172 027, assigned to BASF Aktiengesellschaft (Ludwigshafen, DE), January 9, 2001.
27. A. Fredj, J.P. Johnston, and C.A.J.K. Thoen, Dye transfer inhibiting compositions comprising polymeric dispersing agents, WO Patent 9 402 578, assigned to Procter & Gamble (US), Fredj, A., Johnston, J. P., and Thoen C. A. J. K., February 03, 1994.
28. M. Maldonado-Santoyo, L.C. Cesteros, I. Katime, and S.M. Nuño Donlucas, Miscible blends of poly(vinyl phenyl ketone hydrogenated) and poly(styrene-*co*-4-vinylpyridine), *Polymer*, 45(16):5591–5596, July 2004.
29. X.H. Pen and J.R. Shen, Miscibility and mechanical properties of poly-(styrene-*co*-4-vinylpyridine) and poly(butyl acrylate-*co*-acrylic acid) ionomer blends, *Polym. Int.*, 46(2):106–110, June 1998.
30. E.M. José R. Isasi, C. Cesteros, and I. Katime, Miscibility and specific interactions in blends of poly(2-ethyl-2-oxazoline) with hydroxylated polymethacrylates, *Macromol. Chem. Phys.*, 197(2):641–649, 1996.
31. L.C. Cesteros, J.R. Isasi, and I. Katime, Phase behavior of poly(4-vinylpyridine)/poly(vinyl acetate-*co*-vinyl alcohol) blends, *Macromolecules*, 27(26):7887–7892, December 1994.
32. M. Vivas de Meftahi and J.M.J. Frechet, Study of the compatibility of blends of polymers and copolymers containing styrene, 4-hydroxystyrene and 4-vinylpyridine, *Polymer*, 29(3):477–482, March 1988.
33. J.R. Isasi, L.C. Cesteros, and I. Katime, Miscibility and phase behaviour in poly(4-vinylpyridine)/poly(ethylene-*co*-vinyl alcohol) blends, *Macromol. Rapid Commun.*, 15(11):903–908, 1994.
34. L.G. Parada, E. Meaurio, L.C. Cesteros, and I. Katime, Miscibility behaviour in blends of copolymers of vinyl alcohol with poly(ethyloxazoline), *Macromol. Chem. Phys.*, 199(8):1597–1602, August 1998.
35. X.D. Li and S.H. Goh, Specific interactions and miscibility of ternary blends of poly(2-vinylpyridine), poly(*N*-vinyl-2-pyrrolidone) and aliphatic dicarboxylic acid: Effect of spacer length of acid, *Polymer*, 43(25):6853–6861, 2002.

36. S.-W. Kuo, C.-L. Lin, and F.-C. Chang, The study of hydrogen bonding and miscibility in poly(vinylpyridines) with phenolic resin, *Polymer*, 43(14):3943–3949, June 2002.
37. J.Z. Yi and S.H. Goh, Miscibility of poly(methacrylic acid)/poly(4-vinylpyridine) blends and complexes investigated by high-resolution solid-state 13C NMR, *Polymer*, 42(22):9313–9316, October 2001.
38. J. Wang, M.K. Cheung, and Y. Mi, Miscibility in blends of poly(4-vinylpyridine)/poly(4-vinylphenol) as studied by 13C solid-state NMR, *Polymer*, 42(7):3087–3093, March 2001.
39. S.J. Azhari and M.A. Diab, Thermal degradation and stability of poly-(4-vinylpyridine) homopolymer and copolymers of 4-vinylpyridine with methyl acrylate, *Polym. Degrad. Stabil.*, 60(2-3):253–256, 1998.
40. K.S. Khairou and M.A. Diab, Stability and degradation of poly(4-vinylpyridine) and copolymers of 4-vinylpyridine with methyl methacrylate, *Polym. Degrad. Stabil.*, 44(1):17–20, 1994.
41. A.C. Jandrey, A.P. de Aguiar, M.R.M.P. de Aguiar, and L.C. de Santa Maria, Thermodegradation of poly(2-vinylpyridine-*co*-styrene-*co*-divinylbenzene) and *N*-oxide derivatives, *Thermochim. Acta*, 424(1-2):63–68, December 2004.
42. M.Y. Efremov, E.A. Olson, M. Zhang, and L.H. Allen, Glass transition of thin films of poly(2-vinyl pyridine) and poly(methyl methacrylate): Nanocalorimetry measurements, *Thermochim. Acta*, 403(1):37–41, June 2003.
43. J.R. Ebdon, L. Guisti, B.J. Hunt, and M.S. Jones, The effects of some transition-metal compounds on the flame retardance of poly(styrene-*co*-4-vinyl pyridine) and poly(methyl methacrylate-*co*-4-vinyl pyridine), *Polym. Degrad. Stabil.*, 60(2-3):401–407, 1998.
44. V. Barresi, D.F. Condorelli, C.G. Fortuna, G. Musumarra, and S. Scire, In vitro antitumor activities of 2,6-di-[2-(heteroaryl)vinyl]pyridines and pyridiniums, *Bioorg. Med. Chem.*, 10(9):2899–2904, September 2002.
45. T. Tashiro, Antibacterial and bacterium adsorbing macromolecules, *Macromol. Mater. Eng.*, 286(2):63–87, February 2001.
46. G. Li, J. Shen, and Y. Zhu, A study of pyridinium-type functional polymers. I. Preparation and characterization of soluble pyridinium-type functional polymers, *J. Appl. Polym. Sci.*, 62(13):2247–2255, December 1996.
47. G.J. Li and J.R. Shen, A study of pyridinium-type functional polymers. IV. Behavioral features of the antibacterial activity of insoluble pyridinium-type polymers, *J. Appl. Polym. Sci.*, 78(3):676–684, October 2000.

48. G. Li, J. Shen, and Y. Zhu, Study of pyridinium-type functional polymers. II. Antibacterial activity of soluble pyridinium-type polymers, *J. Appl. Polym. Sci.*, 67(10):1761–1768, March 1998.
49. G.J. Li, J.R. Shen, and Y.L. Zhu, A study of pyridinium-type functional polymers. III. Preparation and characterization of insoluble pyridinium-type polymers, *J. Appl. Polym. Sci.*, 78(3):668–675, October 2000.
50. N. Kawabata and M. Nishiguchi, Antibacterial activity of soluble pyridinium-type polymers, *Appl. Environ. Microbiol.*, 54(10):2532–2535, 1988.
51. N. Kawabata and S. Kawato, Removal of airborne bacteria by filtration using a composite microporous membrane made of a pyridinium-type polymer showing strong affinity with microbial cells, *Epidemiol. Infect.*, 121(2):349–356, October 1998.
52. E.S. Park, H.S. Kim, M.N. Kim, and J.S. Yoon, Antibacterial activities of polystyrene-block-poly(4-vinyl pyridine) and poly(styrene-random-4-vinyl pyridine), *Eur. Polym. J.*, 40(12):2819–2822, December 2004.
53. K. Ando, Fiber for reinforcing rubber products, US Patent 6 875 509, assigned to Asahi Fiber Glass Company, Limited (Tokyo, JP), April 5, 2005.
54. H. Watanabe, S. Makino, and T. Kuroda, Process for producing an adhesive-treated polyester fiber cord, US Patent 6 528 113, assigned to Teijin Limited (Osaka, JP), March 4, 2003.
55. M.B. Chan-Park and S.S. Tan, Thermal graft copolymerization of 4-vinyl pyridine on polyimide to improve adhesion to copper, *Int. J. Adhes. Adhes.*, 22(6):471–475, 2002.
56. J.S. Shih, B. Srinivas, and J.C. Hornby, Water soluble dye complexing polymers, US Patent 6 103 831, assigned to ISP Investments Inc. (Wilmington, DE), August 15, 2000.
57. G. Broze and D. Bastin, Built nonaqueous liquid nonioinic laundry detergent composition containing hexylene glycol and method of use, US Patent 4 873 012, assigned to Colgate-Palmolive Company (New York, NY), October 10, 1989.
58. D.W. Emery, N. Jones, H.d. Ruijter, and A.H.J. Strijbosch, Dye-transfer-inhibiting compositions and particulate detergent compositions containing them, US Patent 6 274 664, assigned to Unilever Home & Personal Care USA, Division of Conopco, Inc. (Greenwich, CT), August 14, 2001.
59. D.S. Caswell, N.Y. Sakkab, A.J. Danneels, K.P.M. Engels, R.A. Murphy, T. Trinh, E.H. Wahl, L. Waegemans, D.V.M. Lecluyse, D.G.J. Perot, K.L. Ignoul, T.D.K. Goetry, S.L. Diersing, and G.M. Frankenbach,

Highly concentrated fabric softener compositions and articles containing such compositions, US Patent 6 958 313, assigned to The Procter & Gamble Company (Cincinnati, OH), October 25, 2005.

60. G.M. Frankenbach, E.S. Baker, R.G. Baker, M.J. Declercq, H.J.M. Demeyere, R.M. Heiden, C.A. Hensley, B.A. Kolb, R.A. Murphy, R.E. Pegoli, T. Trinh, E.H. Wahl, M.R. Weaver, D.L. DuVal, J.H. Shaw, Jr., M. Nogami, and R. Wu, Clear or translucent aqueous fabric softener compositions containing high electrolyte content and optional phase stabilizer, US Patent 7 037 887, assigned to The Procter & Gamble Company (Cincinnati, OH), May 2, 2006.
61. G.A. Olah, Environmentally safe alkylation of aliphatic and aromatic hydrocarbons with olefins using solid HF-equivalent catalysts, US Patent 7 038 099, May 2, 2006.
62. G.A. Olah, T. Mathew, A. Goeppert, G. Rasul, G.K.S. Prakash, and P.M. Esteves, Carbocationic rearrangement cation and protonated of pivaloyl pivalaldehyde in superacid medium: A novel solution equivalent of the mclafferty rearrangement, *J. Am. Soc. Mass Spectrom.*, 15(7):959–965, July 2004.
63. W. Reppe and H. Kröper, Verfahren zur Herstellung von Carbonsäuren, ihren Anhydriden, Estern, Thioestern oder Amiden, DE Patent 765 969, assigned to IG Farbenindustrie Ag (De), June 01, 1953.
64. G. Kiss, Palladium-catalyzed Reppe carbonylation, *Chem. Rev.*, 101 (11):3435–3456, November 2001.
65. H.-C. Cheung, V. Santillan, M.O. Scates, E.C. Sibrel, and G.P. Torrence, Method and apparatus for sequesting entrained and volatile catalyst species in a carbonylation process, US Patent 6 627 770, assigned to Celanese International Corporation (Dallas, TX), September 30, 2003.
66. N. Yoneda, T. Minami, Y. Nakagawa, I. Ohta, A. Yamaguchi, H. Sugiyama, and F. Uemura, Process for the production of carbonyl compound, US Patent 6 066 762, assigned to Chiyoda Corporation (Yokohama, JP), May 23, 2000.
67. C.-J. Tsai, Y.-L. Liu, and H.-C. Tsai, Polymer-supported carbonylation catalyst and its use, US Patent 6 420 304, assigned to China Petrochemical Development Corporation (Taipei, TW), July 16, 2002.
68. F. Hung-Low, G.C. Uzcategui, M.C. Ortega, A.B. Rivas, J.E. Yanez, J. Alvarez, A.J. Pardey, and C. Longo, Hydroesterification and hydroformylation of 1-hexene catalyzed by rhodium complexes immobilized on poly(4-vinylpyridine), *Catal. Today*, 107-108:273–281, October 2005.
69. A.J. Pardey, M. Fernandez, J. Alvarez, C. Urbina, D. Moronta, V. Leon, M. Haukka, and T.A. Pakkanen, Catalysis of the water-gas shift reaction by [Rh(COD)(4-picoline)2]PF6 immobilized on poly(4-vinylpyridine): Characterization of the catalyst and the effect of temperature

under continuous-flow conditions, *Appl. Catal., A*, 199(2):275–283, June 2000.

70. A.J. Pardey, M. Fernandez, J. Alvarez, C. Urbina, D. Moronta, V. Leon, C. Longo, P.J. Baricelli, and S.A. Moya, The reduction of nitrobenzene as catalyzed by poly(4-vinylpyridine)-immobilized [Rh(COD)-(amine)2](PF6) complexes under WGSR conditions, *J. Mol. Catal. A: Chem.*, 164(1-2):225–234, December 2000.
71. A.J. Pardey, M. Fernandez, A.B. Rivas, M.C. Ortega, C. Urbina, D. Moronta, C. Longo, M. Mediavilla, P.J. Baricelli, and S.A. Moya, Water gas shift reaction and nitrobenzene reduction catalysis by tetracarbonyldi-[mu]-chlorodirhodium(i) complex immobilized on poly-(4-vinylpyridine), *Inorg. Chim. Acta*, 329(1):22–30, February 2002.
72. S. Xu, X. Xi, J. Shi, and S. Cao, A homogeneous catalyst made of poly(4-vinylpyridine-*co*-*N*-vinylpyrrolidone)-Pd(0) complex for hydrogenation of aromatic nitro compounds, *J. Mol. Catal. A: Chem.*, 160 (2):287–292, October 2000.
73. X. Xi, Y. Liu, J. Shi, and S. Cao, Palladium complex of poly(4-vinylpyridine-*co*-acrylic acid) for homogeneous hydrogenation of aromatic nitro compounds, *J. Mol. Catal. A: Chem.*, 192(1-2):1–7, February 2003.
74. B.S. Lee, S. Mahajan, and K.D. Janda, Cross-linked poly(4-vinylpyridine/styrene) copolymers as a support for immobilization of ytterbium triflate, *Tetrahedron*, 61(12):3081–3086, March 2005.
75. H.B. Friedrich and N. Singh, The very efficient oxidation of alcohols by poly(4-vinylpyridine)-supported sodium ruthenate, *Tetrahedron Lett.*, 41(20):3971–3974, May 2000.
76. H. El-Hamshary and F.N. Assubaie, Aqueous oxidation of 2,6-dimethylphenol with hydrogen peroxide catalyzed by transition metal complexes supported on polyvinylpyridinium salts, *J. Macromol. Sci., Pure Appl. Chem.*, A41(2):107–114, 2004.
77. C.H. Lu, Developer compositions containing quaternized vinylpyridine polymers, and copolymers, US Patent 4 397 934, assigned to Xerox Corporation (Stamford, CT), August 9, 1983.
78. A. Aoki, M. Nakaya, and T. Miyashita, Photopatterning using a cross-linkable polymer Langmuir-Blodgett film, *Macromolecules*, 31 (21):7321–7327, October 1998.
79. T. Miyashita, M. Nakaya, and A. Aoki, Molecular photopatterning with a two-dimensional network of polymer LB films, *Thin Solid Films*, 327-329:833–836, August 1998.
80. P.K. Bhowmik, R.A. Burchett, H. Han, and J.J. Cebe, Synthesis and characterization of poly(pyridinium salt)s with organic counterion exhibiting both lyotropic liquid-crystalline and light-emitting properties, *Polymer*, 43(6):1953–1958, March 2002.

81. L. Akcelrud, Electroluminescent polymers, *Prog. Polym. Sci.*, 28(6): 875–962, June 2003.
82. E. Vaganova and S. Yitzchaik, Tunable emission in poly(4-vinylpyridine)-based gel, *Acta Polym.*, 49(10–1):636–641, 1998.
83. H. Hong, R. Sfez, S. Yitzchaik, and D. Davidov, Led device based on self assembled poly(4-vinylpyridine) with PVK and PBD as transport layer, *Synth. Met.*, 102(1-3):1217–1217, June 1999.
84. H. Hong, R. Sfez, E. Vaganova, S. Yitzchaik, and D. Davidov, Electrostatically self-assembled poly(4-vinylpyridine-*co*-vinylpyridinium-chloride)-based led, *Thin Solid Films*, 366(1-2):260–264, May 2000.
85. C.T. Lima Luz and F.M.B. Coutinho, The influence of the diluent system on the porous structure formation of copolymers based on 2-vinylpyridine and divinylbenzene-diluent system: I. *n*-heptane/diethylphthalate, *Eur. Polym. J.*, 36(3):547–553, March 2000.
86. C.T. Lima Luz and F.M.B. Coutinho, The influence of the diluent system on the porous structure formation of copolymers based on 2-vinylpyridine and divinylbenzene. Diluent system: II. *n*-heptane/toluene, *Polymer*, 42(11):4931–4938, May 2001.
87. M. Grendze, D.W. McQuigg, J.T. Wyeth, E. Crowe, K.M. Weisheit, and E.F.V. Scriven, Crosslinked, non-swellable ampholytic base resins, US Patent 6 420 439, assigned to Reilly Industries, Inc. (Indianapolis, IN), July 16, 2002.
88. E.S.P. Bouvier, R.E. Meirowitz, and P.D. McDonald, Water-wettable chromatographic media for solid phase extraction, US Patent 6 773 583, assigned to Waters Investments Ltd. (Wilmington, DE), August 10, 2004.
89. N. Fontanals, P. Puig, M. Galia, R.M. Marce, and F. Borrull, New hydrophilic polymeric resin based on 4-vinylpyridine-divinylbenzene for solid-phase extraction of polar compounds from water, *J. Chromatogr. A*, 1035(2):281–284, May 2004.
90. Y. Fan, M. Zhang, and Y.-Q. Feng, Poly(acrylamide-vinylpyridine-*N,N'*-methylene bisacrylamide) monolithic capillary for in-tube solid-phase microextraction coupled to high performance liquid chromatography, *J. Chromatogr. A*, 1099(1-2):84–91, December 2005.
91. N.S. Shaglaeva, S.V. Amosova, L.P. Shaulina, I.P. Golentovskaya, G.A. Zavarzina, R.G. Sultangareev, S.V. Fedorov, and G.A. Pirogova, Copolymerization of divinyl sulfide with vinylpyridines and their derivatives, *Polym. Sci. Ser. B*, 46(5-6):121–125, May–June 2004.
92. H.-A. Tsai and M.-J. Syu, Synthesis and characterization of creatinine imprinted poly(4-vinylpyridine-*co*-divinylbenzene) as a specific recognition receptor, *Anal. Chim. Acta*, 539(1-2):107–116, May 2005.

93. L. Donato, A. Figoli, and E. Drioli, Novel composite poly(4-vinylpyridine)/polypropylene membranes with recognition properties for (s)-naproxen, *J. Pharm. Biomed. Anal.*, 37(5):1003–1008, April 2005.
94. A. Seebach and A. Seidel-Morgenstern, Enantioseparation on molecularly imprinted monoliths–preparation and adsorption isotherms, *Anal. Chim. Acta*, 591(1):57–62, May 2007.
95. Y. Liu, X. Chang, D. Yang, Y. Guo, and S. Meng, Highly selective determination of inorganic mercury(II) after preconcentration with Hg-(II)-imprinted diazoaminobenzene-vinylpyridine copolymers, *Anal. Chim. Acta*, 538(1-2):85–91, May 2005.
96. H.B. Sonmez and N. Bicak, Quaternization of poly(4-vinyl pyridine) beads with 2-chloroacetamide for selective mercury extraction, *React. Funct. Polym.*, 51(1):55–60, April 2002.
97. M. Khajeh, Y. Yamini, E. Ghasemi, J. Fasihi, and M. Shamsipur, Imprinted polymer particles for selenium uptake: Synthesis, characterization and analytical applications, *Anal. Chim. Acta*, 581(2):208–213, January 2007.
98. R.K. Nagarale, G.S. Gohil, and V.K. Shahi, Recent developments on ion-exchange membranes and electro-membrane processes, *Adv. Colloid Interface Sci.*, 119(2-3):97–130, February 2006.
99. M.Y. Kariduraganavar, R.K. Nagarale, A.A. Kittur, and S.S. Kulkarni, Ion-exchange membranes: preparative methods for electrodialysis and fuel cell applications, *Desalination*, 197(1-3):225–246, October 2006.
100. D.M. Stachera and R.F. Childs, Tuning the acid recovery performance of poly(4-vinylpyridine)-filled membranes by the introduction of hydrophobic groups, *J. Membr. Sci.*, 187(1-2):213–225, June 2001.
101. D.M. Stachera, R. F. Childs, A.M. Mika, and J.M. Dickson, Acid recovery using diffusion dialysis with poly(4-vinylpyridine)-filled microporous membranes, *J. Membr. Sci.*, 148(1):119–127, September 1998.
102. A.M. Mika, R.F. Childs, and J.M. Dickson, Ultra-low pressure water softening: A new approach to membrane construction, *Desalination*, 121(2):149–158, March 1999.
103. A.M. Mika, R.F. Childs, and J.M. Dickson, Chemical valves based on poly(4-vinylpyridine)-filled microporous membranes, *J. Membr. Sci.*, 153(1):45–56, February 1999.
104. A.J.B. Kemperman, ed., *Handbook on bipolar membrane technology*, Twente University Press, Enschede, 2000.
105. K.N. Mani, Electrodialysis water splitting technology, *J. Membr. Sci.*, 58(2):117–138, May 1991.
106. E. Da Silva, L. Lebrun, and M. Metayer, Elaboration of a membrane with bipolar behaviour using the semi-interpenetrating polymer networks technique, *Polymer*, 43(19):5311–5320, September 2002.

107. L. Lebrun, E. Da Silva, G. Pourcelly, and M. Métayer, Elaboration and characterisation of ion-exchange films used in the fabrication of bipolar membranes, *J. Membr. Sci.*, 227(1-2):95–111, December 2003.
108. T. Sata, K. Kawamura, and K. Matsusaki, Electrodialytic transport properties of anion-exchange membranes prepared from poly(vinyl alcohol), poly(*N*-ethyl 4-vinylpyridinium salt) and [beta]-cyclodextrin, *J. Membr. Sci.*, 181(2):167–178, January 2001.
109. J.-J. Qin and T.-S. Chung, Development of high-performance polysulfone/poly(4-vinylpyridine) composite hollow fibers for CO_2/CH_4 separation, *Desalination*, 192(1-3):112–116, May 2006.
110. J.-J. Shieh, T.-S. Chung, R. Wang, M.P. Srinivasan, and D.R. Paul, Gas separation performance of poly(4-vinylpyridine)/polyetherimide composite hollow fibers, *J. Membr. Sci.*, 182(1-2):111–123, February 2001.
111. S. Suryanarayan, A.M. Mika, and R.F. Childs, Gel-filled hollow fiber membranes for water softening, *J. Membr. Sci.*, 281(1-2):397–409, September 2006.
112. Y. Li, M.J. Yang, and Y. She, Humidity sensitive properties of crosslinked and quaternized poly(4-vinylpyridine-*co*-butyl methacrylate), *Sens. Actuators, B*, 107(1):252–257, May 2005.
113. Y. Sakai, Y. Sadaoka, and M. Matsuguchi, Humidity sensors based on polymer thin films, *Sens. Actuators, B*, 35(1-3):85–90, September 1996.
114. Y. Sakai, Y. Sadaoka, and M. Matsuguchi, A humidity sensor using cross-linked quaternized polyvinylpyridine, *J. Electrochem. Soc.*, 136 (1):171–174, 1989.
115. Y. Sakai, Y. Sadaoka, M. Matsuguchi, Y. Kanakura, and M. Tamura, Humidity sensor using polytetrafluoroethylene-graft-quaternized-polyvinylpyridine, *J. Electrochem. Soc.*, 138(8):2474–2478, August 1991.
116. S. Wu, Y. Zhu, F. Li, and J. Shen, Humidity sensing properties of the vinylpyridine-butyl acrylate-styrene copolymers, *J. Appl. Polym. Sci.*, 74(8):1992–1996, November 1999.
117. M.J. Yang, Y. She, and Y. Li, Humidity sensitive properties of quaternized poly(4-vinylpyridine-*co*-butyl methacrylate), *J. Mater. Sci. Lett.*, 21(19):1477–1479, October 2002.
118. Q. Yang, L. Wang, W. Xiang, J. Zhou, and J. Li, Grafting polymers onto carbon black surface by trapping polymer radicals, *Polymer*, 48 (10):2866–2873, May 2007.
119. Y. Li, L. Hong, Y. Chen, H. Wang, X. Lu, and M. Yang, Poly(4-vinylpyridine)/carbon black composite as a humidity sensor, *Sens. Actuators, B*, 123(1):554–559, April 2007.
120. A. Liu and J.-I. Anzai, A poly(4-vinylpyridine) derivative bearing Os-(5,6-dmphen)2Cl (5,6-dmphen=5,6-dimethyl-1,10-phenanthroline):

A novel electrochemical indicator for detecting DNA hybridization, *Mater. Sci. Eng., C*, 24(4):503–505, June 2004.

121. S. Poyard, C. Martelet, N. Jaffrezic-Renault, S. Cosnier, and P. Labbe, Association of a poly(4-vinylpyridine-*co*-styrene) membrane with an inorganic/organic mixed matrix for the optimization of glucose biosensors, *Sens. Actuators, B*, 58(1-3):380–383, September 1999.
122. M. Guenther, D. Kuckling, C. Corten, G. Gerlach, J. Sorber, G. Suchaneck, and K.-F. Arndt, Chemical sensors based on multiresponsive block copolymer hydrogels, *Sens. Actuators, B*, 126(1):97–106, September 2007.
123. J.-F. Brunelli and S. Fouquay, Acrylic copolymers as additives for inhibiting paraffin deposition in crude oils, and compositions containing same, US Patent 6 218 490, assigned to CECA S.A. (Puteaux, FR), April 17, 2001.
124. K.N. Bakeev, J.-C. Chuang, T. Winkler, M.A. Drzewinski, and D.E. Graham, Method for preventing or retarding the formation of gas hydrates, US Patent 6 281 274, assigned to ISP Investments Inc. (Wilmington, DE), August 28, 2001.
125. K.N. Bakeev, J.-C. Chuang, T. Winkler, M.A. Drzewinski, and D.E. Graham, Corrosion inhibition during transport of water and a hydrocarbon through a pipeline, US Patent 6 432 355, assigned to ISP Investments Inc. (Wilmington, DE), August 13, 2002.
126. M.L. Stephens and H.F. Efner, Cement composition and process therewith, US Patent 6 124 383, assigned to Phillips Petroleum Company (Bartlesville, OK), September 26, 2000.
127. F.J. Boden, R.P. Sauer, I.L. Goldblatt, and M.E. McHenry, Polar grafted polyolefins, methods for their manufacture, and lubricating oil compositions containing them, US Patent 6 686 321, assigned to Castrol Limited (GB), February 3, 2004.
128. Y. Abed, Z. Arrar, B. Hammouti, A. Aouniti, S. Kertit, and A. Mansri, An electrochemical study of the action of a poly(vinylpyridine) derivative as inhibitor for corrosion of iron in 1 M H_2SO_4, *J. Chim. Phys. Phys.- Chim. Biol.*, 96(8):1347–1355, 1999.
129. A. Chetouani, K. Medjahed, K.E. Benabadji, B. Hammouti, S. Kertit, and A. Mansri, Poly(4-vinylpyridine isopentyl bromide) as inhibitor for corrosion of pure iron in molar sulphuric acid, *Prog. Org. Coat.*, 46 (4):312–316, June 2003.
130. A. Chetouani, K. Medjahed, K.E. Sid-Lakhdar, B. Hammouti, M. Benkaddour, and A. Mansri, Poly(4-vinylpyridine-poly(3-oxide-ethylene) tosyle) as an inhibitor for iron in sulphuric acid at 80 °C, *Corros. Sci.*, 46(10):2421–2430, October 2004.
131. C.A. Stone and K. Dunn, Vinylpyridine: An unusual cause of a chemical burn, *Burns*, 22(2):150–151, March 1996.

132. X. Peng and J. Shen, Preparation and biodegradability of polystyrene having pyridinium group in the main chain, *Eur. Polym. J.*, 35(9): 1599–1605, September 1999.
133. N. Kawabata, W. Sakakura, and Y. Nishimura, Static suppression of tomato bacterial wilt by bacterial coagulation using a new functional polymer that coagulates bacterial cells and is highly biodegradable, *Biosci. Biotechnol. Biochem.*, 69(3):537–543, March 2005.
134. N. Kawabata, H. Kishimoto, T. Abe, T. Ikawa, K. Yamanaka, H. Ikeuchi, and C. Kakimoto, Control of tomato bacterial wilt without disinfection using a new functional polymer that captures microbial cells alive on the surface and is highly biodegradable, *Biosci. Biotechnol. Biochem.*, 69(2):326–333, February 2005.

8

Poly(vinylimidazole)

8.1 Monomers

Imidazole is produced by the *Radziszewski* reaction*. A 1,2-carbonyl compound is condensed with an aldehyde and ammonia (1). The reaction is carried out in water at 50–100°C (2). Several routes for the synthesis of imidazole derivates have been described (3). 1-Vinylimidazole (VI) is produced by a base catalyzed addition of acetylene to imidazole. Synonyms for VI are simply vinylimidazole, *N*-vinylimidazole, or 1-ethenyl-1H-imidazole. Vinylimidazole related monomers are shown in Figure 8.1.

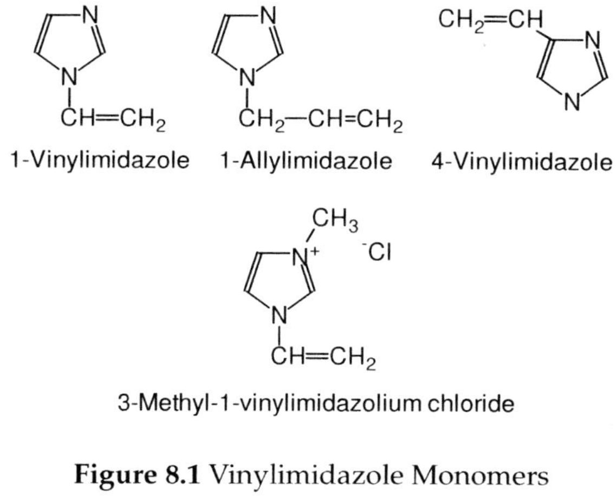

Figure 8.1 Vinylimidazole Monomers

*Bronisław Radziszewski 1838–1914

VI is used in copolymers in the production of cationic polymers. Alkyl imidazoles are used as hardeners for epoxy resins and urethanes. Quaternary imidazolinium derivates are functional as algicide and have anti-bactericidal properties. VI can be quaternized in the 3-position to yield, a variety of quaternized monomers. These monomers are shown in Table 8.1. These monomers are used for

Table 8.1 Quaternized Vinylimidazoles (4)

Monomer
3-Methyl-1-vinylimidazolium chloride
3-Benzyl-1-vinylimidazolium chloride
3-*n*-Dodecyl-1-vinylimidazolium bromide
3-*n*-Octadecyl-1-vinylimidazolium chloride

copolymers in detergent compositions.

8.1.1 Comonomers

A number of compounds are suitable as comonomers for VI in the preparation of copolymers. If the comonomer has hydrophilic properties, an especially high amount of comonomers can be introduced into the copolymer with VI without markedly decreasing water solubility. Comonomers for VI described in this chapter are shown in Table 8.2.

Table 8.2 Comonomers for 1-Vinylimidazole

Comonomer	Examples of Use
N-Vinyl-2-pyrrolidone	Dye transfer inhibitors (DTI)s
Vinyl acetate (VA)	Printing Applications
Hydroxyalkyl acrylate	Adhesive compositions
Vinylmethylacetamide	Adhesive compositions
Acrylonitrile	Chelating hydrogels
Divinylbenzene	Crosslinking agent in imprinted polymer catalysts
N,N'-Divinylethyleneurea	Crosslinking agent in dye transfer inhibitors

In order to maintain the water solubility the amount of hydrophobic monomers in the VI copolymers has to remain low. However, hydrophobic comonomers in the VI copolymers can contribute to a decrease in the permeability of gases.

8.2 Polymerization and Fabrication

8.2.1 Solution Polymerization

Poly(1-vinylimidazole) (PVI) and copolymers with VI can be obtained by radical polymerization of VI or copolymerization with a suitable comonomer. The polymerization can be performed in an aqueous solution. In this case, water-soluble initiators, such as 2,2′-azobis-(2-amidinopropane)dihydrochloride, c.f. Figure 8.2, are used (5). Also, the preparation of copolymers of VI and VA have been described.

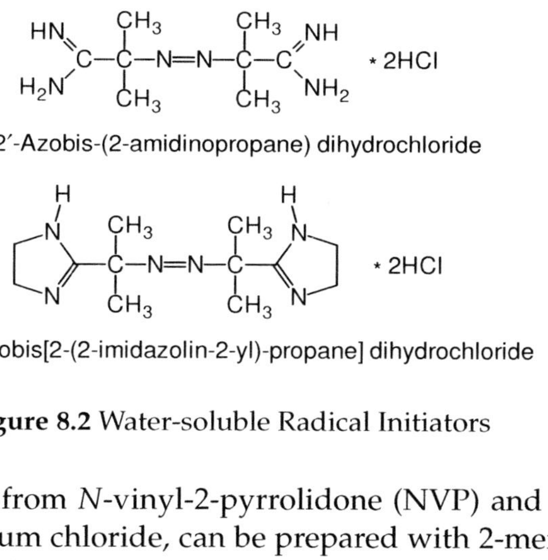

Figure 8.2 Water-soluble Radical Initiators

A copolymer from *N*-vinyl-2-pyrrolidone (NVP) and 3-methyl-1-vinylimidazolium chloride, can be prepared with 2-mercaptoethanol and 2,2′-azobis-(2-amidinopropane)dihydrochloride in aqueous solution at 65°C. Similarly a copolymer from NVP and VI is prepared using 2,2′-azobis[2-(2-imidazolin-2-yl)propane]dihydrochloride (4).

Copolymers from NVP and VI are prepared in aqueous solution using 2,2′-azobis(2-methylbutyronitrile) in isopropanol as initiator (6). The mixture is then stirred at 85°C for 1 *h*. The reaction is completed at 60°C by adding *tert*-butyl hydroperoxide and sodium bisulfite. 2-mercaptoethanol can be added as a chain transfer agent (CTA).

8.2.2 *Precipitation Polymerization*

PVI or copolymers containing VI can be manufactured by a radical polymerization in solvents wherein the monomers dissolve and the PVI or the VI-containing copolymer precipitate. Examples for such solvents are butanone and isopropanol. In order to increase the yield of the polymerization, a nonpolar solvent can be added in amounts that the monomer remains dissolved but the precipitation of PVI or VI-containing copolymers takes place to an increased extent. For this purpose the combination of butanone and hexane is suitable.

Copolymers of VI with vinylmethylacetamide have been described that are prepared according to the precipitation method (5). Copolymers of VI and NVP, strongly crosslinked with *N,N'*-divinylethyleneurea, can be synthesized by radical polymerization with 2,2'-azobis-(2-amidinopropane)dihydrochloride. As precipitating agent, poly(ethylene glycol) (PEG) or poly(oxyethylene) (POE)-nonylphenyl ether is used (7). A yellowish microporous powder with an average particle size of 450 μm is obtained.

8.2.3 *Grafting*

VI and acrylic acid (AA) can be grafted by γ-radiation on to poly(propylene) (PP) (8). Due to the absence of any reactive functional groups, PP exhibits poor moisture absorption and poor dye uptake. These problems can be overcome by a surface modification.

The surface of poly(tetrafluoroethylene) (PTFE) and several other polymers (9–11) can be modified by plasma treatment and UV-induced grafting of 4-vinylpyridine (4-VPy), 2-vinylpyridine (2-VPy), or, VI. On to the modified surface, copper or nickel can be deposited by electroless plating (12, 13).

Poly(ethylene terephthalate) (PET) fibers can be grafted in *N,N*-dimethylformamide (DMF) with VI by 2,2'-azobisisobutyronitrile (AIBN) as radical initiator (14). The grafting reaction changes the properties of the PET fibers, such as intrinsic viscosity, water absorption capacity, fiber diameter and mechanical properties.

8.2.4 *Organic-inorganic Hybrid Materials*

A vinylimidazole silica hybrid can be prepared by performing a free radical addition reaction of vinylimidazole and 3-(methacryloxy)-propyl trimethoxysilane, and then hybridizing with tetramethoxy silane (TMOS) by a sol-gel technique (15).

The radical copolymerization is performed in tetrahydrofuran (THF) using AIBN as initiator. In the second step, THF, TMOS and hydrochloric acid is added to the reaction mixture to perform the condensation reaction. The reaction product is precipitated by an excess of *n*-hexane. The hybrid polymer exhibits an excellent thermal stability.

8.3 Properties

PVI shows a high degree of oxygen permeability. Commercial copolymers of NVP and vinylimidazole are often yellow in color. When crude vinylimidazole is flash distilled under vacuum, substantially colorless polymers can be obtained (16).

8.4 Applications

8.4.1 *Lithographic Printing*

Lithography is a printing technique. It is used to print text or artwork on to paper or on to another suitable material. In fact, the modern techniques of lithography utilize photographic techniques. Therefore, it is basically not correct to differentiate between lithography and photolithography. Photolithography is employed in the manufacture of semiconductor devices (17). Photolithography is also addressed as microlithography and nanolithography. It is similar to the conventional lithography used in book printing. Lithographic techniques consist of the following main steps:

- Preparation of the substrate
- Coating with a photoresist film
- Selective irradiation
- Development and etching to form a resist pattern on the substrate

- Final steps as rinsing and drying.

Photoresists include a negative photoresist in which unexposed portions to the active ray or radiation are dissolved and removed in development, and a positive photoresist, in which exposed portions to the active ray or radiation are dissolved and removed in development. Negative or positive photoresists are selected and used according to the intended purpose.

8.4.1.1 *Protective Top Coats*

Water-soluble polymers are used as temporary coatings on organic light sensitive substrates in the manufacture of lithographic printing plates. The polymers should protect the substrate from aerial oxygen during storage exposure and during the time between exposure and development.

Water-soluble layers containing poly(vinyl alcohol) and PVI or a copolymer of VI exhibit favorable oxygen impermeability properties and an improved adhesion to organic substrates (5). In contrast to many other additives to poly(vinyl alcohol) layers, large amounts of PVI can be used without entailing disadvantages for the oxygen impermeability.

During the developing step of the light sensitive systems, the layers are removed residue-free with water or aqueous developing solutions so that no negative properties, such as impaired ink receptivity are entailed if the processed layers are used as printing molds.

8.4.1.2 *Anti-reflection Coatings*

In the photolithographic formation of a resist pattern, it is understood that multiple interference of light occurs in a photoresist film, and that the width of the resist pattern varies with a varying thickness of the photoresist film. The multiple interference of light occurs because irradiated light of single wavelength coming into the photoresist film formed on a substrate interferes with reflected light from the substrate, and the quantity of absorbed light energy varies in the thickness direction of the photoresist film. The variation in the thickness of photoresist film affects the width of the resulting resist pattern after development, thereby deteriorating the dimensional precision of the resist pattern.

A composition for lithographic anti-reflection coating with a balanced compatibility and a conventional photoresist composition includes a copolymer of vinylimidazole and NVP. Further, a fluorine-containing surfactant is added, such as perfluorooctanoic acid, or perfluorooctylsulfonic acid (18).

Perfluorooctylsulfonic acid is advantageous as it has a high inhibitory activity against interference, a high solubility in water, and is easy to adjust the pH of the resulting composition. On the other hand, for safety and health reasons, perfluorooctanoic acid is preferred.

8.4.2 *Printing Inks*

Ink compositions for use in ink jet printers with an improved water fastness can be prepared by the addition of PVI (19). PVI is synthesized by a radical polymerization in aqueous solution. Ammonium persulfate is used a radical initiator and 2-mercaptoethanol is used to adjust a weight average molecular weight of 800–30,000 *Dalton*. Detailed prescriptions of how to prepare ink compositions are given in the literature (19).

8.4.3 *Dye Transfer Inhibitors*

One of the most persistent and troublesome problems arising during conventional laundering operations is the tendency of some colored fabrics to release their dyes into the laundering solutions. These so-called fugitive dyes are then transferred on to other fabrics oftentimes having colors different from the fugitive dyes. If, for example, white laundry is washed together with colored textiles, the white laundry is stained.

The dye transfer problem can be fixed by complexing or by absorbing the fugitive dyes before they have the opportunity to become attached to other fabric articles in the laundering solution (20).

Water-soluble copolymers containing VI, 4-VPy *N*-oxide units, NVP, or related monomers are suitable as DTIs in detergents for laundry applications (6). Common comonomers are NVP and VI, to form binary copolymers (21). Copolymers from VI and NVP are very compatible with other ingredients of the laundry formulation, such as clays in that the dye transfer inhibiting properties of the polymers are not adversely affected by the presence of clays formulated therewith. In addition, the softening performance of the clays is maintained in presence of the dye transfer inhibiting copolymers.

Polymers from quaternized vinylimidazoles, such as 3-methyl-1-vinylimidazolium chloride have been also proposed as DTIs (4). Likewise, copolymers of VI and NVP can be partially quaternized with dimethyl sulfate to result a terpolymer with moieties of VI, 1-vinyl-3-methylimidazolium methosulfate and NVP. In addition, benzyl chloride is a suitable agent for quaternization (22).

Grafted proteins improve the primary and secondary detergency and the soil release properties of phosphate-free and low-phosphate detergent and cleaner formulations. Unlike globular proteins, which are usually readily soluble in water, scleroproteins, such as keratin, elastin, fibroin and collagen are in general not soluble in water. However, these proteins can be at least be partially degraded so that they become soluble in water. Copolymers prepared in the presence of water-soluble proteins or protein hydrolyzates exhibit at least partial degradability.

Water-soluble crosslinked copolymers from NVP and VI can be prepared by the addition of *N,N′*-divinylethyleneurea as crosslinking agent and 2-mercaptoethanol as CTA by radical copolymerization (23). Care must be taken to avoid gelation. Crosslinked copolymers bind small amounts of dye distinctly more strongly than water-soluble DTIs. Thus, these polymers have a great advantage on use by comparison with the non-crosslinked water-soluble products.

Crosslinked water-insoluble polymers exhibit not only excellent dye transfer inhibition, but anti-fading benefits as well. A synergistic dye transfer inhibition action can be obtained if water-insoluble systems are used in combination with water-soluble systems (24). Crosslinked copolymers of this type are also suitable for absorbing heavy metals from wastewaters or beverages (25).

Polymeric DTIs in laundry detergents can be identified by pyrolysis gas chromatography (26). The identification runs via the monomers. A maximum of monomer yield is obtained at a pyrolysis temperature around 650–700°C.

8.4.4 *Adhesive Compositions*

Compact disks (CD)s and digital versatile discs (DVD)s include two polycarbonate substrates bonded together through an adhesive composition. One or both of the polycarbonate discs may contain data. To enable the data to be read by an optical reader, at least one of the polycarbonate discs has a metal-coated surface, e.g., a 50 *nm* coating of aluminum.

One problem in selecting a suitable adhesive composition for the discs is the potential for polycarbonate to outgas which makes it difficult for many adhesives to maintain the integrity of an initially formed bond between the adhesive and the surfaces of the discs.

If debonding occurs it can be difficult or impossible for the optical reader to read the data. In addition, many adhesive compositions contain chemical species that corrode the metal coating. The rate of corrosion and debonding tends to increase as humidity and temperature increase.

A suitable adhesive composition for CDs includes an adhesive polymer from *N*-vinyl caprolactam (VCL), NVP, or VI, and an AA ester monomer, such as 2-ethylhexyl acrylate (EHA) or isooctyl acrylate (27). In the first step, the monomers, e.g., EHA and VCL are blended with a photoinitiator, such as 2,2-dimethoxy-1,2-diphenylethane-1-one. The mixture is then partially photopolymerized with ultraviolet light to form a syrup.

In the next step, the syrup is coated under exclusion of oxygen at a desired thickness. During or after the coating process the syrup is further exposed to energy to complete the polymerization and to crosslink the adhesive composition. Then the film is laminated to the surface of a polycarbonate substrate.

Alternatively, the adhesive composition may be applied directly to a surface of the polycarbonate substrate followed by polymerization of the adhesive composition or a release treated surface can be laminated to the adhesive composition.

When the adhesive composition is provided in the form of a film, it is preferably sandwiched between two release liners where one of the release liners has a release coating that exhibits a lesser degree of adhesion to the adhesive film relative to the release coating of the second release liner.

Several test procedures to elucidate the performance of the adhesive have been described, including (27):

- Accelerated aging test methods
- Peel strength test methods
- Cleavage test methods
- Optical clarity test methods
- DVD performance test method.

8.4.5 Lubricating Additives

Grafted copolymers of nitrogenous, heterocyclic monomers with poly(olefin)s have been proposed for use in lubricating oils as viscosity index improving agents and as dispersants for keeping the insoluble materials in the crankcase of an internal combustion engine in suspension. VI can be grafted on to poly(olefin)s. Other monomers suitable for grafting are 1-allylimidazole, *N*-methyl-*N*-vinylacetamide and 4-VPy. In lubricating oils, these grafted poly(olefin)s act as an viscosity index improver (28,29).

The preparation of a grafted poly(olefin) in a pilot plant is performed under fairly similar conditions as those in a laboratory. The poly(olefin) in mineral oil solution is heated under inert atmosphere up to 170–190°C. Then around 1% VI of the poly(olefin) is added over period of 1 *min*. After thoroughly mixing for 20 *min*, 0.20% di-*tert*-butyl peroxide initiator is added within 30 *min*. The resulting mixture is allowed to react for another 30–60 *min*. The preparation is also possible in a twin screw extruder.

In addition, diblock copolymers consisting of styrene and poly(isoprene) blocks have been used as base polymers for grafting of VI (30). Before grafting, the poly(isoprene) block is hydrogenated to remove most of its original ethylenic unsaturation. A series of organic peroxides have been demonstrated to be useful to initiate grafting in the extruder.

Inhibitors may be added to limit the degree of crosslinking of the poly(olefin). Limiting the degree of crosslinking will reduce the viscosity increase resulting from the grafting reaction and provide a final grafted poly(olefin) as well which has the benefit of improved shear stability. Suitable inhibitors are hindered phenols, such as octadecyl-3,5-di-*tert*-butyl-4-hydroxyhydrocinnamate or hydroquinone (28).

8.4.6 *Additives for Electrolytes*

Rechargeable lithium ion batteries are widely used as small batteries with high energy densities. It is important to inhibit the reaction of the anode with an electrolyte. When the potential of the anode is going down to a very base potential, the anode is likely to react with the electrolyte. This can considerably influence the performance of the battery, especially battery capacity, service time, and cycling characteristics.

For nonaqueous solvents for lithium ion secondary batteries, organic solvents, such as ethylene carbonate, dimethyl carbonate, methyl ethyl carbonate, γ-butyrolactone, methyl propionate, butyl propionate and ethyl propionate are used.

To inhibit reactions of the anode with the electrolyte, a monomer, which is capable of anionic polymerization, is added. This monomer forms a protective film on the surface of the anode when charging. Examples of such anionic polymerizable monomers are isoprene, styrene, 2-VPy, VI, *n*-butyl acrylate, ethyl acrylate, methyl methacrylate, NVP, ethyl cinnamate, methyl cinnamate, ionone, and myrcene (31).

8.4.7 *Sensors*

8.4.7.1 *Oxygen Sensor*

Polymeric porphyrin cobalt complexes as shown in Figure 8.3 can act as synthetic carriers for oxygen. In contrast to hemoglobin they exhibit a lower tendency for irreversible oxidation by O_2. The cobalt complexes are immobilized on copolymers. The oxygen affinity depends strongly on the nature of the polymeric ligand, as demonstrated with copolymers of 4-vinylimidazole and VI and 4-VPy and 2-VPy (32).

Copolymers of 2,2,3,3,4,4,5,5-octafluoropentyl methacrylate and VI or copolymers of octyl methacrylate and VI absorb oxygen in a rapid reaction (33,34). The oxygenated complexes show a maximum of light absorption at around 550 *nm*. Membranes based on these

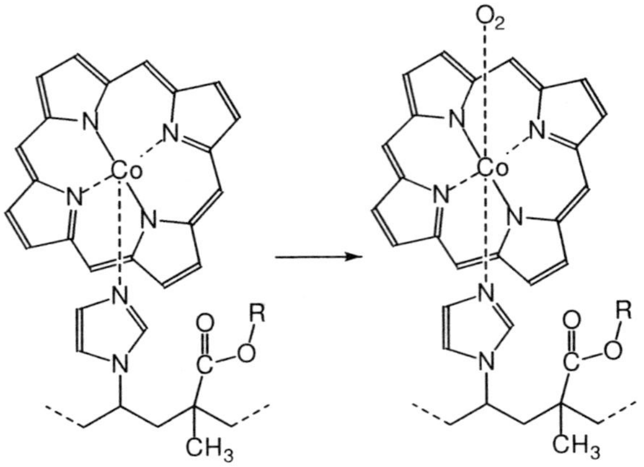

Figure 8.3 Polymeric Porphyrin Cobalt Complexes (33)

copolymer complexes can serve as oxygen sensitive components in optical sensors for oxygen.

8.4.8 Enzyme Related Technology

8.4.8.1 Enzyme Immobilization

Immobilized Catalysts. Immobilized enzymes have a particular relevance in food technology. The immobilization can be achieved by adsorption to insoluble materials, entrapment in polymeric matrix, encapsulation, or grafting on to an insoluble support. The method of adsorption to a solid support is easy to perform and may result in reversibly immobilized enzymes.

Glucose oxidase (GOD) (35), yeast invertase (36), or α-amylase (37) can be immobilized on to a polymeric support by adsorption. To poly(1-vinylimidazole-*co*-ethylene glycol dimethacrylate (EGDM)) hydrogel beads, copper salts are added to form copper chelates via the pendent imidazole moieties. The adsorption of invertase on to these polymeric metal chelates has been studied. Polymers that are not bearing metal chelates show only a weak adsorption of invertase, whereas polymers with copper chelates exhibit a significant adsorption capacity.

The storage stability of immobilized enzymes is in general better than the storage stability of the free enzymes. For example, the yeast invertase loses all of its activity within 30 days in solution. The adsorbed yeast invertase loses 20% of its activity during the same period (36). The most important advantage of immobilization is the repeated use of enzymes.

The enzymes can be desorbed by treatment with a competitive chelating agent, i.e., ethylenediamine tetraacetic acid (EDTA). The matrix can be repeatedly used for loading with invertase. The adsorption capacity for invertase does not change during 10 successive adsorption/desorption cycles.

Enzyme Electrodes. Amperometric biosensors consist of a graphite electrode with an immobilized enzyme. The enzyme is immobilized in a polymer by a coupling agent. The polymer contains a complex for the mediation of the electric potential. The enzyme performs the appropriate reactions in presence of the analyte by which the electric potential is established. The quantity of the analyte can thereby be determined. The first enzyme electrode was described by Clark and Lyons (38).

In contrast to the common sense that suggests that enzymes would be operational only in aqueous systems, biosensors for nonaqueous environments are available (39). Solvents, such as acetonitrile, acetone, 2-butanol, chloroform, hexane, and THF are suitable for organic phase amperometric biosensors.

The analysis of the primary amine histamine, c.f. Figure 8.4 is important in clinical and food chemistry.

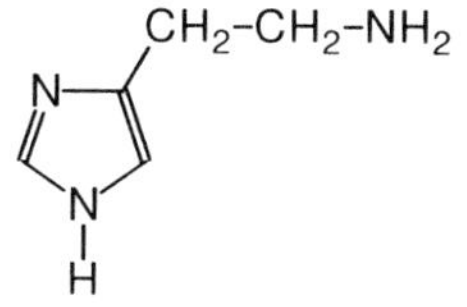

Figure 8.4 Histamine

For example, histamine is produced in fish products by the bacterial decarboxylation of histidine. If the fish products are not properly stored histamine may accumulate to toxic levels.

Histamine can be analyzed by chromatographic techniques and by biosensors using histamine oxidase, methylamine dehydrogenase, and histamine dehydrogenase. Histamine dehydrogenase catalyzes the oxidative deamination of histamine to imidazole, acetaldehyde, and ammonia. This enzyme shows its highest activity toward histamine among other amines. It is thermally stable and does not need molecular oxygen as an electron acceptor.

Biosensors based on amine oxidase (AO) have been compared. One is based on adsorbed AO on graphite electrodes. Here, the detection is based on a direct electron transfer mechanism. The second design is based on Os(4, 4′−dimethylbipyridine)$_2$Cl$_2$ complexed with PVI (40). Both electrode designs can detect histamine in the μmol range. However, the complex based electrodes show superior characteristics with regard to stability, selectivity, and linear range.

Histamine dehydrogenase can be immobilized in PVI complexed with Os(4, 4′−dimethylbipyridine)$_2$Cl$_2$ (41). The enzyme is coupled to the polymer with PEG diglycidyl ether.

To a glassy carbon electrode, a solution containing the enzyme, the polymer, and the coupling agent is deposited and dried (42). The histamine is detected using a flow injection analysis system, with the electrode as detector. A detection limit for histamine of 100 $p\,mol$ could be reported.

Several other biosensors utilize enzymes immobilized in PVI according to the principles described in this section. The devices are summarized in Table 8.3.

Chiral Stationary Phases. Human serum albumin can be immobilized on quaternized PVI coated silica anion exchangers to give chromatographic chiral stationary phases (43, 44). PVI was synthesized by radical polymerization with AIBN as initiator in presence of methanol and quaternized with methyl iodide (45). Enantiomeric separations have been demonstrated with racemic mixtures of tryptophan, warfarin, and oxazepam.

8.4.8.2 Artificial Enzymes

Enzymes as catalysts in industrial processes offer the advantages of mild conditions, reduced side reactions, and specificity towards

Table 8.3 Biosensor Electrodes with Poly(1-vinylimidazole) Metal Complexes

Enzyme	Analytes	References
Tyrosinase	Phenols	(39)
Amine oxidase	Histamine	(40, 41)
Quinohemoprotein alcohol dehydrogenase	Alcohols	(46, 47)
Alcohol oxidase, horseradish peroxidase	Alcohols	(48)
Glycerol dehydrogenase	Glycerol	(49)
Glucose oxidase	Glucose	(50, 51)
Horseradish peroxidase	Hydrogen peroxide	(52)
L-Glutamate oxidase, horseradish peroxidase[a]	*L*-Glutamate	(53)
Cellobiose dehydrogenase, oligosaccharide dehydrogenase[a]	Sugars	(54)

[a] Dual enzyme system

substrates. The disadvantages of enzymes in comparison to classical chemical catalysts are their minor stability in organic solvents and their expense. Tailor made polymeric catalysts are more durable and more resistant to harsh environments than biological molecules.

Molecular imprinting is a promising technique for the synthesis of such artificial enzyme catalysts. Molecular imprints, in contrast to biological molecules, are highly crosslinked materials. They are tolerant to conditions that denature most biological molecules.

In simplest form, a functional polymer, such as vinylimidazole is assembled around a template molecule, e.g., *p*-nitrophenyl phosphate and copolymerized with divinylbenzene (DVB) as crosslinking agent. After the removal of the template the resulting polymers are efficiently catalyzing the hydrolysis of *p*-nitrophenyl acetate (55). The template polymerization technique has several other interesting issues as reviewed in the literature (56).

However, in general, it is not easy to find the optimal template molecule. Thus, the development of a molecular imprinted catalyst (MIC) by the trial-and-error approach can be time consuming. Combinatorial techniques have been developed to optimize MICs (57).

Another approach uses molecular modelling software (58). These techniques involve the construction and screening of a virtual library of intermediates of an enzyme-catalyzed reaction. The goal is to identify the most stable intermediate which should provide an optimal catalytic performance (59). According to the theoretical results the MIC is then synthesized.

The concept pointed out in general has been exemplified for the transesterification reaction between *p*-nitrophenyl acetate and hexanol (59,60). According to the optimization results, a MIC was prepared by copolymerizing 4-vinylimidazole and itaconic acid with trimethylpropanol trimethacrylate in the presence of *p*-nitrophenyl acetate. The MIC synthesized in this way exhibited a substrate specificity with a 6.5 fold preference for *p*-nitrophenyl acetate over *p*-nitrophenyl salicylate.

In the same way an artificial phosphotriesterase, an artificial creatininase, and an artificial hydantoinase, was synthesized by the copolymerization of 4-vinylimidazole and zinc methacrylate or cobalt methacrylate (61,62).

Pores in the matrix can be imparted by using either methanol as porogen (59) or by polymerizing in a silica matrix, which is subsequently dissolved by hydrofluoric acid (63).

8.4.8.3 Biofuel Cells

Biofuel cells consist of a two-electrode set modified by biocatalytic enzymes to specifically oxidize and reduce substrates respectively. For implantable biofuel cells these substrates are available in vivo. Most biofuel cells are simple in design and easy to miniaturize (64–66). Certain requirements are necessary for implantable biofuel cells.

- No harmful products should be produced by the electrode reactions
- The naturally available molecules should be suitable for the electrode reactions to get electrical power.

A practical design of an implantable, membrane-less and biocompatible biofuel cell consists of a GOD-based anode and a laccase-based cathode (67). The electron transfer to or from the biocatalytic active sites is mediated by polymer bound or entrapped redox complexes.

Laccase is a naturally occurring enzyme, for example in fruits. It is responsible for the development of the brownish color of peeled fruits. The efficiency of laccase depends on its origin (68) and on the pH. Both components of the fuel, glucose, and O_2 are found in body fluids. GOD oxidizes glucose in presence of oxygen to gluconolactone or gluconic acid.

Several redox polymers and complexes for the mediation of the reduction of O_2 have been developed (67,69,70). The polymer either is made up of vinylimidazole units or is a copolymer of vinylimidazole and acrylamide. To the backbone, osmium bis-(1,10-phenanthroline) is coordinated via two imidazole moieties. The structure of the polymer complex is shown in Figure 8.5.

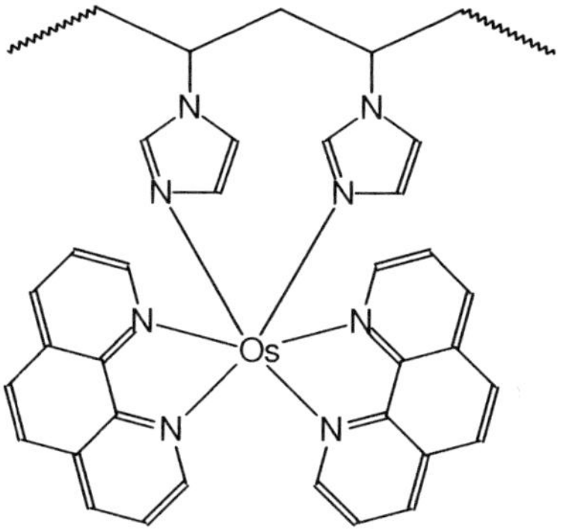

Figure 8.5 Electron Transfer Mediating Polymer Complex (67)

The polymer also serves to immobilize the enzyme. Thereby the enzyme must retain its activity. A systematic study of various copolymers, using acrylonitrile (AN) as the main component, reveals that VI as comonomer, using GOD as enzyme, is most useful in this aspect (71).

The redox electrode is prepared in the following manner. Graphite disc electrodes with a diameter of 6 *mm* are coated with solutions containing the laccase or GOD enzyme, the redox polymer and POE bis-glycidyl ether as a crosslinking agent. The films are dried for 24 *h* (67).

The maximum theoretical electromotive force calculated from thermodynamic considerations of the glucose oxidation and oxygen reduction, is approximately 1 V at physiological pH. Under physiological conditions at a pH of 7.4 a maximum power density of 16 $\mu W\,cm^{-2}$ at a cell voltage of 0.25 V is obtained. At a lower pH value of 5.5 the maximum power density is 40 $\mu W\,cm^{-2}$ at 0.4 V. The maximum power density decreases at still lower pH.

8.4.9 *Protein Purification*

An important issue in biotechnology is the separation and purification of the target product from the fermentation broth. A wide variety of combinations of several different steps, such as precipitation, centrifugation, filtration, and various chromatographic techniques are used for the recovery and for the purification of products manufactured by biotechnology (72).

8.4.9.1 *Immobilized Metal Affinity Chromatography*

Immobilized metal ion affinity chromatography (IMAC) was introduced in 1975 (73). The IMAC purification process is based on the employment of a chelating matrix loaded with soft metal ions, such as Cu^{2+} or Ni^{2+}.

Electron donating groups on the surface of proteins, especially the imidazole side chain of histidine, can bind to the non-coordinated sites of the loaded metal. The interaction between the electron donor group with the metal can be made reversible by lowering the pH or by displacement with imidazole. Thus, a protein possessing electron donating group, such as histidine can be purified by reversible interactions of the metal complex and the protein (74,75).

Besides imidazole, polymers that contain the imidazole group can be used. Copolymers of VCL and VI can be used as displacers because they undergo complex formation with Cu^{2+}, Ni^{2+}, Co^{2+}, and Zn^{2+} ions (76,77). These complexes can form larger complexes with proteins containing certain moieties, such as histidine, methionine and tryptophane.

VCL renders the copolymer with thermosensitivity. The copolymer precipitates nearly quantitatively from an aqueous solution by increasing the temperature to 48°C (78). A displacer copolymer poly(VI-VCL) is more efficient in comparison to a monomeric displacer of the same chemical nature, i.e., imidazole, probably due to the multiple interaction of the imidazole groups within the same macromolecule with a single Cu^{2+} ion.

8.4.9.2 Affinity Precipitation Techniques

In IMAC certain drawbacks have been encountered. The scaling up can cause problems. Sometimes a sample pretreatment is compulsory to prevent the clogging of the column. To circumvent these deficiencies, new techniques have been developed, such as affinity precipitation and affinity phase extraction.

In affinity precipitation, certain ligands are coupled to a water-soluble polymer. The polymers used for affinity precipitation are hydrophilic and water-soluble but change reversibly to hydrophobic and water-insoluble when the conditions, such as pH, temperature, and ionic strength are changed.

A polymer from *N*-isopropylacrylamide (NIPAAm) is a thermoresponsive polymer with an upper critical solution temperature of about 32°C in water. Poly(NIPAAm) bears no reactive groups that can be used for coupling of an affinity ligand. The reactive groups needed are preferably introduced by copolymerization of NIPAAm with VI. The copolymer can be loaded with Cu^{2+} or Ni^{2+}, the former being more efficient in the performance.

The method was demonstrated with a histidine-tagged lactate dehydrogenase. For the affinity precipitation, a metal complexed copolymer solution is mixed with the crude liquid containing the enzyme. The polymer-enzyme complex is precipitated by adding NaCl solution and incubating in a water bath at 35°C for 10 *min* and centrifuged. The target product is then released from the complex by adding EDTA (79).

Bacillus stearothermophilus was purified by a related technique, i.e., affinity partitioning. Affinity partitioning was performed in an aqueous two-phase polymer system formed by dextran and a copolymer of VCL and VI. The enzyme moves preferentially into the copolymer phase in the presence of copper ions. Finally, the precipitation of the copolymer at 45°C results in an aqueous solution of the purified enzyme (80).

The principle of separation by metal affinity interactions can also be applied to the separation of single-stranded nucleic acids by interactions involving purines (81).

8.4.10 *Hydrogels*

8.4.10.1 *Metal Recovery*

The concentration and separation of trace metals are important for solving wastewater problems, such as the removal of toxic heavy metals and the recovery of valuable or radioactive metals.

PVI hydrogels can bind many metal ions (82). The binding is less complete with Pb^{2+} Ca^{2+} and Na^{+} ions. The absorption of Hg^{2+} from aqueous solutions by PVI hydrogel particles was studied as a function of pH, counter ion, and cation concentration (83). An essential removal of Hg^{2+} ions can be achieved at pH of 2.

Copolymers of AN and VI have been adapted for the recovery of uranium ions. The hydrogels are synthesized by irradiating AN and VI solutions to doses up to 14 $k\,Gy$ in a ^{60}Co$-\gamma$-source. Insoluble network structures are formed. The samples prepared after irradiation are immersed into DMF to extract monomers and soluble fractions. The gels are treated with hydroxylamine to convert the nitrile group into amidoxime groups as shown in Eq. 8.1.

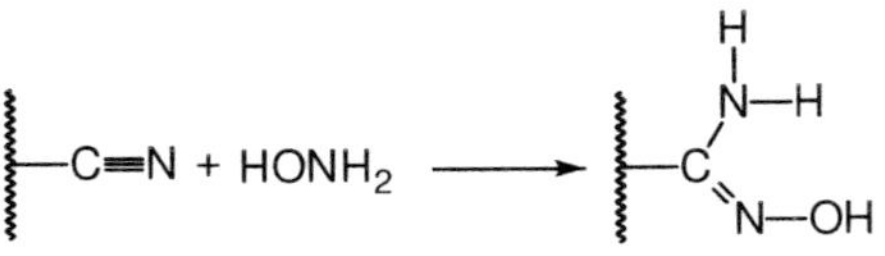

(8.1)

The amidoximated hydrogels from AN and VI are very suitable chelating systems for the adsorption of UO_2^{2+} ion from aqueous solutions (84, 85). Further, crosslinked poly(1-vinylimidazole-*co*-acrylic acid) and crosslinked poly(1-vinylmidazole-*co*-2-acrylamido-2-methyl-1-propane sulfonic acid) are suitable for the retention of copper ions and uranium ions and various other ions (86). The retention of uranyl ions for the latter resin is close to 100% at pH 5.0. The regeneration of the resins can be achieved by alkaline treatment (87).

Copolymers from EGDM with VI are prepared as beads with an average diameter of 150–200 μm. These beads are suitable for the adsorption of heavy metal ions. The regeneration of the chelating beads can be easily achieved with 0.1 M HNO_3 (88).

8.4.10.2 *Drug Delivery*

When hydrogels are used as carriers in drug delivery systems, the release of the drugs from a swollen drug polymer matrix can be controlled either by the rate of hydrolysis reaction or by the diffusion of the drug. Normally, drug delivery by hydrolysis proceeds with a sufficiently high rate in strongly alkaline medium. However, for practical applications, it is necessary that the hydrolysis takes place at a sufficient rate at a pH of around 7.

However, it is possible to catalyze the hydrolysis by alkaline moieties bound to the polymer. The imidazole catalyzed hydrolysis of *p*-nitrophenyl esters from hydrogels has been demonstrated (89). The esters form a charge transfer complex with the nucleophilic catalytic group.

Model copolymers from 2-hydroxyethyl methacrylate, VI and 2,4-dinitrophenyl-*p*-vinyl benzoate as hydrolyzable unit, as well as other monomers, such as *N*-methacryloyl-*L*-histidine were synthesized (89). The polymer is crosslinked with small amounts of EGDM. The mechanism of the hydrolysis reaction is shown in Figure 8.6.

8.4.11 Composite Membranes

8.4.11.1 *Fuel Cell Membrane*

Nafion® (DuPont) is a sulfonated PTFE. It has many uses such as drying gases, as ion exchanging membrane at the chloro alkali electrolysis, and as a proton exchanging membrane in fuel cells. The Nafion® membrane for direct methanol fuel cell can be modified by polymerizing VI in the pores in order to reduce the methanol permeability (90).

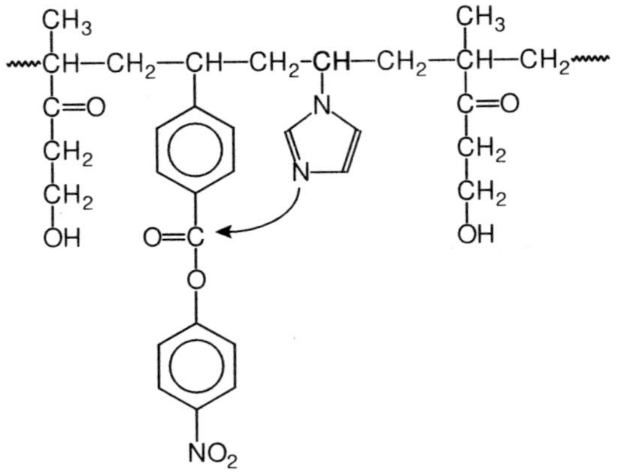

Figure 8.6 Drug Delivery by Catalyzed Hydrolysis through Anchimeric Effect (89)

8.4.11.2 Oxygen Binding Membrane

A membrane with a thickness in the nanometer range uses a copolymer of vinylimidazole and octyl methacrylate. A cobalt phthalocyanine complex is incorporated in the polymer. In solution, this complex shows reversible oxygen absorption, even at low temperatures.

In the so prepared membranes, the permeability coefficient of oxygen is more than twentyfold greater than that of nitrogen and increases with lower upstream oxygen pressure (91). This behavior is explained by the concentration of the oxygen in the membrane being increased by the cobalt phthalocyanine complex.

Also, membranes made from cobalt porphyrins attached to copolymers from vinylimidazole and fluoropentyl methacrylate or octyl methacrylate exhibit a high oxygen permeability (34,92,93).

8.4.12 Drug Uses

8.4.12.1 Drug Release Matrices

Oral drug forms with a delayed release of the active ingredient have an improved patient compliance owing to a reduced frequency of intake, a reduction in side effects owing to the avoidance of plasma level peaks, more uniform levels of the medicinal substance in the blood, and the avoidance of local irritations.

Coated slow-release formulations consist of cores containing the medicinal substance. The coating film is insoluble in water but has semipermeable pores through which the medicinal substance diffuses. It is also possible to achieve control and prolongation of the release by embedding the medicinal substance in a matrix.

Embedding the medicinal substance in a matrix offers the particular advantages of simple and low cost production and high drug safety because dose-dumping effects cannot occur. In other words, the occurrence of high plasma concentrations due to incorrect intake, e.g., chewing instead of swallowing, will not be a problem.

The ancillary substances, such as hydroxypropyl cellulose, hydroxypropyl methyl cellulose, alginic acid or alginates, and xanthan, have technical disadvantages with regard to use. These disadvantages arise because the release of medicinal substance could be dependent on the pH or ionic strength. Further, using these ancillary substances may result in tablets with unsatisfactory properties, such as low hardness or inhomogeneity.

More advantageous is the use of copolymers consisting of hydrophilic components, such as NVP, and further related monomers, such as VCL, VI, and acrylic esters, such as stearyl acrylate (94).

8.4.12.2 Bile Acid Sequestrants

Bile acids are precursors to bile salts and are derived from cholesterol. Following digestion, bile acids can be passively absorbed in the jejunum or, in the case of conjugated primary bile acids, reabsorbed by active transport in the ileum. Bile acids, which are not reabsorbed, are deconjugated and dehydroxylated by bacterial action in the distal ileum and large intestine. Reabsorption of bile acids from the intestine conserves the lipoprotein cholesterol in the bloodstream.

Conversely, the blood cholesterol level can be diminished by reducing reabsorption of the bile acids. Bile acids can be reabsorbed by oral administration of compounds that sequester the bile acids but cannot themselves be absorbed. The sequestered bile acids consequently are excreted.

Certain polymers based on 3-(methacrylamido)propyltrimethylammonium chloride or 2-trimethylammoniumethyl acrylate chloride with crosslinking agents, such as *N,N'*-methylene bismethacrylamide or DVB, have been proposed as ion exchange sequestering agents for bile acids. Other comonomers are *n*-butyl methacrylamide, 2,3,4,5,6-pentafluorostyrene, 2-VPy, or VI (95).

8.4.12.3 *Pyrogen Retention Membranes*

Pyrogens, such as lipopolysaccharide complexes, cause fever in animals and humans after intravenous injection in extremely small amounts. For this reason, infusion media contaminated with pyrogens, must be rendered pyrogen-free prior to use.

A further area of application in which the removal of pyrogens is important is kidney dialysis using high-flux membranes. Here, the dialyzate must either be freed of pyrogens, using a separate membrane unit for example, to prevent pyrogens from being filtered back through the wall of the dialysis membrane, thus entering the blood of the dialysis patient. It is also possible to use a dialysis membrane that itself does not pass pyrogens, thus also preventing the transition of pyrogens into the blood.

Semipermeable membranes for pyrogen retention have been developed that comprise an additive to the base polymer made from a copolymer of NVP and a vinylimidazole compound (96).

8.4.12.4 *Disinfection Agents*

A disinfecting solution for contact lenses comprises the combination of a polymeric biguanide and a quaternized vinylimidazole polymer (97). Unlike hard contact lenses, soft-type contact lenses have a tendency to bind and concentrate fluids, environmental pollutants, water impurities, as well as antimicrobial agents and other active ingredients commonly found in lens-care solutions. In most instances, the low levels of the ingredients in lens-care solutions do not lead to eye tissue irritation when used properly.

Nevertheless, especially due to the inherent binding action of protein deposits to soft-lens materials, some disinfecting agents and preservatives tend to build up on lens surfaces and may become concentrated to potentially hazardous levels. Thus, when released they could cause corneal inflammation and other eye tissue irritation.

Poly(hexamethylene biguanide) (PHMB) and related polymers have been shown to be effective biocidal agents. PHMB, however, is less effective against fungi than bacteria. In contrast, quaternized polymers are effective against fungi and less effective against bacteria.

The combination of a biguanide polymer and a copolymer from NVP and quaternized vinylimidazole has been found to result in a strong antimicrobial activity across the entire range of microorganisms typically associated with ophthalmic preservation and disinfection.

8.4.13 Cosmetic Compositions

Copolymers containing VI are used in several cosmetic formulations. Examples of uses are summarized in Table 8.4.

Table 8.4 Cosmetic Compositions (98)

Hair treatment	Skin treatment
Hair rinses and lotions	Liquid soaps
Treatment fluids for damaged ends	Body lotions
Equalizing agents for permanent waves	Shaving lotions
Hot oil treatment preparations	Face lotions
Conditioners	Cosmetic lotions
Setting lotions	
Hairsprays	

The major requirements of such cosmetic compositions are (98):

- No irritant or toxic effects on skin or hair
- Good feeling on and adhesion to the skin or hair
- Good compatibility with other cosmetic substances
- Prevention of electrostatic charging of the hair
- Good wet-compatibility.

8.4.13.1 Cosmetic Hair Preparations

Cationic polymers are used to give the hair good cosmetic properties such as softness, a pleasant feel, and easy disentangling. Aqueous preparations comprising a copolymer-based on VCL, NVP and VI are useful in cosmetic skin preparations and cosmetic hair preparations (98).

Advantageous copolymers are those which include NVP, 3-methyl-1-vinylimidazolium chloride, and 3-methyl-1-vinylimidazolium methyl sulfate. The polymerization is carried out as usual by a radical initiator. The Fikentscher K values of the polymers should be 50–300. The molecular weight can be adjusted by CTAs.

The solvents used conventionally in cosmetics are ethanol and isopropanol. When the free radical solution polymerization is carried out in such alcoholic solutions, only low-molecular weight materials can be obtained. Therefore, in alcoholic solution polymerization, bifunctional monomers, in small amounts to prevent gelation, such as *N,N′*-divinylurea are added (99). The same type of copolymers is suitable as matrix copolymers for solid controlled release in oral pharmaceutical or cosmetic preparations (94).

It is advantageous to formulate hair compositions that have a high viscosity and are in a thickened liquid form to spread well, such as a styling, care cream, or gel. These properties are appreciated by consumers since the composition does not run down the forehead, the nape of the neck, the face, or into the eyes. However, the introduction of cationic polymers into thickeners often leads to problems of fluidization and of loss of clarity, and resultant cosmetic performance levels are sometimes insufficient for care products.

A special class of thickeners has been proposed that is more compatible to the cationic polymers. This is a terpolymer composed of methacrylic acid (MA), methyl acrylate, and a non-ionic urethane macromonomer, which is the product of reaction of a oligo oxethylated phenol and α,α-dimethyl-*m*-isopropenyl-benzyl-isocyanate (100).

The terpolymer is stable in electrolytic media and it has very good thickening power at a pH equal to or above 5.5. A good level of viscosity can be achieved in high concentrations of alcohol.

8.4.13.2 Conditioning Agents in Shampoos

The main purpose of shampoos for hair is to free the hair from dirt. In addition to this cleansing effect, modern shampoos perform conditioning functions as well. The conditioning effect is achieved by adding conditioning agents to the shampoo composition.

Examples of conditioning agents in shampoos are cationic polymers and silicones. Silicones have the disadvantage that they are generally insoluble in water and so must be stabilized in the shampoo formulation by means of dispersants. In addition, silicones show a strong accumulation effect. They are fixed to the hair and are not completely removed by washing. After a certain time, the hair feels unpleasantly heavy.

Cationic polymers with a high charge density have a greater affinity for the hair. They can be employed without addition of auxiliary dispersing agents in shampoo formulations with anionic surfactants. Copolymers of MA and 3-methyl-1-vinylimidazolium chloride are used as active ingredients in cosmetic hair formulations such as shampoos (101). A typical shampoo formulation contains among other ingredients, sodium lauryl ether sulfate, cocamidopropyl betaine, and a copolymer containing 3-methyl-1-vinylimidazolium chloride.

8.5 Suppliers and Commercial Grades

Major suppliers and tradenames are shown in Table 8.5.

Table 8.5 Examples for Commercially Available 1-Vinylimidazole Polymers

Tradename	Producer	Remarks
Basotronic PVI	BASF	Quaternized PVI for electroless copper plating
Sokalan HP 56	BASF	NVP/VI copolymer
Luviquat FC 370	BASF	NVP/methylvinylimidazolium chloride copolymer (70:30)
Luvitec VPMA 91	BASF GSF Chemicals	VI/NVP copolymer (1:9)

8.6 Safety

VI is readily absorbed through the skin within a short time and cause intoxication. Excitation and convulsion are common features with high doses (2). *N*-Methyl-2-vinylimidazole and *N*-butyl-2-vinylimidazole showed a sensitizing effect on the skin of guinea pig, in addition to corrosive action. Essentially no toxicological information about the polymer is available.

Tradenames appearing in the references are shown in Table 8.6.

Table 8.6 Tradenames in References

Tradename Description	Supplier
Aerosil® Fumed Silica (21)	Degussa AG
Ajax® Detergent (22)	Colgate-Palmolive
Alcalase® Proteolytic enzyme, detergent (21)	Novo Industries A/S
Atlas® G 2612 Poly(oxyethylene) (25) propylene glycol stearate (97)	ICI (Atlas Powder Co. Corp.)
Avicel® Microcristalline cellulose (24)	FMC Corp.
Brij® (Series) Ethoxylated fatty alcohols (97,98)	ICI Surfactants
Brij® 30 Poly(oxyethylene) (4) lauryl ether (98)	ICI Surfactants
Brij® 35 Poly(oxyethylene) (23) lauryl ether (97)	ICI Surfactants
Carezyme Cellulase enzyme for detergent usage (24)	Novo Nordisk A/S
Celluzyme Cellulase enzyme for detergent usage (24)	Novo Nordisk A/S
Ceteareth-25 INCI International Nomenclature of Cosmetic Ingredients, Poly(oxyethylene) cetyl ether (98)	INCI Name
Cosmocil® CQ Hexamethylene biguanide polymer (97)	Zeneca, Inc.
Decalin® Decahydronaphthalene (28,29)	DuPont

Table 8.6 (cont.)

Tradename Description	Supplier
Dequest® 2016	Monsanto Co.
1-Hydroxyethylidene-1,1-diphosphonic acid tetrasodium salt (97)	
Dequest® 2046	Monsanto Co.
Pentasodium(ethylenediamine)tetramethylenephosphonate (6,22)	
Desmodur® N100	Bayer AG
Aliphatic solvent free HDI biuret poly(isocyanate) (5)	
Ebecryl® (Series)	Cytec Industries (UCB)
Urethane acrylate (5)	
EF-101	Tohkem Products Corp.
Perfluorooctylsulfonic acid (18)	
EF-201	Tohkem Products Corp.
Perfluorooctanoic acid (18)	
Esperase®	Novozymes A/S Corp.
Proteolytic enzyme, detergent (21)	
FN-base	Genencor
Proteolytic enzyme, detergent (21)	
Fungamyl	Novozymes A/S Corp.
Amylolitic enzyme for detergent usage (24)	
Kollidon® CL	BASF AG
Poly(vinylpyrrolidone), crosslinked, super-disintegrant in tablets (94)	
Kollidon® VA 64	BASF AG
Vinylpyrrolidone/vinyl acetate (60:40) copolymer (18)	
Kyro™ EOB	Procter & Gamble
Condensation product of higher alcohols with moles ethylene oxide (21)	
Laureth® 4	INCI Name
INCI International Nomenclature of Cosmetic Ingredients, poly(oxyethylene) (4) lauryl ether (98)	
Lipase P Amano	Amano Pharmaceutical Co., Ltd.
Lipase enzyme for detergent usage (21)	
Lipolase®	Novo Industries A/S
Lipase enzyme for detergent usage (24)	
Luvimer® 100P	BASF AG
tert-Butylacrylate/ethylacrylate/methacrylic acid copolymers (101)	
Luviquat FC 370	BASF AG
Vinylpyrrolidone/methylvinylimidazolium chloride copolymer (70:30) (97,100)	

Table 8.6 (cont.)

Tradename Description	Supplier
Luviquat® PQ 11 PN	BASF AG
Quaternized copolymer of vinylpyrrolidone (VP) and dimethylaminoethylmethacrylate (DMAEMA) in aqueous solution, hair care polymer (101)	
Luviset® CAN	BASF AG
Terpolymer of vinyl acetate, crotonic acid and vinyl neodecanoate (101)	
Luviskol® VBM	BASF AG
Vinylpyrrolidone, *tert*-butylacrylate, methacrylic acid copoplymers (101)	
Luvitec® VPMA 91	BASF AG
Vinylimidazole with vinylpyrrolidone (1:9) copolymer (18)	
M1 Lipase	Ibis
Lipase enzyme for detergent usage (24)	
Maxacal®	Gist-Brocades N.V
Proteolytic enzyme (20)	
Maxatase®	Gist-Brocades N.V
Proteolytic enzyme (20, 21)	
Merquat® 100	Calgon Corp.
Poly(dimethyldiallyammonium chloride), antimicrobial polymer (97)	
Miranol®	Rhodia Inc. Corp.
Alkylaspartic acid, ampholytic detergent (97)	
Myrj® 52	ICI Americas Inc.
Poly(oxyethylene) (40) stearate (97)	
Nekal® BX	BASF AG
Sodium alkyl naphthalene sulfonate, surfactant (5)	
Neodol® 23-6.5	Shell
Condensation product of C_{12}–C_{13} linear alcohol with 6.5 moles of ethylene oxide (21)	
Neodol® 45-4	Shell
Condensation product of C_{14}-C_{15} linear alcohol with 4 moles of ethylene oxide (21)	
Neodol® 45-7	Shell
Condensation product of C_{14}-C_{15} linear alcohol with 7 moles of ethylene oxide (21)	
Neodol® 45-9	Shell
Condensation product of C_{14}-C_{15} linear alcohol with 9 moles of ethylene oxide (21)	

Table 8.6 (cont.)

Tradename Description	Supplier
Optimase	MKC
Proteolytic enzyme, detergent (21)	
Pluronic® (Series)	BASF AG
Ethylene oxide/propylene oxide block copolymer, defoamers (21)	
Poloxamine 1107	BASF-Wyandotte Corp.
Poly(oxypropylene)/poly(oxyethylene) block copolymer adduct of ethylene diamine (97)	
Polyquaternium® 10	Nalco Chemical Comp.
2-(2-Hydroxy-3-(trimethylammonium)propoxy)ethyl cellulose ether chloride, cationic cellulose derivative (101)	
Polyquaternium® 7	Nalco Chemical Comp.
N,*N*-Dimethyl-*N*-2-propenyl-2-propene-1-aminium chloride, polymer with acrylamide copolymer (101)	
Polyquaternium® 1	Onyx Corp.
[4-Tris(2-hydroxyethyl)ammonio]-2-butenyl-ω-[tris(2-hydroxyethyl)ammonio]dichloride, antimicrobial agent (97)	
ProJet™ Fast Black 2	Zeneca, Inc.
Black pigment (19)	
Rewopol® NLS 28	REWO Chemicals
Solution of sodium lauryl sulfate in water, developer component (5)	
Sartomer 355	Cray Valley
Ditrimethylol propane tetraacrylate (5)	
Savinase®	Novo Nordisk A/S
Proteolytic enzyme for detergent usage (20, 21, 24)	
Shellflex® 371	Shell
Processing oil (30)	
Softlan®	Colgate-Palmolive
Fabric softener (22)	
Sokalan® HP (Series)	BASF AG
Water-soluble homo- and copolymers of vinylpyrrolidone, vinylimidazole and nonionic monomers (6, 22)	
Tergitol® 15-S (Series)	Union Carbide Corp.
Ethoxylated C11-15-secondary alcohols, surfactant (21)	
Termamyl® (Series)	Novo Industries A/S
α-Amylase for detergent usage (24)	
Tetronic® (Series)	BASF-Wyandotte Corp.
Propoxylated ethylenediamine-poly(ethylene glycol) adduct, surfactant (21, 97)	

Table 8.6 (cont.)

Tradename Description	Supplier
Texapon® 842	Henkel
Sulfuric acid, mono(2-ethylhexyl) ester, sodium salt, Developer component (5)	
Triton® X (Series)	Union Carbide Corp. (Rohm & Haas)
Poly(alkylene oxide), nonionic surfactants (21)	
TSMR-8800	Tokyo Ohka Kogyo Co., Ltd.
Cresol novolak resin with a naphthoquinone diazide compound, positive photoresist (18)	
Tween® (Series)	Uniqema
Ethoxylated fatty acid ester surfactants (97)	
Ultrahold® 8	BASF AG
N-tert-Butylacrylamide/ethylacrylate/acrylic acid terpolymer (101)	

References

1. B. Radziszewski, Über die Constitution des Lophins und verwandter Verbindungen, *Ber.*, 15:1493–1496, June 1882.
2. K. Ebel, H. Koehler, A.O. Gamer, and R. Jäckh, "Imdidazole and derivates," in *Ullmanns Encyclopedia of Industrial Chemistry*, Vol. 17, pp. 493–501. Wiley VCH, Weinheim, 6th edition, 2003.
3. G.P. Speranza and W.-Y. Su, Method for the preparation of imidazoles, US Patent 4 927 942, assigned to Texaco Chemical Co. (White Plains, NY), May 22, 1990.
4. J. Detering, C. Schade, W. Trieselt, and J. Tropsch, Use of vinylpyrrolidone and vinylimidazole copolymers as detergent additives, novel polymers of vinylpyrrolidone and of vinylimidazole, and preparation thereof, US Patent 5 846 924, assigned to BASF Aktiengesellschaft (Ludwigshafen, DE), December 8, 1998.
5. S.P. Pappas, H. Baumann, U. Dwars, C.M. Savariar-Hauck, and H.-J. Timpe, Overcoat for light-sensitive materials comprising (1-vinylimidazole) polymer or copolymer, US Patent 6 649 323, assigned to Kodak Polychrome Graphics LLC (Norwalk, CT), November 18, 2003.
6. D. Boeckh, S. Stein, A. Funhoff, J.A. Lux, and H.-U. Jäger, Use of water-soluble copolymers comprising *N*-vinylimidazole units as color

transfer inhibitors in detergents, US Patent 6 172 027, assigned to BASF Aktiengesellschaft (Ludwigshafen, DE), January 9, 2001.

7. S. Stein, D. Boeckh, C. Schade, and G. Rossler, Powdery porous polymers containing *N*-vinylimidazol units, process for their preparation and their use, US Patent 6 025 317, assigned to BASF Aktiengesellschaft (Ludwigshafen, DE), February 15, 2000.
8. H.F. Naguib, R.O. Aly, M.W. Sabaa, and S.M. Mokhtar, Gamma radiation induced graft copolymerization of vinylimidazole-acrylic acid onto polypropylene films, *Polym. Test.*, 22(7):825–830, October 2003.
9. Z.H. Ma, H.S. Han, K.L. Tan, E.T. Kang, and K.G. Neoh, Thermally induced surface graft copolymerization with concurrent lamination of polyaniline films under atmospheric conditions, *Int. J. Adhes. Adhes.*, 19(5):359–365, October 1999.
10. A.K.S. Ang, E.T. Kang, K.G. Neoh, K.L. Tan, C.Q. Cui, and T.B. Lim, Low-temperature graft copolymerization of 1-vinyl imidazole on polyimide films with simultaneous lamination to copper foils–effect of cross-linking agents, *Polymer*, 41(2):489–498, January 2000.
11. Y.S. Lin, H.M. Liu, and H.T. Chen, Surface modification of polyimide films by argon plasma for copper metallization on microelectronic flex substrates, *J. Appl. Polym. Sci.*, 99(3):744–755, February 2006.
12. G.H. Yang, E.T. Kang, and K.G. Neoh, Electroless deposition of copper and nickel on poly(tetrafluoroethylene) films modified by single and double surface graft copolymerization, *Appl. Surf. Sci.*, 178(1-4): 165–177, July 2001.
13. G.H. Yang, C. Lim, Y.P. Tan, Y. Zhang, E.T. Kang, and K.G. Neoh, Electroless deposition of nickel on fluoropolymers modified by surface graft copolymerization, *Eur. Polym. J.*, 38(11):2153–2160, November 2002.
14. H.I. Unal, C. Inegollu, and O. Sanli, Graft copolymerization of *N*-vinylimidazole on poly(ethylene terephthalate) fibers in a swelling solvent using azobisisobutyronitrile as initiator, *Turk. J. Chem.*, 27(3): 403–415, 2003.
15. J.-C. Yang, Y.-S. Hong, T.-C. Chang, and Y.-T. Wang, Method of preparing a hybrid of polyvinylimidazole and silica, US Patent 6 566 456, assigned to Chung-Shan Institute of Science & Technology (Tao-Yuan, TW), May 20, 2003.
16. S.L. Kopolow, D. Patel, Y.T. Kwak, M. Tallon, D.K. Hood, L. Senak, and J.M.C. Kittrick, Substrate for color inkjet printing, US Patent 6 641 259, assigned to ISP Investments Inc. (Wilmington, DE), November 4, 2003.
17. H.J. Levinson, *Principles of Lithography*, Vol. 146 of *SPIE PM*, SPIE Press, Bellingham (WA), 2nd edition, 2005.

18. K. Wakiya, N. Kubota, S. Yokoi, and M. Kobayashi, Composition for lithographic anti-reflection coating, and resist laminate using the same, US Patent 6 416 930, assigned to Tokyo Ohka Kogyo Co., Ltd. (Kanagawa, JP), July 9, 2002.
19. J.A. Bates, J.A. Belmont, J.E. Johnson, and J.C. Smith, Ink compositions and method for generating images produced therefrom, US Patent 5 958 999, assigned to Cabot Corporation (Boston, MA), September 28, 1999.
20. C.V. White, Soil release polymer in detergent compositions containing dye transfer inhibiting agents to improve cleaning performance, US Patent 5 451 341, assigned to The Procter & Gamble Company (Cincinnati, OH), September 19, 1995.
21. A. Busch, A.C. Convents, and P.J. Leeuwen, Detergent compostions inhibiting dye transfer comprising copolymers of *N*-vinylimidazole and *N*-vinylpyrrolidone, US Patent 5 710 118, assigned to The Procter & Gamble Company (Cincinnati, OH), January 20, 1998.
22. D. Bockh, H.-U. Jäger, J.A. Lux, S. Stein, J. Detering, and C. Schade, Use of quaternized polymerizates containing units of vinyl imidazol as a color fixing and color transfer inhibiting additive to detergent post-treatment agents and detergents, US Patent 6 165 969, assigned to BASF Aktiengesellschaft (Ludwigshafen, DE), December 26, 2000.
23. C. Schade, J. Detering, S. Stein, D. Boeckh, and H.-U. Jäger, Water soluble crosslinked copolymers, their preparation and their use, US Patent 5 962 613, assigned to BASF Aktiengesellschaft (Ludwigshafen, DE), October 5, 1999.
24. P.J. Van Leeuwen, A.C. Convents, A. Busch, T.L. Cachet, and C.E.A. Joos, Laundry detergent composition comprising substantially water-insoluble polymeric dye transfer inhibiting agent, US Patent 5 912 221, assigned to Procter & Gamble Company (Cincinnati, OH), June 15, 1999.
25. C. Schade and K.-H. Schneider, Agglomerated, finely divided, crosslinked vinylimidazole copolymers, preparation thereof, and use thereof, US Patent 5 804 662, assigned to BASF Aktiengesellschaft (Ludwigshafen, DE), September 8, 1998.
26. T. Uchiyama, A. Kawauchi, and D.L. DuVal, Quick identification of polymeric dye transfer inhibitors in laundry detergents by pyrolysis-gas chromatography/mass spectrometry, *J. Anal. Appl. Pyrolysis*, 45 (2):111–119, May 1998.
27. R.E. Bennett, G.S. Bennett, M.A. Craton, and S.-H. Chou, Polycarbonate articles and adhesive composition therefor, US Patent 6 815 035, assigned to 3M Innovative Properties Company (St. Paul, MN), November 9, 2004.

28. F.J. Boden, R.P. Sauer, I.L. Goldblatt, and M.E. McHenry, Polar grafted polyolefins, methods for their manufacture, and lubricating oil compositions containing them, US Patent 6 300 289, assigned to Castrol Limited (Wiltshire, GB), October 9, 2001.
29. F.J. Boden, R.P. Sauer, I.L. Goldblatt, and M.E. McHenry, Polar grafted polyolefins, methods for their manufacture, and lubricating oil compositions containing them, US Patent 6 686 321, assigned to Castrol Limited (GB), February 3, 2004.
30. R.J. Sutherland, Dispersant viscosity index improvers, US Patent 6 083 888, assigned to Shell Oil Company (Houston, TX), July 4, 2000.
31. R. Shimizu, Non-aqueous electrolyte battery, US Patent 6 291 107, assigned to NEC Moli Energy Corp (Yokohama, JP), September 18, 2001.
32. Y. Suzuki, H. Nishide, and E. Tsuchida, Membranes of the picket fence cobalt porphyrin complexed with poly(vinylimidazole and -pyridine)s: Selective optical response to oxygen, *Macromolecules*, 33(7): 2530–2534, April 2000.
33. S. Röösli, E. Pretsch, W.E. Morf, E. Tsuchida, and H. Nishide, Selective optical response to oxygen of membranes based on immobilized cobalt(II) porphyrins, *Anal. Chim. Acta*, 338(1-2):119–125, February 1997.
34. B.Q. Shentu, Z.X. Weng, and H. Nishide, Preparation and facilitated oxygen transport of poly(vinyl imidazole-*co*-fluoroalkyl methaerylate)-cobaltporphyrin complexed membrane, *Chem. J. Chin. Univ.-Chin.*, 24(3):547–550, March 2003.
35. N. Pekel, B. Salih, and O. Güven, Activity studies of glucose oxidase immobilized onto poly(*N*-vinylimidazole) and metal ion-chelated poly(*N*-vinylimidazole) hydrogels, *J. Mol. Catal. B: Enzym.*, 21(4-6): 273–282, February 2003.
36. B. Osman, A. Kara, L. Uzun, N. Besirli, and A. Denizli, Vinyl imidazole carrying metal-chelated beads for reversible use in yeast invertase adsorption, *J. Mol. Catal. B: Enzym.*, 37(1-6):88–94, December 2005.
37. A. Kara, B. Osman, H. Yavuz, N. Besirli, and A. Denizli, Immobilization of α-amylase on Cu^{2+} chelated poly(ethylene glycol dimethacrylate-*N*-vinyl imidazole) matrix via adsorption, *React. Funct. Polym.*, 62 (1):61–68, January 2005.
38. J. Clark, L. C. and C. Lyons, Electrode systems for continuous monitoring in cardiovascular surgery, *Ann. New York Acad. Sci.*, 102:29–45, 1962.
39. E.I. Iwuoha, M.R. Smyth, and M.E.G. Lyons, Organic phase enzyme electrodes: Kinetics and analytical applications, *Biosens. Bioelectron.*, 12(1):53–75, 1997.

40. M. Niculescu, I. Frebort, P. Pec, P. Galuszka, B. Mattiasson, and E. Csoregi, Amine oxidase based amperometric biosensors for histamine detection, *Electroanalysis*, 12(5):369–375, March 2000.
41. T.J. Ohara, R. Rajagopalan, and A. Heller, "wired"enzyme electrodes for amperometric determination of glucose or lactate in the presence of interfering substances, *Anal. Chem.*, 66(15):2451–2457, August 1994.
42. K. Takagi and S. Shikata, Flow injection determination of histamine with a histamine dehydrogenase-based electrode, *Anal. Chim. Acta*, 505(2):189–193, March 2004.
43. M.C. Millot, S. Servagent-Noinville, N.L. Taleb, M.H. Baron, M. Revault, and B. Sébille, Structural changes of human serum albumin immobilized on chromatographic supports: A high-performance liquid chromatography and Fourier-transform infrared spectroscopy study, *J. Chromatogr. B*, 753(1):101–113, March 2001.
44. M.C. Millot, N.L. Taleb, and B. Sébille, Binding of human serum albumin to silica particles by means of polymers: A liquid chromatographic study of the selectivity of resulting chiral stationary phases, *J. Chromatogr. B*, 768(1):157–166, February 2002.
45. N.L. Taleb, M.-C. Millot, and B. Sébille, Enantioselectivity properties of human serum albumin immobilized on anion-exchangers based on polyvinylimidazole-coated silica: Effect of protein loading on separation properties, *J. Chromatogr. A*, 776(1):45–53, July 1997.
46. M. Niculescu, T. Erichsen, V. Sukharev, Z. Kerenyi, E. Csoregi, and W. Schuhmann, Quinohemoprotein alcohol dehydrogenase-based reagentless amperometric biosensor for ethanol monitoring during wine fermentation, *Anal. Chim. Acta*, 463(1):39–51, July 2002.
47. A. Vilkanauskyte, T. Erichsen, L. Marcinkeviciene, V. Laurinavicius, and W. Schuhmann, Reagentless biosensors based on co-entrapment of a soluble redox polymer and an enzyme within an electrochemically deposited polymer film, *Biosens. Bioelectron.*, 17(11-12):1025–1031, December 2002.
48. A.R. Vijayakumar, E. Csoregi, A. Heller, and L. Gorton, Alcohol biosensors based on coupled oxidase-peroxidase systems, *Anal. Chim. Acta*, 327(3):223–234, July 1996.
49. M. Niculescu, S. Sigina, and E. Csoregi, Glycerol dehydrogenase based amperometric biosensor for monitoring of glycerol in alcoholic beverages, *Anal. Lett.*, 36(9):1721–1737, 2003.
50. J.J. Fei, Y.H. Wu, X.B. Ji, J. Wang, S.S. Hu, and Z.Q. Gao, An amperometric biosensor for glucose based on electrodeposited redox polymer/glucose oxidase film on a gold electrode, *Anal. Sci.*, 19(9): 1259–1263, September 2003.

51. J. Fei, K. Wu, F. Wang, and S. Hu, Glucose nanosensors based on redox polymer/glucose oxidase modified carbon fiber nanoelectrodes, *Talanta*, 65(4):918–924, February 2005.
52. S. Gaspar, I.C. Popescu, I.G. Gazaryan, A. Gerardo Bautista, I.Y. Sakharov, B. Mattiasson, and E. Csoregi, Biosensors based on novel plant peroxidases: A comparative study, *Electrochim. Acta*, 46 (2-3):255–264, November 2000.
53. A. Belay, A. Collins, T. Ruzgas, P. T. Kissinger, L. Gorton, and E. Csöregi, Redox hydrogel based bienzyme electrode for l-glutamate monitoring, *J. Pharm. Biomed. Anal.*, 19(1-2):93–105, February 1999.
54. M. Tessema, T. Larsson, T. Buttler, E. Csoregi, T. Ruzgas, M. Nordling, S.-E. Lindquist, G. Pettersson, and L. Gorton, Simultaneous amperometric determination of some mono-, di-, and oligosaccharides in flow injection and liquid chromatography using two working enzyme electrodes with different selectivity, *Anal. Chim. Acta*, 349(1-3): 179–188, August 1997.
55. Y. Kawanami, T. Yunoki, A. Nakamura, K. Fujii, K. Umano, H. Yamauchi, and K. Masuda, Imprinted polymer catalysts for the hydrolysis of *p*-nitrophenyl acetate, *J. Mol. Catal. A: Chem.*, 145(1-2):107–110, September 1999.
56. S. Połowiński, Template polymerisation and co-polymerisation, *Prog. Polym. Sci.*, 27(3):537–577, April 2002.
57. T. Takeuchi, A. Seko, J. Matsui, and T. Mukawa, Molecularly imprinted polymer library on a microtiter plate. High-throughput synthesis and assessment of cinchona alkaloid-imprinted polymers, *Instrum. Sci. Techol.*, 29(1):1–9, January 2001.
58. S.A. Piletsky, K. Karim, E.V. Piletska, C.J. Day, K.W. Freebairn, C. Legge, and A.P.F. Turner, Recognition of ephedrine enantiomers by molecularly imprinted polymers designed using a computational approach, *Analyst*, 126(10):1826–1830, 2001.
59. Z. Meng, T. Yamazaki, and K. Sode, A molecularly imprinted catalyst designed by a computational approach in catalysing a transesterification process, *Biosens. Bioelectron.*, 20(6):1068–1075, December 2004.
60. Z.H. Meng and K. Sode, The molecular reaction vessels for a transesterification process created by molecular imprinting technique, *J. Mol. Recognit.*, 18(3):262–266, May–June 2005.
61. Z.H. Meng, T. Yamazaki, and K. Sode, Enhancement of the catalytic activity of an artificial phosphotriesterase using a molecular imprinting technique, *Biotechnol. Lett.*, 25(13):1075–1080, July 2003.
62. Z.H. Meng, T. Yamazaki, and K. Sode, Construction of artificial cyclic amide amidohydrolases using molecular imprinting technique, *Biotechnol. Lett.*, 27(7):471–475, April 2005.

63. K. Lettau, A. Warsinke, A. Laschewsky, K. Mosbach, E. Yilmaz, and F.W. Scheller, An esterolytic imprinted polymer prepared via a silica-supported transition state analogue, *Chem. Mater.*, 16(14):2745–2749, July 2004.
64. N. Mano and A. Heller, A miniature membraneless biofuel cell operating at 0.36 V under physiological conditions, *J. Electrochem. Soc.*, 150(8):A1136–A1138, August 2003.
65. H.H. Kim, N. Mano, X.C. Zhang, and A. Heller, A miniature membrane-less biofuel cell operating under physiological conditions at 0.5 V, *J. Electrochem. Soc.*, 150(2):A209–A213, February 2003.
66. S.C. Barton, J. Gallaway, and P. Atanassov, Enzymatic biofuel cells for implantable and microscale devices, *Chem. Rev.*, 104(10):4867–4886, October 2004.
67. F. Barrière, P. Kavanagh, and D. Leech, A laccase-glucose oxidase biofuel cell prototype operating in a physiological buffer, *Electrochim. Acta*, 51(24):5187–5192, July 2006.
68. S.C. Barton, M. Pickard, R. Vazquez-Duhalt, and A. Heller, Electroreduction of O_2 to water at 0.6 V (SHE) at pH 7 on the wired Pleurotus ostreatus laccase cathode, *Biosens. Bioelectron.*, 17(11-12):1071–1074, December 2002.
69. S. Tsujimura, K. Kano, and T. Ikeda, Electrochemical oxidation of NADH catalyzed by diaphorase conjugated with poly-1-vinylimidazole complexed with Os(2,2′-dipyridylamine)$_2$Cl, *Chem. Lett.*, 31 (10):1022–1023, October 2002.
70. S. Timur, Y. Yigzaw, and L. Gorton, Electrical wiring of pyranose oxidase with osmium redox polymers, *Sens. Actuators, B*, 113(2):684–691, February 2006.
71. T. Godjevargova, R. Nenkova, and V. Konsulov, Immobilization of glucose oxidase by acrylonitrile copolymer coated silica supports, *J. Mol. Catal. B: Enzym.*, 38(2):59–64, February 2006.
72. R.K. Scopes, ed., *Protein Purification. Principles and Practice*, Springer Advanced Texts in Chemistry, Springer Verlag, New York, 3rd edition, 1994.
73. J. Porath, J. Carlsson, I. Olsson, and G. Belfrage, Metal chelate affinity chromatography, a new approach to protein fractionation, *Nature*, 258:598–599, 1975.
74. P.S. Nelson, T.-T. Yang, S.R. Kain, and T.H. Smith, Method for purification of recombinant proteins, US Patent 6 242 581, assigned to Clontech Laboratories, Inc. (Palo Alto, CA), June 5, 2001.
75. G.S. Chaga, Twenty-five years of immobilized metal ion affinity chromatography: Past, present and future, *J. Biochem. Bioph. Methods*, 49(1-3):313–334, October 2001.

76. A.E. Ivanov, I.Y. Galaev, S.V. Kazakov, and B. Mattiasson, Thermosensitive copolymers of *N*-vinylimidazole as displacers of proteins in immobilised metal affinity chromatography, *J. Chromatogr. A*, 907(1-2): 115–130, January 2001.
77. A.E. Ivanov, S.V. Kazakov, I.Y. Galaev, and B. Mattiasson, Thermosensitive copolymer of *N*-vinylcaprolactam and 1-vinylimidazole: Molecular characterization and separation by immobilized metal affinity chromatography, *Polymer*, 42(8):3373–3381, April 2001.
78. P. Arvidsson, A.E. Ivanov, I.Y. Galaev, and B. Mattiasson, Polymer versus monomer as displacer in immobilized metal affinity chromatography, *J. Chromatogr. B*, 753(2):279–285, April 2001.
79. A. Kumar, A.A.M. Khalil, I.Y. Galaev, and B. Mattiasson, Metal chelate affinity precipitation: Purification of (His)6-tagged lactate dehydrogenase using poly(vinylimidazole-*co*-*N*-isopropylacrylamide) copolymers, *Enzyme Microb. Tech.*, 33(1):113–117, July 2003.
80. B. Mattiasson, M.B. Dainyak, and I.Y. Galaev, Smart polymers and protein purification, *Polym. Plast. Tech. Eng.*, 37(3):303–308, 1998.
81. S. Balan, J. Murphy, I. Galaev, A. Kumar, G.E. Fox, B. Mattiasson, and R.C. Willson, Metal chelate affinity precipitation of RNA and purification of plasmid DNA, *Biotechnol. Lett.*, 25(13):1111–1116, July 2003.
82. B.L. Rivas, H.A. Maturana, M.J. Molina, M.R. Gomez-Anton, and I.F. Pierola, Metal ion binding properties of poly(*N*-vinylimidazole) hydrogels, *J. Appl. Polym. Sci.*, 67(6):1109–1118, February 1998.
83. M.J. Molina, M.R. Gomez-Anton, B.L. Rivas, H.A. Maturana, and I.F. Pierola, Removal of Hg(II) from acid aqueous solutions by poly(*N*-vinylimidazole) hydrogel, *J. Appl. Polym. Sci.*, 79(8):1467–1475, February 2001.
84. N. Pekel, N. Sahiner, and O. Güven, Development of new chelating hydrogels based on *N*-vinyl imidazole and acrylonitrile, *Radiat. Phys. Chem.*, 59(5-6):485–491, November 2000.
85. N. Pekel and O. Güven, Separation of uranyl ions with amidoximated poly(acrylonitrile/*N*-vinylimidazole) complexing sorbents, *Colloids Surf., A*, 212(2-3):155–161, January 2003.
86. B.L. Rivas, M. Jara, and E.D. Pereira, Preparation and adsorption properties of the chelating resins containing carboxylic, sulfonic, and imidazole groups, *J. Appl. Polym. Sci.*, 89(10):2852–2856, September 2003.
87. B.L. Rivas, E. Pereira, M. Jara, and C. Esparza, Resins with the ability to bind copper and uranyl ions, *J. Appl. Polym. Sci.*, 99(3):706–711, February 2006.

88. A. Kara, L. Uzun, N. Besirli, and A. Denizli, Poly(ethylene glycol dimethacrylate-*N*-vinyl imidazole) beads for heavy metal removal, *J. Hazard. Mater.*, 106(2-3):93–99, January 2004.
89. R.N. Karmalkar, M.G. Kulkarni, and R.A. Mashelkar, Pendent chain linked delivery systems: I facile hydrolysis through anchimeric effect, *J. Controlled Release*, 42(2):185–193, November 1996.
90. B.C. Bae, H.Y. Ha, and D. Kim, Preparation and characterization of nafion/poly(1-vinylimidazole) composite membrane for direct methanol fuel cell application, *J. Electrochem. Soc.*, 152(7):A1366–A1372, 2005.
91. N. Preethi, H. Shinohara, and H. Nishide, Reversible oxygen-binding and facilitated oxygen transport in membranes of polyvinylimidazole complexed with cobalt-phthalocyanine, *React. Funct. Polym.*, 66 (8):851–855, August 2006.
92. H. Nishide, A. Kato, and E. Tsuckida, High oxygen-binding affinity of poly(4-vinylimidazole-*co*-octylmethacrylate)-cobaltporphyrin complex: Effect of hydrogen-bond at the imidazole residue, *Mol. Cryst. Liquid Cryst.*, 342:249–254, 2000.
93. B. Shentu, H. Shinohara, and H. Nishide, High oxygen permeation and persistent oxygen-carrying in a poly(vinylimidazole-*co*-fluoroalkyl methacrylate) cobaltporphyrin membrane, *Polym. J.*, 33(10): 807–811, 2001.
94. H. Meffert and F. Ruchatz, Use of *N*-vinyllactam or *N*-vinylamine containing copolymers as matrix for producing solid pharmaceutical and cosmetic presentations, US Patent 6 436 440, assigned to BASF Aktiengesellschaft (Ludwigshafen, DE), August 20, 2002.
95. W.H. Mandeville, III and S.R. Holmes-Farley, Process for removing bile salts from a patient and compositions therefor, US Patent 6 060 517, assigned to GelTex Pharmaceuticals, Inc. (Waltham, MA), May 9, 2000.
96. T. Niklas, F. Wechs, and A. Nothdurft, Shaped objects for pyrogen retention and processes for their manufacture, US Patent 6 632 361, assigned to Membrana GmbH (Wuppertal, DE), October 14, 2003.
97. D.J. McCanna, S.E. Maier, D.J. Heiler, S.P. Spooner, and E. Xia, Compositions comprising polyquaterniums in combination with polymeric biguanides for disinfecting contact lenses, US Patent 6 153 568, November 28, 2000.
98. P. Hossel, K. Sperling, and V. Schehlmann, Aqueous compositions and their use, US Patent 6 191 188, assigned to BASF Aktiengesellschaft (Ludwishafen, DE), February 20, 2001.
99. R. Blankenburg and A. Sanner, Soluble copolymers for hair cosmetics, US Patent 5 635 169, assigned to BASF Aktiengesellschaft (Ludwigshafen, DE), June 3, 1997.

100. C. Dupuis, Cosmetic composition containing a cationic polymer and an acrylic terpolymer, and use of this composition for the treatment of keratinous material, US Patent 6 214 326, assigned to L'Oreal S.A. (Paris, FR), April 10, 2001.
101. R. Dieing, P. Hossel, and A. Sanner, Use of cationic copolymers of unsaturated acids and *N*-vinylimidazolium salts in cosmetic hair formulations, US Patent 6 355 231, assigned to BASF Aktiengesellschaft (Ludwigshafen, DE), March 12, 2002.

9

Poly(vinylpyrrolidone)

9.1 Monomers

N-Vinyl-2-pyrrolidone (NVP) is a valuable and useful fine chemical. Due to its unique physical properties, such as water solubility, high polarity, low toxicity, chemical stability, and cation activity, it has been widely applied in the manufacture of adhesives, paints, textiles, foods and personal medicines. The homopolymers or copolymers thereof have improved film strength, dye compatibility, rigidity, and adhesion.

Precursors for NVP are 2-pyrrolidone and *N*-(β-hydroxyethyl)-2-pyrrolidone. 2-Pyrrolidone is obtained from 1,4-butanediol via γ-butyrolactone and treatment with ammonia. *N*-(β-hydroxyethyl)-2-pyrrolidone is obtained from γ-butyrolactone and 2-aminoethanol.

There are several routes to produce NVP in industrial scale. NVP is commonly manufactured by reacting 2-pyrrolidone with acetylene by a *Reppe* Reaction. However, acetylene is difficult to handle.

Consequently, alternative methods have been proposed for the production of NVP. One process starts with *N*-(β-hydroxyethyl)-2-pyrrolidone as raw material by reacting with thionyl chloride to form *N*-β-chloroethyl-2-pyrrolidone. In the next step, hydrogen chloride is eliminated. In another route, by the reaction with acetic anhydride the ester is formed, followed by removing the acetic anhydride to obtain NVP.

N-(β-hydroxyethyl)-2-pyrrolidone can be directly dehydrated to result in NVP without the formation of intermediates involved. Dehydration catalysts are active aluminum, or oxides, such as zirconium oxide, thorium oxide, cerium oxide, zinc oxide and chromium oxide (1). The addition of water reduces the amount of *N*-ethyl-2-

pyrrolidone, which is a by-product (2). The synthesis routes to NVP are shown in Figure 9.1.

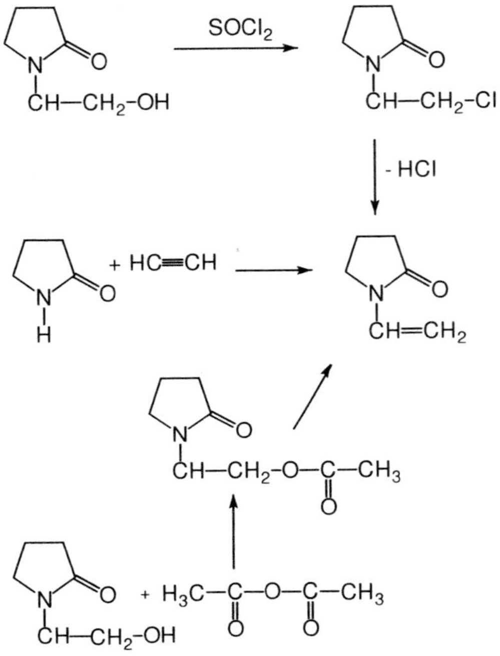

Figure 9.1 Synthesis Routes to *N*-Vinyl-2-pyrrolidone (1)

A large number of applications require a NVP which comprises less than 0.1% by weight of impurities. In principle, such a degree of purity can be achieved by fractional distillation or by a multistage, fractional crystallization (3, 4). NVP copolymerizes readily with several other monomers. Comonomers for NVP described in this chapter are shown in Table 9.1.

The monomer 3-(2-aminoethyl)-α-aminoethyl-*N*-vinyl-2-pyrrolidone (AEAEVP) is synthesized by the Michael Addition reaction from 1-vinyl-3-(*E*)-ethylidene pyrrolidone (EVP) and ethylenediamine as shown in Figure 9.2. The product is a white, needle-shaped crystalline solid having a melting point of 59–61°C.

Table 9.1 Comonomers for NVP

Comonomer	Examples of Use
Vinyl acetate	Laundry detergents
N,N'-dimethylaminopropyl acrylamide	Laundry detergents
N-3,3-Dimethylaminopropyl methacrylamide	Personal care applications
Vinylimidazole	Photosensitive resin compositions
2-methylene-1,3-dioxepane	Biodegradable copolymers
1-Vinyl-3-(*E*)-ethylidene pyrrolidone	Crosslinking agent for proliferous polymers
3-(2-Aminoethyl)-α-aminoethyl-*N*-vinyl-2-pyrrolidone	Metal ion complexing agent
N,N'-divinylimidazolidin-2-one	Crosslinking agent for clarifying resins
Maleic anhydride	Coatings for rust removal

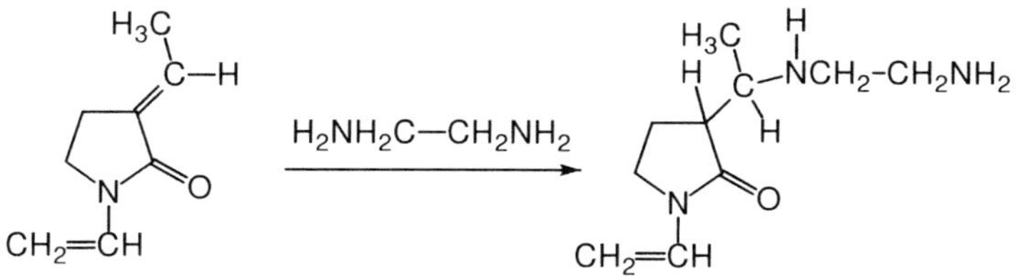

Figure 9.2 Synthesis of 3-(2-Aminoethyl)-α-aminoethyl-*N*-vinyl-2-pyrrolidone (5)

9.2 Polymerization and Fabrication

9.2.1 Homopolymerization

NVP is polymerized via free radical polymerization techniques (6). The polymerization in organic solvents, for example in alcoholic solution, leads to a poly(*N*-vinyl-2-pyrrolidone) (PNVP) of low molecular mass, since the organic solvents may act as chain transfer agents (CTA)s. To prepare highly concentrated aqueous solutions of the polymer, however, the major part of the organic solvents has to be distilled off.

The polymerization of NVP in aqueous solution is originally carried out in the presence of hydrogen peroxide as initiator (7). In this case, the molecular weight of the PNVP depends on the hydrogen peroxide concentration. Low molecular weight types result from high hydrogen peroxide concentrations, and vice versa. Hydrogen peroxide is used advantageously together with redox systems, i.e., iron(II) salts or copper (II) salts.

Highly concentrated aqueous solutions of PNVP cannot be prepared directly because it is impossible to control the exothermic reaction there. Further, hydrogen peroxide effects an unwanted grafting effect in that it increases the molecular weight. Instead of hydrogen peroxide, organic peroxides are used beneficially.

9.2.1.1 Colorless Polymers

Conventional NVP polymers and copolymers easily color. This tendency is remarkable in the case of low-molecular weight polymers. The coloring is explained by lower molecular weights of the polymers require relatively large amount of initiator that easily cause side reactions. The coloring of a NVP polymer can be suppressed, by (8):

1. Adjusting the pH with a carbonate or hydrogen carbonate when polymerizing *N*-vinylpyrrolidone by using hydrogen peroxide
2. Using 2,2′-azobisisobutyronitrile (AIBN) and a hydroperoxide together in polymerizing NVP

3. Using *tert*-butyl peroxypivalate and a chelating buffer together
4. Using NVP obtained by a gas phase dehydration reaction rather than NVP obtained by the *Reppe* process in combination with special initiators, such as 2,2′-azobis-(2-methylpropionate) or *tert*-amyl peroxy-2-ethyl hexanoate.

2-Pyrrolidone or *N*-(2-hydroxyethyl)pyrrolidone is the raw material for the synthesis of NVP. It is usually derived from γ-butyrolactone. The polymerization of NVP is easily hindered when the γ-butyrolactone content in NVP exceeds a certain amount. When acetylene is used as a raw material for the synthesis of NVP, by-products are formed by the methylation of NVP, which hinders the polymerization of NVP (9).

The polymerization to achieve the homopolymer of NVP is conducted in aqueous medium. Care must be taken to chose the correct initiator system. When an organic peroxide and a sulfite are employed as a redox initiator, the resulting polymers show little coloration and smell, and contain only a small amount of 2-pyrrolidone as a by-product (10).

9.2.2 *Copolymers*

9.2.2.1 *Copolymers with Vinyl acetate (VA)*

Copolymers containing NVP and VA can be prepared by radical polymerization with 2,5-dimethyl-2,5-di-(*tert*-butylperoxy)hexane (which is Lupersol® 101) in isopropanol as solvent at around 80°C. After completion, isopropanol is stripped while water is added. Then the mixture is pressurized and a post heating up to 130°C is initiated. In this way, the residual level of NVP can be reduced down to 50 *ppm* (11).

9.2.2.2 *Proliferous Copolymers*

Proliferous Copolymers with VA. For the proliferous polymerization, no radical initiator is used. The monomers, namely NVP and VA, small amounts (0.8–1.2%) of crosslinking agent, such as ethylidene-vinylpyrrolidone, and water, are held in a sealed stirred reactor under a nitrogen atmosphere at about 80°C until the reaction mixture pops, i.e., proliferous polymerization occurs. In this stage, solid white particles appear.

When the reaction is complete, after about 2–5 *h*, the mixture is left for an additional hour, then cooled, and discharged. The crosslinked copolymer, is washed with water, filtered, and dried in vacuo at 70°C. The yield is about 70–95% (12). The proliferous copolymers find use for pharmaceutical tablets, which exhibit rapid dissolution and disintegration of a drug therein. The materials show a reduced hygroscopicity.

Proliferous Copolymers with AEAEVP. The synthesis of AEAEVP has been shown already in Figure 9.2, page 295. This monomer, when present in small amounts in copolymers is effective in completing and removing traces of heavy metal ions, such as copper and iron ions from aqueous solutions.

The comonomer may be generated *in situ* during the course of proliferous polymerization of NVP, or obtained independently of the polymerization, and added to the NVP monomer to form the polymerization reaction mixture.

Excess EVP present in the reaction product of the Michael Addition reaction may be used as crosslinking agent in the polymerization process. EVP undergoes a proliferous polymerization in the presence of diethylenetriamine, or dimethylethylenediamine.

9.2.2.3 *Compositional Homogeneous Copolymers*

Copolymers from NVP and *N*-3,3-dimethylaminopropyl methacrylamide (DMAPMA) (13) or diallyldimethylammonium chloride (14) find use in hair styling compositions. Since human hairs are negatively charged, the copolymer of NVP and DMAPMA will form complexes with hairs and form a uniform thick film on keratin. That is the reason why a copolymer of NVP and DMAPMA is used in hair care products (13).

In general, by feeding the monomers at the begin of the copolymerization, the composition of the copolymer changes in the course of conversion because of the unequal reactivity of the comonomers. For hair styling compositions, it has been demonstrated that compositional homogeneous copolymers have a superior performance over copolymers prepared in a conventional manner because they provide an improved curl retention, i.e., hold during use.

In comparison to NVP, DMAPMA is consumed faster during copolymerization. This unequal rate of incorporation can be compensated by a special feeding schedule of DMAPMA in the course of copolymerization (15). To set up the feeding schedule, measurements of the residual monomer composition are performed during copolymerization. According to the results, the feeding schedule is adjusted in several iterative experiments until the ratio of residual monomers is approximately constant over the whole range of copolymerization.

9.2.2.4 *Biodegradable Copolymers*

The radical copolymerization of NVP using 2-methylene-1,3-dioxepane (MDO) as comonomer runs via a highly special mechanism, which consists of ring opening accompanied by the rearrangement of the MDO moiety. The mechanism of copolymerization is shown in Figure 9.3.

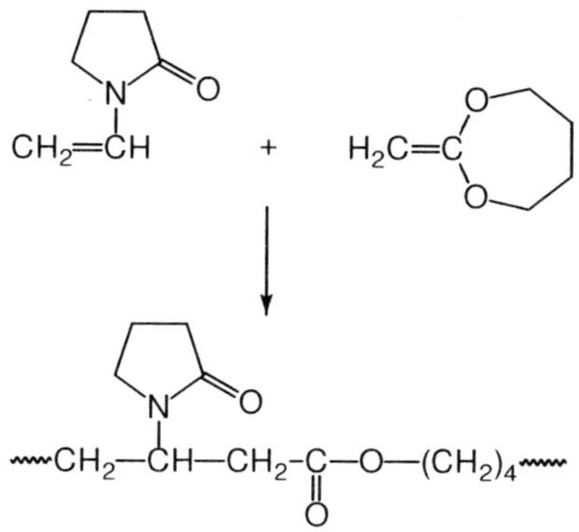

Figure 9.3 Copolymerization of NVP with 2-Methylene-1,3-dioxepane (16)

The copolymerization can be carried out either as solution polymerization in isopropanol as solvent, or as precipitation polymerization using *n*-heptane (16). In solution polymerization, as radical initiator 2,2′-Azobis(2-methylbutyronitrile) (Vazo® 67, DuPont) is used, whereas in precipitation polymerization *tert*-butyl peroxypivalate (Lupersol® 11, Atochem) is used. The copolymer is hydrolytic degradable both in acid and in alkaline medium, and biodegradable as well.

9.2.2.5 Block Copolymers

A_nB block copolymers can be prepared from precursor polymers, i.e., prepolymers, by linking the precursor polymers together. The A block is formed from a monofunctional hydroxyl terminated PNVP prepolymer. The B block is formed from a linear poly(urethane) (PU) with isocyanate groups at the ends OCN−PU−NCO.

A hydroxyl terminated PNVP prepolymer is synthesized by the radical polymerization of NVP with AIBN as initiator and isopropoxyethanol as a CTA. After polymerization, the isopropoxyethanol is partially evaporated in vacuo. The polymer is precipitated from concentrated solution with cold ether (17). Variants of the hydroxyl terminated PNVP prepolymer have been described. For example, the prepolymer can be cain extended by the esterification with citric acid. Further, hydroxy terminated copolymers made from NVP and methyl acrylate (MA), or the usage of a carboxylic acid functionalized CTA, such as 2-carboxyethoxy-2-propane have been described.

The B block prepolymer is a PU made from diisocyanatodiphenyl methane (MDI) and poly(tetramethylene oxide). MDI is used in stoichiometric excess to ensure isocyanate endgroups. Tin dilaurate is a suitable catalyst. Alternatively, the B block prepolymer can be functionalized with amide or imide groups.

Eventually, the PVP/PU/PVP block copolymer is synthesized by the reaction of the prepolymers, namely the A block prepolymer in dimethylacetamide solution and the B block prepolymer in dioxane solution, respectively, at 60°C in dry atmosphere.

9.2.2.6 Hydrogels

Hydrogels are of interest for medical applications, such as drug delivery and in separation science. Most hydrogels are prepared by irradiation techniques, although chemical crosslinking is feasible. Some hydrogels exhibit pH-responsive properties.

PNVP hydrogels can be prepared by crosslinking with γ-ray irradiation (18). Possible applications are in the medical field, such as wound dressing (19). In addition, hydrogels for wound dressings have been made from a mixture of aloe vera, poly(vinyl alcohol) (PVA) and PNVP by freezing and thawing, followed by γ-ray irradiation (20).

Copolymers of NVP and MA that are crosslinked with *N,N'*-methylenebisacrylamide (MBA) are hydrogels, suitable for the extraction of various metal ions, such as copper, nickel, or cadmium ions (21). For the same purpose, NVP was grafted on to various cellulose derivatives (22).

Hydrogels consisting of PNVP and polysaccharides, such as κ-carrageenan or agar have been prepared by γ-ray irradiation (23–25). The gel strength and gel fraction attain a maximum up to a certain irradiation dose.

Semi-interpenetrating networks based on hydrogen bonding between crosslinked PNVP and linear poly(acrylic acid) (PAA), have been prepared in three steps: In the first step, a linear PAA was prepared by free radical polymerization using AIBN. Then a crosslinked porous PNVP bead was obtained by the radical crosslinking copolymerization of NVP and MBA in suspension. The porous PNVP bead was then soaked with an aqueous solution of PAA and stabilized by multiple freezing thawing steps (26). The hydrogel exhibits a pH sensitive swelling behavior in the pH range of 2.25 to 4.00.

Thermosensitive terpolymers from *N*-propylacrylamide, *N*-isopropylacrylamide (NIPAAm), and NVP have been tested as a possible liquid embolic agent. The material gels in aqueous solution at body temperature (27).

Hydrogel coatings from PNVP can be formed on surfaces of poly(ethylene terephthalate) by depositing the polymer on the surface and irradiating with an electron beam (28). The films swell by a factor of 7 and the swollen films are stable against shear stress and variation of the pH.

9.3 Properties

PNVP is a hygroscopic, amorphous polymer. It is delivered as a white, free flowing powder, or in aqueous solution in molecular weights ranging from 10 *k Dalton* to 3 *M Dalton*. It is soluble in water and most polar solvents (29). PNVP is considered physiologically inert.

It can be plasticized with water and common organic plasticizers. Due to its high polarity, it forms complexes with phenols and carboxylic acids, as well as with anionic dyes and inorganic salts. The high polarity imparts good adhesive properties. The glass transition temperature of PNVP increases with its molecular weight up to 180°C.

9.3.1 *Fikentscher K Value*

Traditionally for NVP polymers, the molecular weight is often expressed as the Fikentscher K value. The Fikentscher K value is an empirical molecular weight value based on dilute solution viscosity η_{rel} and the concentration c.

$$\begin{aligned} \frac{\log \eta_{rel}}{c} &= \frac{K_0 + 75K_0^2}{1 + 1.5K_0 c} \\ K &= K_0 \times 10^3. \end{aligned} \tag{9.1}$$

For NVP polymers, the relation of the Fikentscher K value to the molecular weight is summarized in Table 9.2.

Table 9.2 Relationship of Viscosity, K Value, and Approximate Molecular Weight for Poly(*N*-vinyl-2-pyrrolidone) (30)

Viscosity[a] /[*cSt*]	Concentration /[%]	K value Range	M_η /[*Dalton*]	M_w /[*Dalton*]
7	20	13–19	10,000	12,000
25	20	26–34	40,000	55,000
50	10	50–62	220,000	400,000
400	10	80–100	630,000	1,280,000
7000	10	115–125	1,450,000	2,800,000

[a] Viscosity in water

9.3.2 Miscible Blends

Due to its capability of forming hydrogen bonds, PNVP is miscible with a variety of other polymers. Polymers that are miscible with PNVP are summarized in Table 9.3.

Table 9.3 Polymers for Miscible Blends with Poly(*N*-vinyl-2-pyrrolidone)

Polymer	References
Collagen	(31)
Chitosan	(32,33)
Cellulose acetate	(34)
Cellulose acetate hydrogen phthalate	(35)
Cellulose diacetate	(36)
Poly(vinyl alcohol)	(37,38)
Poly(*L*-lactide)	(39)
Poly(methyl methacrylate), atactic	(40)
Poly(methyl methacrylate), syndiotactic	(40)
Poly(vinylphenol)	(41,42)
Poly(styrene-*co*-vinylphenol)	(43)
Oligo(ethylene glycol)	(44)
Poly(*N*-phenyl-2-hydroxytrimethylene amine)	(45)
Poly(bisphenol A hydroxyether)	(46)
Poly(hydroxyether sulfone)	(47)
Poly(ethersulfone) (PES)[a]	(48)
PES[b]	(49)

[a] with poly(1-vinylpyrrolidone-*co*-styrene)
[b] with poly(1-vinylpyrrolidone-*co*-acrylonitrile)

For poly(methyl methacrylate) (PMMA), the kind of tacticity is governing the miscibility with PNVP. Based on the observation of two glass transition temperatures, atactic PMMA as well as syndiotactic PMMA is miscible with PNVP, whereas isotactic PMMA is immiscible or partially miscible.

The immiscibility is probably due to a stronger interaction among isotactic methyl methacrylate segments because its ordination and molecular packing contribute to form a rigid domain (40).

9.3.3 *Optical Properties*

PNVP films show photoluminescence (PL). When irradiated with UV light, a response to the ultraviolet excitation in the region of approximately 440 *nm* is observed (50). Hybrid films of PNVP and silica with a high thermal stability up to 250°C have been prepared. Doping with europium effects a white PL.

9.4 Special Additives

9.4.1 *Antioxidants*

The heat resistance and storage stability of PNVP can be improved by adding antioxidants as shown in Table 9.4. The stability is measured in the following way: To 50 g of PNVP with an initial Fikentscher K value of 88, 0.05 g of antioxidant is added. Both compounds are dissolved in 200 *g* water or methanol. The solution is cast on to a Teflon sheet, then dried at 90°C in vacuo, and then pulverized. The powder is kept at 50°C in air, and the K value is measured every week. The durability of the K value stability is the number of the days at which a change of 5% is obtained.

Before the concept of durability of the K value stability (51) was developed, the stability of the samples was tested by keeping them for 2 *h* at 120°C in air (52). After this treatment, the change in the K value was measured. This experimental concept seems to be less sensitive in comparison to the method where the samples are kept at elevated temperature as long as the K value becomes unstable.

9.5 Applications

Polymers and copolymers based on NVP are widely used in various fields, such as medicines, cosmetics, pressure-sensitive adhesives or adhesives, paints, dispersants, inks, and electronic parts, because these materials are excellent for biocompatibility, safety, and hydrophilicity. In addition, crosslinked products of the polymers are useful as water absorbent resins for various uses requiring water absorption and water retention, for example, disposable diapers (8). In detail, applications include (29):

Table 9.4 Antioxidants for PNVP (51)

Compound	Stability[a] /[*d*]
No additive	14
Thiourea	70
Dimethylthiourea	56
Thioacetamide	49
Hydroquinone	56
4-Benzoyloxy-2,2,6,6-tetramethylpiperidine	49
N,N'-di-*sec*-butylphenylenediamine	42
2-Mercaptobenzimidazole	42
Triphenyl phosphite	35
Thiourea 2-mercaptobenzimidazole	196
Thioacetamide 2-mercaptobenzimidazole	175
4-Benzoyloxy-2,2,6,6-tetramethylpiperidine-2-mercaptobenzimidazole	175
N,N'-di-*sec*-butylphenylenediamine triphenyl phosphite	119
Hydroquinone triphenyl phosphite	168
Thiourea sodium hydroxide (pH 3.5)	21
Thiourea succinic acid (pH 5.5)	56
Thiourea sodium hydroxide (pH 11.6)	70

[a] Days at 50°C after the K value changed by 5%

- Dye transfer inhibition in detergents
- Complexing agent in photoprocessing
- Protective colloids in emulsion polymerization
- Ink jet printing
- Hollow fiber membranes
- Oil field applications (53)
- Lithographic printing, including negative photoresist technology (54),
- Adhesive technology
- Combustible ceramic binders (55)
- Tablet binder formulations
- Dispersants, thickeners and stabilizers.

9.5.1 Medical Devices

9.5.1.1 Biomedical Polymers

Polymers used to create medical devices are typically chosen for their bulk properties. However, it is often desirable for the surfaces of such medical devices to possess different properties than that of the bulk polymer. For example, it may be desirable for a polymer surface to have a different level of compatibility with other polymers or tissues, surface energy, etc., than that of the bulk polymer.

Block copolymers have been used to modify PU surfaces, which are important biomedical polymers used in implantable devices such as artificial hearts, cardiovascular catheters, pacemaker lead insulation, etc. Such block copolymers have been used to enhance antimicrobial properties, lubricity, barrier properties, anticoagulant properties, etc. PU/PNVP block copolymers can be made from hydroxyl terminated PNVP prepolymers and isocyanate terminated PU prepolymers (17).

A copolymer of NIPAAm with NVP as hydrophilic comonomer exhibits thermosensitive properties. Aqueous solutions of the copolymer undergo a thermal induced phase transition and precipitate or gel with heating. The gels show a significant contraction at storage. However, the collapsed gels have still a high water content. Due to these properties, the materials arse suitable for biomedical application as an injectable implant material (56).

9.5.1.2 *Plasma Substitutes*

For first aid in cases of physiological shock, it is frequently necessary to use plasma substitutes, since blood stocks are in most cases not immediately available. Furthermore, it is necessary to be able to replace plasma without taking any risks as to compatibility, for example with respect to the blood group and the Rhesus factor. Pure salt solutions, such as physiological sodium chloride solution are not suitable for the treatment of shock, because their residence time in the vascular system is insufficient.

PNVP solutions are used extensively as plasma substitute. Because of their similarity to peptides, PNVPs exhibit good compatibility, and providing they have the appropriate medium molecular weight, their colloidal osmotic pressure resembles to that of blood plasma.

However, if the molecular weight is too high, they are not completely excreted and deposit in the reticuloendothelium. In contrast, if the polymer has a low molecular weight, it shows a good excretion via the kidneys, but the colloidal osmotic pressure and hence the effectiveness of the substitute, is low. PNVP is an inert polymer, since it is produced by free radical polymerization and so its polymeric backbone consists of carbon-carbon bonds.

The manufacture of biodegradable PNVP by linking NVP blocks by means of peptide units or analogously by means of sugar units is expensive and laborious. However, the polymer can be functionalized with hydroxyl endgroups. These endgroups can be linked together with dicarboxylic acid derivates or aliphatic diisocyanates to form a block copolymer (57).

With the introduction of urethane groups, the biodegradability of PNVP is enhanced. The number of degradable bonds, and the type of bond, can be controlled by the type of the carboxylic acid derivatives or the isocyanate. Examples of linking agents are oxalic acid dichloride, adipic acid dichloride, succinic acid dichloride, and hexamethylene diisocyanate. The block polymers exhibit good colloidal osmotic effects and are biodegradable. The low-molecular weight compounds resulting from such a degradation are easily excreted through the kidneys.

9.5.1.3 *Wound Dressing*

Copolymers from NVP and allyl alcohol bear pendant hydroxyl functionalities (58). These copolymers can be linked to isocyanate terminated PU prepolymers. The PU is chosen to take up iodine to give a complex use as an antimicrobial coating on a medical article, as an antimicrobial foam sponge, or wound dressing.

Iodine is a well-known germicide with an activity against a wide range of bacteria and viruses. Some polymeric materials form complexes with iodine. These complexes, termed iodophors, are used commercially with a sponge or brush for germicidal cleaning or scrubbing. PNVP iodophors are extensively used in germicidal preparations.

The PU block copolymer combines the iodine complexing property of PNVP with the excellent physical properties of PU. The PNVP, being covalently bonded in the polymer chain, does not leach out when contacted with water.

Delivery of iodine is accordingly slower so that less iodine is needed in the iodophor. Because iodine delivery is slower, the levels of released iodine is sufficient for antimicrobial activity so that the danger of iodine toxicity is substantially eliminated.

9.5.1.4 *Surface Stabilizers*

Nanoparticulate compositions are particles consisting of a poorly soluble therapeutic or diagnostic agent that are adsorbed on to the surface of a surface stabilizer. The surface stabilizers physically adhere to the surface of the nanoparticulate drug, but do not chemically react with the drug. A copolymer of NVP and VA can be used as a surface stabilizer for nanoparticulate compositions (59). Auxiliary surface stabilizers include non-ionic and ionic surfactants.

9.5.1.5 *Drug Delivery Systems*

The earliest drug delivery systems, of the 1970s were based on polymers formed from lactic acid (60).

The release of the drugs medications from the polymer device traditionally has been diffusion-controlled. However, biodegradable polymer systems have been developed that degrade into biologically harmless compounds, setting free the drug. Another more recent issue is to develop polymers that are accumulating in various regions of the body where the drug is needed, together with the not yet released drug. PNVP has a comparatively long mean resident time, when injected intravenously, as exemplified with mice (61). The pharmacokinetic characteristics are shown in Table 9.5.

Table 9.5 Mean Residence Times of Polymers in Blood (61)

Polymer	*M* /[*Dalton*]	Residence Time /[*min*]
Poly(ethylene glycol)	5,000	78.9
Poly(ethylene glycol)	12,000	139.5
PNVP		278.8
Poly(acrylamide)		166.3
Poly(dimethylacrylamide)		79.7
Poly(vinyl alcohol)		262.5
Dextran	10,400	97.6

The polymers show a specific accumulation in various organs. The target where the polymer should accumulate can be tailored by modification of the polymer. For example, poly(1-vinylpyrrolidone-*co*-styrene) accumulates preferably in the liver, but poly-(1-vinylpyrrolidone-*co*-vinyl laurate) accumulates preferably in the spleen, 20 times higher than PNVP and poly(1-vinylpyrrolidone-*co*-styrene) (62).

In solid drugs, the molecular mobility below the glass transition temperature is an important parameter for storage stability (63). The change of the enthalpy in the course of storage can serve to estimate the storage stability. Based on enthalpy relaxation measurements, it has been suggested that an amorphous pharmaceutical solid must be stored at least fifty degrees below the glass transition temperature in order to reduce the molecular mobility of the material to such an extent that it is physically stable for a significant period of time. Absorbed water depresses the glass transition temperature.

This issue must be considered in the prediction of stability. In the case of lactose, it has been pointed out that there is a critical humidity for crystallization in the course of long-term stability. The presence of PNVP generally increases the critical relative humidity for crystallization (64).

For common hydrophobic and hydrophilic pharmaceutical materials, such as magnesium stearate or lactose, it has been confirmed that intrinsic adhesion forces of PNVP toward these materials are dependent on the relative humidity. Therefore, in order to adjust the adhesion properties of PNVP, humidity control is an important issue in the development of drug delivery systems based on PNVP (65).

The addition of drugs to PNVP may influence the glass transition temperature caused by a plasticizing effect. It was attempted to predict the plasticizing effect by solubility parameters when various drugs were added to PNVP. However, the correlation between the solubility parameters and the plasticizing effect of drugs turned out to be very poor. In contrast, infrared spectroscopy showed the interaction between the drugs and PNVP and proved to be extremely useful with regard to the plasticizing effect of various drugs (66).

The interaction of drugs with the PNVP host has been elucidated with molecular modelling techniques (67). Experimental methods for the characterization of the interaction of drugs with the polymer are differential scanning calorimetry, fourier transform infrared spectroscopy, and X-ray diffraction (68).

The solubility of drugs can be considerably enhanced when coprecipitated with the polymer to form microspheres (69). In this way, the crystallinity of the drug may be changed. For example, for a leukotriene biosynthesis inhibitor, MK-0591, c.f. Figure 9.4, the extent of crystallization inhibition increases with the molecular weight of PNVP. For comparable molecular weights, the homopolymer is more effective in the crystallization inhibition of the drug than a VA copolymer (70).

A copolymer from NVP and dimethylmaleic acid, or dimethylmaleic acid anhydride, respectively, has a high renal-targeting capability as a drug carrier. The relationship between the molecular weight of the copolymer and its renal accumulation has been evaluated in mice. A molecular weight in the range of 6–8 *k Dalton* shows the highest renal accumulation (71,72).

Mucoadhesive Materials. Mucins are glycoproteins secreted from various types of cell membranes as mucus. These proteins have sev-

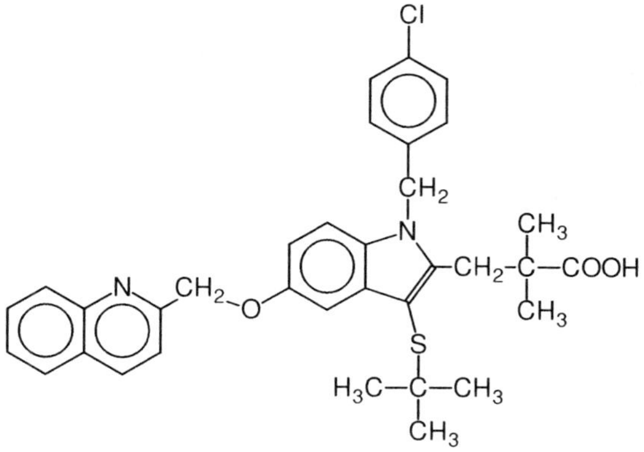

Figure 9.4 3-(1-(4-Chlorobenzyl-3-(*tert*-butylthio)-5-(quinolin-2-ylmethoxy)indol-2-yl))-2,2-dimethyl propanoic acid (MK-0591)

eral functions in epithelial cells including cytoprotection, extravasation during metastases, maintenance of luminal structure, and signal transduction (73). Intestinal mucus is believed to be the main barrier to drug absorption. The mucus mainly affords protection for the underlying epithelial cells (74).

Mucoadhesive materials are useful in a wide variety of applications, particularly in pharmaceutical compositions. Mucoadhesive pharmaceutical compositions can provide prolonged and improved coating and protection of the mouth, esophagus, oropharynx, or stomach to inhibit irritation and accelerate healing of inflamed or damaged tissue. The sustained coating provides a matrix to deliver therapeutic agents to mucosal tissues at higher concentrations for higher efficacy, lower side effects, and sustained release of the active agent (75).

Inter-polymer complexes have been prepared by a template polymerization technique of acrylic acid (AA) using PNVP as a template polymer (76, 77). The inter-polymer complexes are formed by hydrogen bonds between the carboxyl groups of PAA and the carbonyl groups of PNVP.

The adhesive force and the drug release rate can be controlled by changing the ratio of PNVP and PAA. Copolymers from NVP and methacrylic or AA crosslinked with MBA are potentially useful in ophthalmic drug delivery, such as pilocarpine or chloramphenicol (78).

9.5.2 Laundry Detergents

PNVP is used in laundry detergents to complex vagrant dyes that are released from colored fabrics during the wash process. It prevents their redeposition on other garments. In powder detergent formulations, NVP polymers can simply be dry blended during the manufacture of powder detergents. In general, the dye complexation efficiency of the polymer increases with increasing molecular weight of the polymer. Powdered laundry detergents that contain PVP, non-ionic surfactants, and anionic surfactants can be readily formulated either by conventional spray drying techniques or by agglomeration.

However, when PNVP is incorporated into an aqueous detergent composition containing non-ionic surfactants, stability problems are often encountered, especially when higher molecular weight PNVP is used. The instability of an aqueous detergent composition manifests itself by phase separation upon standing. Stable formulations can be prepared with copolymers of NVP and VA in combination with certain amounts of anionic and non-ionic surfactants (79).

Stable detergent formulations can be made that contain greater than 1% by weight copolymer if the weight ratio of anionic to non-ionic surfactant is above about 4:1. Other dye transfer inhibitors, mostly binary copolymers based on NVP with various other comonomers are shown in Table 9.6.

9.5.3 Adhesives

From vinylpyrrolidone/vinyl acetate copolymers, transparent, flexible, and oxygen permeable films can be made. The materials adhere to glass, plastics, and metals. For these reasons, the copolymers are suitable in the usage as hot melt adhesives and other adhesive applications (29). The water solubility, film hardness, adhesive strength, and glass transition temperature increases with increasing content of NVP.

Blends of high molecular weight PNVP with a short-chain liquid poly(ethylene glycol) (PEG) show pressure-sensitive adhesion in a

Table 9.6 Comonomers for *N*-Vinyl-2-pyrrolidone in Polymers for Dye Transfer Inhibitors (80)

Comonomers
N,N'-Dimethylaminopropyl acrylamide (DMAPMA)
N,N'-Dimethylaminoethyl methacrylate (DMAEMA)
4-Vinylpyridine
N,N'-Dimethylaminopropyl methacrylamide quaternized with diethyl sulfate
N,N'-Dimethylaminoethyl methacrylate quaternized with diethyl sulfate
2-Tri(methylammonium)ethyl methacrylate chloride
3-Methyl-1-vinylimidazolium chloride
Diallyldimethylammonium chloride (14)
1-Vinylimidazole
1-Vinylimidazole *and* *N,N'*-dimethylaminoethyl methacrylate
1-Vinylimidazole *and* 3-methyl-1-vinylimidazolium chloride *and* ethyl acrylate

narrow range of composition, around 36% PEG (81–83). PEG and water behave as tackifiers in the blends with glassy PNVP. PEG is a stronger tackifier than water.

Hot melt pressure-sensitive adhesives that are in addition cured by photo crosslinking have been developed. The base material is a mixture from copolymers of NVP, *n*-butyl acrylate (BA) with small amounts of AA (84). The composition contains tackifiers such as modified rosin resins. The photoinitiator consists of acetophenone or benzophenone derivatives, which are modified with polymerizable vinyl groups and incorporated in the copolymers. The polymerizable vinyl groups are introduced into the photoinitiator by esterification with AA or methacrylic acid (85). An example is shown in Figure 9.5.

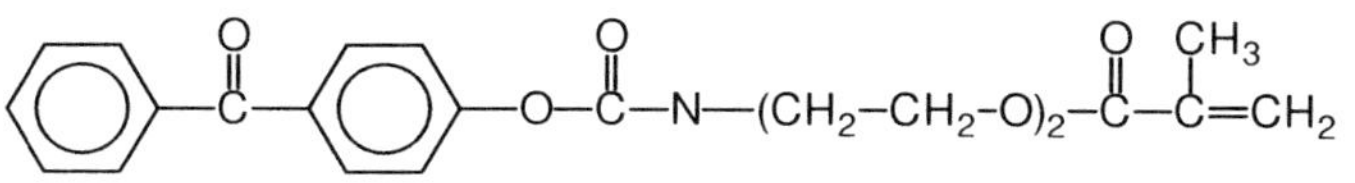

Figure 9.5 Photoinitiators with Pendent Vinyl Groups (85)

The polymeric photoinitiator is made by radical copolymerization of the functional monomeric photoinitiator with 2-ethylhexyl acrylate, BA, AA and tetrahydrofurfuryl-2-acrylate. The amount of monomeric photoinitiator relative to the other monomers is 0.1–4%. The crosslinking of the polymer occurs preferably by means of a chemical grafting reaction of the photoinitiator with a spatially adjacent polymer chain.

Pressure-sensitive adhesive tapes exhibit both high adhesion and high cohesion values. They adhere strongly to automotive paints and to rubber or plastic foam layers (86). The use of NVP in the copolymers improves the adhesive properties and the electric properties of the adhesives.

9.5.4 Membranes

9.5.4.1 Membranes for Reverse Osmosis

A reverse osmosis membrane can be produced by wetting a porous membrane with a mixture of NVP, maleic anhydride (MA), and maleimide and irradiation of the wetted membrane with a laser in order to polymerize the monomers (87). The membrane substrate consists of a poly(sulfone) (PSf) polymer. The resulting membrane exhibits a high salt rejection and improved chlorine stability.

9.5.4.2 Ultrafiltration Membranes

Poly(acrylonitrile) is a favored membrane matrix material because of its thermal stability up to 130°C and its resistance to many organic solvents. It is widely used commercially as ultrafiltration (UF) membrane. PAN membranes are chemically modified reactions involving the nitrile groups in order to obtain membranes with an improved permeation behavior. Other common techniques are surface graft polymerization using hydrophilic monomers, such as 2-hydroxyethyl methacrylate, acrylic acid, methacrylic acid, or *N*-vinyl-2-pyrrolidone (88).

PES is essentially a hydrophobic material. For membrane applications it can be mixed with poly(1-vinylpyrrolidone-*co*-styrene) copolymers. The copolymers contain NVP 59–92% (48,89). Membranes prepared from the miscible blends of PES and the NVP-containing copolymer show a better performance in solute rejection and water flux than those prepared from PES or a mixture of PES and PNVP. Membranes prepared from immiscible blends of PES and the NVP-containing copolymer show a worse solute rejection in comparison to the other membranes.

PSf forms homogeneous mixtures with poly(1-vinylpyrrolidone-*co*-acrylonitrile) that contain a content of acrylonitrile of 2–16% (49). UF membranes can be produced from miscible blends of PSf and poly(1-vinylpyrrolidone-*co*-acrylonitrile) copolymers (90). The copolymers are synthesized via radical copolymerization (91).

9.5.4.3 *Fuel Cell Membranes*

Membranes of PNVP and chitosan (CS) are suitable for direct methanol fuel cell applications (92). Relevant properties of these materials are shown in Table 9.7. The mechanical properties of the blends are greatly affected by UV irradiation, but the extent of change is smaller in the blend than in pure CS (93).

Table 9.7 Properties of Membranes for Fuel Cell Applications (92)

Membrane	Water uptake /[%]	Methanol uptake /[%]	Methanol permeability [a] /[$cm^2 s^{-1}$]	Proton conductivity /[$S\ cm^{-1}$]
CS/PNVP	[b]	1.8	9.2E-8	0.019
CS/PNVP[c]	52.1	0.11	7.3E-8	0.024
Nafion 117	33.3	9.32	21.6E-8	0.086

[a] At 30°C, with 50 % methanol
[b] Highly swollen in water
[c] Crosslinked CS/PNVP

Blends of sulfonated poly(ether ether ketone) with PNVP exhibit a reduced methanol uptake. The methanol permeability decreases while the high proton conductivity is maintained (94). When the content of PNVP is lowered below 20% in the blends, a great reduction of methanol uptake and the methanol permeability is observed.

The family of membranes for fuel cells includes PNVP modified with PVA/poly(2-acrylamido-2-methyl-1-propane sulfonic acid) (95) and composites of sulfonated poly(styrene) microspheres, PNVP and poly(vinylidene fluoride) (96).

9.5.4.4 *Pervaporation Membranes*

The pervaporation separation technique has been used for the dehydration of tetrahydrofuran (THF) from its aqueous mixtures by blend membranes of PNVP and CS. The membranes are physically blended and crosslinked with glutaraldehyde as well as with sulfuric acid in a methanol/sulfuric acid mixture bath to enhance their selectivity and mechanical strength properties (97). PVA/PNVP blends can be independently crosslinked:

1. Chemically by using glutaraldehyde
2. Photochemically by using 4,4′-diazostilbene-2,2′-disulfonic acid disodium salt.

In this way, interpenetrating polymer network membranes can be fabricated (98). Pervaporation studies of THF/water and THF/methanol mixtures showed that the membranes are excellent in the dehydration of THF, but much less efficient for the separation of THF/methanol mixtures.

High performance pervaporation membranes for the selective removal of ethanol from ethyl *tert*-butyl ether consist of blends of cellulose acetate and poly(1-vinylpyrrolidone-*co*-vinyl acetate) (99).

9.5.4.5 *Hollow Fiber Membranes*

An asymmetrical, microporous, hollow fiber membrane can be made from a mixture of PSf and PNVP dissolved in an aprotic solvent. The physical morphology of the hollow fiber membrane, i.e., the asymmetric microporous wall, is rapidly formed by passing the polymeric dope mixture through an outer annular orifice of a tube-in-orifice spinneret, while simultaneously passing a precipitating solution through the central tube of the spinneret.

The emerging hollow fiber travels downward for about 2–10 *m* before submersion into a quenching bath. The asymmetrical, microporous, hollow fiber membrane is biocompatible and suitable for use in, for example, dialysis, hemodialysis, ultrafiltration, and water filtration applications (100).

On PSf-based hollow fiber membranes the PNVP can be fixed by a treatment with γ-ray irradiation. The treatment fixes the PNVP to a great extent which is then suitable for dialysis membranes (101).

PNVP modified PSf membranes give a lower protein adsorption from a plasma solution than PSf and other surface modified membranes (102). It is believed that the hydrophilic surface of membranes without ionic groups causes the suppression of platelet adhesion.

9.5.5 *Cleaning Compositions*

PNVP is used in a variety of cleaning compositions. These compositions cover a range of applications from household cleaners to industrial applications for the deflocculation of beverages.

9.5.5.1 *Soil*

Copolymers of NVP and *N*-vinylimidazole, together with copolymers composed from NVP and dialkylaminoalkyl acrylate or methacrylate (in alkaline aqueous solution), are used in cleaning compositions (103, 104). The compositions show an improved stability against bacterial contamination.

Various other ingredients are added to the formulation such as perfumes, suds controlling agents, chelating agents, builders, etc. The cleaning compositions are specially designed for cleaning hard surfaces soiled with greasy stains or burnt sticky food residues that are typically found in kitchens.

9.5.5.2 *Clarifying Agents for Wine and Beer*

In the production of wine, fruit must, and similar products displaying instability with regard to color, aroma and taste is a crucial feature. These properties are dependent on the concentration of phenolic constituents and the content of heavy metal ions. Thus, the controlled adjustment of the concentration phenolic constituents and heavy metal ions is a critical issue for the production of final products with stable sensory properties.

Phenolic substances can be bound by complexing, adsorption, or precipitation. Suitable products are native or modified proteins such as gelatine, caseinate, dried blood, or potassium caseinate. In addition, crosslinked polymers based on NVP are used. The formulations are added to must or wine that contain turbid suspended substances as well. After treatment, they are filtered off with the suspended substances. The filtration properties are improved in the formulations described below in detail.

Formulations that can be easily handled are blends made from cellulose, potassium caseinate and a copolymer from NVP and 2% *N,N′*-divinylimidazolidin-2-one as crosslinking agent. Likewise, copolymers from *N*-vinylimidazole, NVP and *N,N′*-divinylethyleneurea can be used (105).

A formulation is prepared by suspending, 80 *kg* of cellulose in a solution of 10 *kg* of potassium caseinate in 400 *l* of water at 60°C. The suspension is atomized at approximately 180°C at 50 *bar* via a single-component nozzle. Thus, the mixture contains 90 *kg* of dry product. To the mixture containing 90 *kg* dry product, 10 *kg* of polymer, prepared from 98% of NVP with 2% crosslinking agent in the form of *N,N′*-divinylimidazolidin-2-one, are added dry in the atomization zone.

With this formulation white wine can be treated before filling into bottles. White wine was treated with 0.1% of the formulation over a period of 30 *min*. The formulation was re-suspended twice in the course of the treatment and was then separated off by means of a single-medium filter using a cellulose bed and charged into bottles having volumes of 0.5 liters. The treatment decreases the browning.

9.5.5.3 *Rust Removal*

Copolymers constituted from NVP and maleic acid are useful for rust removal from metallic surfaces. These copolymers may be formed by hydrolysis of a precursor copolymer of NVP and MA. Copolymers with K values of 20–50 are suitable for use. The copolymer is applied in the form of an aqueous solution containing 35–60% polymer (106, 107). The rusty surface is coated with the aqueous solution and allowed to dry. The rust becomes incorporated into the layer and the layer containing the rust detaches itself from the surface.

9.5.6 *Oil Field Applications*

Polymers and copolymers containing NVP have found applications in oil field industries, in particular as:

- Kinetic inhibitors against gas hydrate formation
- Fluid loss control agents
- Drilling muds.

9.5.6.1 *Gas Hydrate Inhibitors*

The formation of gas hydrates in a pipeline during the transport of oil and natural gas is a serious problem, especially in areas with a low temperature in the winter season or in the sea. Gas hydrate formation occurs when water temperatures are low and if no special steps are taken. Anti-freeze compounds, such as glycol or methanol, can be added to minimize the gas hydrate formation. However, large quantities of these compounds are required to be effective. Anti-freeze agents are addressed as thermodynamic inhibitors.

Another method to prevent the formation is gas hydrates that consist of kinetic inhibitors that prevent the nucleation reaction. For such a purposes several types of polymers have been used. A terpolymer of NVP, *N*-vinyl caprolactam and a quaternized comonomer, such as dialkylaminoalkyl methacrylamide, dialkyldialkenyl ammonium halide and a dialkylaminoalkyl acrylate or methacrylate is suitable to prevent the formation of gas hydrates (108).

The polymers have a molecular weight in the range of 500 to 2500 *Dalton*, and are polymerized in solution, such as 2-butoxyethanol. The composition includes ethylene glycol as carrier solvent (109).

9.5.6.2 *Drilling Muds*

A copolymer composed from NVP, itaconic acid, acrylamide (AAm) and 2-acrylamido-2-methyl-1-propane sulfonic acid was tested both as freshwater and saltwater mud. Drilling muds containing the copolymer show an excellent tolerance to salt and to high temperatures (110).

9.5.6.3 *Enhanced Oil Recovery*

Secondary or tertiary recovery of oil utilizes polymer thickened aqueous drive fluids. Aqueous solutions of hydrophobically associating polymers exhibit enhanced viscosification, reduced salt sensitivity, and other rheological properties that are useful in chemically enhanced oil recovery processes. The polymers consist of water-soluble groups, such as AAm and NVP and a salt of AA, whereas the water-insoluble moiety is a higher alkyl acrylamide (53).

9.5.7 *Photoresist Resin Compositions*

The photosensitive resin compositions described subsequently are useful in the manufacture of color cathode ray tubes. Matrix materials for the water-soluble photosensitive resin compositions include blends of PVA and PNVP, copolymers of NVP and VA (111), or PNVP, copolymers of NVP and vinylimidazole (112).

Due to the incompatibility between poly(*N*-vinyl-2-pyrrolidone) and PVA, copolymers of NVP and VA are preferred (111). The VA units improve the adhesion of the coating solution to the glass video display panel. The polymers are not photosensitive as such, but exhibit a particularly high oxygen permeability, which is advantageous for the crosslinking reaction of the actual photosensitive composition.

Azido compounds are used as photosensitive components, c.f. Figure 9.6. The tetraalkylammonium salts have a greater solubility in water in comparison to the disodium salts, which improves the video image color contrast (111).

Another concept uses a photosensitive polymer. The polymer is added in amounts of ca. 10% based on PNVP, or the copolymer, to an aqueous solution of PNVP (112). The precursor of the photosensitive polymer is synthesized by the radical copolymerization of *N,N*-dimethylacrylamide and diacetone acrylamide in aqueous solution using 2,2′-azobis[2-(2-imidazolin-2-yl)propane]-dihydrochloride, c.f. Figure 9.7, as a radical initiator. This is an azo-type initiator, which does not contain nitrile groups. It has a half-life time of 10 h at 44°C in aqueous solution, with an energy of activation of 108.0 $kJ\ mol^{-1}$.

To the aqueous polymer solution, sodium 4-azidobenzaldehyde-2-sulfonate is added that reacts under alkaline conditions with the pendent acetone groups as shown in Figure 9.8.

The photo crosslinkable composition is formed by mixing an aqueous solution of 3% PNVP with an aqueous solution of the photo crosslinkable polymer so that the ratio of PNVP to photo crosslink-

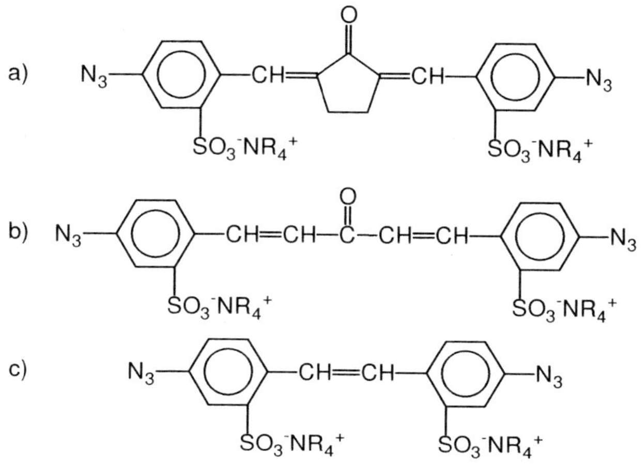

Figure 9.6 Photosensitive Components for Photo Crosslinking
a) Di(tetraalkylammonium)-2,6-di(*p*-azidobenzal)-cyclopentanone-2,2′-disulfonate, b) Di(tetraalkylammonium)-2,2′-(1,5-penta-3-one-1,4-dienyl)-bis-(5-azidobenzene-disulfonate), c) Di(tetraalkylammonium)-4,4′-diazidostilbene-2,2′-disulfonate

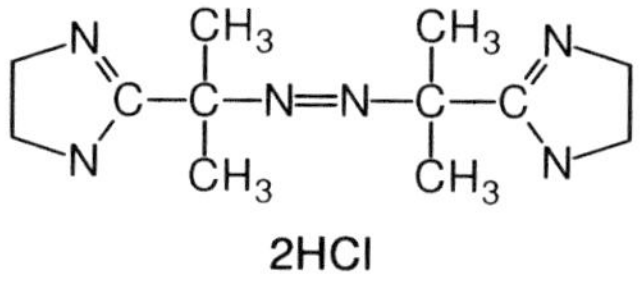

Figure 9.7 2,2′-Azobis[2-(2-imidazolin-2-yl)propane]dihydrochloride

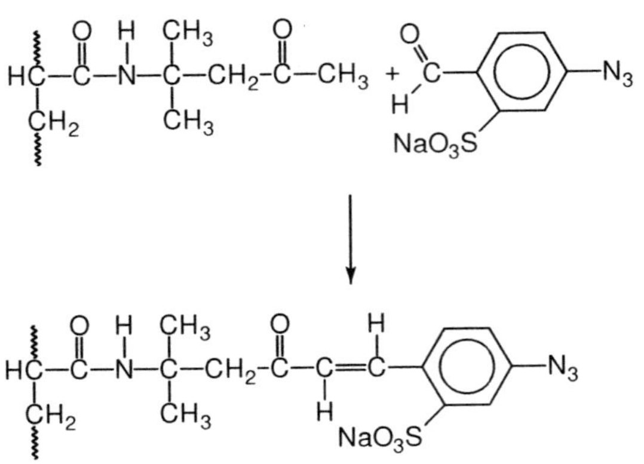

Figure 9.8 Formation of a Photo Crosslinkable Polymer (112)

able polymer is 10:1. The solution is then coated on a substrate. In the next step, the composition is dried to form the photoresist layer, which is exposed through a shadow mask. The layer is exposed to light having a wavelength of 300–400 *nm*.

The term gap exposure means that there is a space between mask and photosensitive layer. This is in contrast to contact exposure, where the mask is placed in direct contact with the photoresist layer. It is possible to perform gap exposure with a gap spacing between mask and layer being 10 *mm* and obtain a satisfactory photo cured pattern showing reciprocity law failure characteristics. The reciprocity law states that the degree of exposure is equal to the product of intensity and the duration of exposure. Thus, the same degree of exposure should result from reducing the duration of exposure and increasing light intensity.

However, the reciprocity law does not always hold, particularly in extreme situations of either duration or light intensity. This behavior is addressed as reciprocity failure. The reason for reciprocity failure in the present situation may be explained as follows:

When the azido groups in the photoresists layer are excited upon exposure, nitrenes are generated. They either react with themselves or with the polymer and crosslinking occurs to cause photocuring. However, in the presence of oxygen or water, the intended crosslinking reaction competes with the side reactions with oxygen or water.

Therefore, the crosslinking is restrained. The matrix polymer, PNVP has a high oxygen permeability, so during gap exposure, aerial oxygen is efficiently taken into the photoresist layer and reacts with the internal nitrenes to restrain the crosslinking reaction.

Thus, the photocuring reaction takes place in the areas of high illumination intensity, i.e., in the centers of the beam passing holes. In contrast, the photocuring reaction is restrained in the area of low illumination intensity, i.e., in the peripheral edges of beam passing holes. Therefore, the resulting photo cured pattern shows smaller dots than the beam passing holes in the shadow mask. This phenomenon prevents the undesired docking effect. Docking refers to an overlap of patterns that are created in different cycles of exposure.

9.6 Suppliers and Commercial Grades

Table 9.8 Examples for Commercially Available Poly(*N*-vinyl-2-pyrrolidone)

Tradename	Producer	Remarks
Luvitec®	BASF AG	Homopolymers
Sokalan®	BASF AG	Homopolymer
Luviquat® FC 370	BASF AG	NVP/methylvinylimidazolium chloride copolymer (70:30) (113)
Luviquat® PQ 11	BASF AG	copolymers of NVP/*N*,*N*′-dimethylaminoethyl methacrylate, quaternized with diethyl sulfate (114)
Luvitec® VPMA 91	BASF AG	1-Vinylimidazole (VI)/NVP copolymer (1:9) (115)
Luvitec® VP155K18P	BASF AG	VI/NVP copolymer (104)
S-630	GAF Corp.	VA/NVP copolymer (40:60) (54)
Ganex® P-904	ISP	Butylated PNVP (116)
Ganex® V-516	ISP	NVP/1-hexadecene copolymer (116)
Gaffix® VC-713	ISP	Terpolymer of *N*-vinyl-2-pyrrolidone, *N*-vinyl caprolactam and *N*,*N*′-dimethylaminoethyl methacrylate (108)

PNVP is sold either in powdery form or in aqueous solutions. The aqueous solutions appear colorless to slightly yellowish. Grades with different molecular weight, characterized by their K values, are available. Examples for commercially available grades and tradenames are shown in Table 9.8.

9.7 Safety

In subchronic studies, inhaled NVP was found to be haemotoxic, hepatotoxic and irritant to the nose. In long-term studies with rats over a period for 2 years, survival was unaffected. However, reduced body weight gain, haemotoxicity, increased liver weight, hepatocellular carcinomas, necrosis, reparative hyperplasia, adenomas and adenocarcinomas of the nasal cavity, and squamous cell carcinomas of the larynx were observed (117). Certain studies reveal that the upper respiratory tract and the liver are the main targets for NVP toxicity (118).

9.8 Environmental Impact and Recycling

9.8.1 *Biodegradable Polymers*

Copolymers of NVP with MDO are biodegradable. The polymers are used in commercial applications where hydrolytic and biodegradability is an important requirement. Accordingly, these polymers are particularly advantageous as binders in products, such as fish feed. The copolymer also is useful as a dispersant in systems where PNVP itself can be used (16).

Tradenames appearing in the references are shown in Table 9.9.

Table 9.9 Tradenames in References

Tradename Description	Supplier
Aculyn™ (Series) hydrophobically-modified poly(acrylate) (14)	Rohm and Haas
Aerosil® Fumed Silica (59)	Degussa AG

Table 9.9 (cont.)

Tradename Description	Supplier
Aerosol® OT	Cytec Industries, Inc.
Sodium sulfosuccinic acid dioctyl ester (59)	
Amberlite® (Series)	Rohm & Haas
Ion exchangers based on poly(styrene) (57)	
Antarox® WA-1	Rhone-Poulenc, Inc.
p-Phenylphenol propylene oxide/ethylene oxide (116)	
Arlacel® C	ICI
Sorbitan sesquioleate (116)	
Arlacel® P-100	ICI
Polyhydroxy stearic acid (116)	
Biosoft® D-62	Stepan
Sodium alkyl benzene sulfonate (79)	
Borchigen® BN911	Bayer AG
Modified polyester resin (116)	
Borchigen® SN88	Bayer AG
Poly(urethane) oligomer (116)	
Carbopol® (Sseries)	Lubrizol Advanced Materials, Inc.
Poly(acrylate) (14, 59)	
Carbowax® (Series)	Union Carbide Corp.
Poly(ethyleneoxide glycol) (PEG) (58, 59)	
Chemax® DT-30	Chemax Inc.
Poly(oxyethylene) 30 tallow diamine (116)	
Chemeen® C-15	Chemax Inc.
Cocoamine (116)	
Crodestas® F-110	Croda, Inc.
Mixture of sucrose stearate and sucrose distearate (59)	
Crodestas® SL-40	Croda, Inc.
Sucrose stereate ? (59)	
Darvan® C	RT Vanderbilt Co., Inc.
Solution of poly(methacrylic acid) as ammonium salt (55)	
Dequest® 2010	Monsanto Co.
1-Hydroxyethylidene-1,1-diphosphonic acid (etidronic acid) (103)	
Disperbyk® 110	BYK Chemie GmbH
Saturated polyester with acidic groups (116)	
Disperbyk® 180	BYK Chemie GmbH
Alkylol ammonium salt of a block copolymer with acidic groups, dispersant (116)	

Table 9.9 (cont.)

Tradename Description	Supplier
Disperbyk® 182 Dispersant (116)	BYK Chemie GmbH
Dobanol® 23-3 C_9–C_{13} EO 3 nonionic surfactant (103)	Shell
Dow Corning® (Series) Silicone Products (14)	Dow
Dowanol® TPM Tripropylene glycol methyl ether (116)	Dow
Duponol® Sodium monododecyl sulfate (59)	DuPont
Dymel® 152A 1,1-Difluoroethane, Aerosol propellent (15)	DuPont
EDDS® Ethylenediamine *N,N'*-disuccinic acid, biodegradable chelating agent (103)	Palmer Research Laboratories
EF-101 Perfluorooctylsulfonic acid (115)	Tohkem Products Corp.
EF-201 Perfluorooctanoic acid (115)	Tohkem Products Corp.
Empimin® LV 33 Octyl sulphate (103)	Albright and Wilson
Ethodumeens® (Series) Ethoxylated amine surfactant (116)	Akzo Nobel
Ethoxy 3389 Ethoxylated triethanolamine (116)	Ethox Chemical Co.
Exxsol® D130 Isoparaffinic hydrocarbon solvents (the number refers to approxiumate flashpoint in°C) (116)	Exxon
Finsolv® PG-22 Dipropylene glycol dibenzoate (116)	Finetex Co.
Frescolat® MGA Menthone glycerol acetal, coolant in cosmetic formulations (75)	Haarmann and Reimer
Gaffix® VC-713 (Dimethylamino)ethyl methacrylate / vinylcaprolactam / *N*-vinyl-2-pyrrolidinone terpolymer (108)	GAF
Gafquat® (Series) Copolymer of vinylpyrrolidone and dimethylaminoethyl methacrylate quaternized with diethyl sulfate (103, 104)	ISP

Table 9.9 (cont.)

Tradename Description	Supplier
Ganex® P-904	ISP
Butylated poly(vinyl pyrrolidone) (116)	
Ganex® V-516	ISP
Vinylpyrrolidone/1-hexadecene copolymer (116)	
Germaben® II	ISP
4-Hydroxybenzoic acid methyl ester, mixture with *N*-[1,3-bis(hydroxymethyl)-2,5-dioxo-4-imidazolidinyl]-N,N′-bis(hydroxymethyl)urea, 1,2-propanediol and propyl 4-hydroxybenzoate (15)	
Glydant®	Glyco Chemicals, Inc.
Antimicrobial agent (14)	
Hypermer® LP-6	ICI Surfactants
Dispersing agent (116)	
Igepon® AC-78	GAF
Sodium cocoyl isothionate (15)	
Isachem® AS	Enichem
Alkyl sulphate/branched alcohol (103)	
Isofol® 12	Condea Chemie GmbH
2-Butyl octanol, sud controlling agent (103)	
Isofol® 16	Condea Chemie GmbH
2-Hexyl decanol, sud controlling agent (103)	
Isopar® (Series)	Exxon
Isoparaffinic solvent (116)	
Jambu®	Takasago Perfumery Co., Ltd.
Salivating agents (75)	
Jeffamine® (Series)	Huntsman Petrochemical Corp.
Amine capped polyalkoxylene glycol (103)	
Kathon®	Rohm & Haas
Mixture or 5-chloro-2-methyl-4-isothiazoline-3-one and 2-methyl-4-isothiazoline-3-one, bactericide (14)	
Ketoprofen	Wyckoff
2-(3-Benzoylphenyl)propanoic acid (59)	
Kollidon® VA 64	BASF AG
Vinylpyrrolidone/vinyl acetate (60:40) copolymer (59, 115)	
Lewatit® (Series)	Lanxess Deutschland GmbH
Divinylbenzene/styrene copolymer ion exchange resins (57)	
Lupersol® 101	Arkema, Inc.
2,5-Dimethyl-2,5-di(t-butylperoxy)hexane (11)	

Table 9.9 (cont.)

Tradename Description	Supplier
Lupersol® 11 *tert*-Butyl peroxypivalate (11, 16)	Arkema, Inc.
Luviksol® VA64 Vinyl pyrrolidone/vinyl acetate copolymer (79)	BASF AG
Luviksol® VA73W Vinyl pyrrolidone/vinyl acetate copolymer (79)	BASF AG
Luvimer® 100P *tert*-Butylacrylate/ethylacrylate/methacrylic acid copolymers (114)	BASF AG
Luviquat FC 370 Vinylpyrrolidone/methylvinylimidazolium chloride copolymer (70:30) (113)	BASF AG
Luviquat® PQ 11 PN Quaternized copolymer of vinylpyrrolidone (VP) and dimethylaminoethylmethacrylate (DMAEMA) in aqueous solution, hair care polymer (114)	BASF AG
Luviset® CAN Terpolymer of vinyl acetate, crotonic acid and vinyl neodecanoate (114)	BASF AG
Luviskol® VA 64 50% Solution of a copolymer of vinylpyrrolidone and vinylacetate (60:40) in water (79)	BASF AG
Luviskol® VA 73 W 50% Solution of a copolymer of vinylpyrrolidone and vinylacetate (70:30) in water (79, 103)	BASF AG
Luviskol® VBM Vinylpyrrolidone, *tert*-butylacrylate, methacrylic acid copoplymers (114)	BASF AG
Luvitec® VPMA 91 Vinylimidazole with vinylpyrrolidone (1:9) copolymer (115)	BASF AG
Mackazoline® C Cocoyl hydroxyethyl imidazoline (116)	McIntyre Group Ltd.
Merguard® Methyldibromo glutaronitrile, cosmetic ingredient (14)	Nalco (Calgon Corp.)
Methocel® Methylcellulose (14)	Dow
MGA Menthone glycerol acetal, coolant in cosmetic formulations (75)	Haarmann and Reimer
Miranol® Alkylaspartic acid, ampholytic detergent (14)	Rhodia Inc. Corp.

Table 9.9 (cont.)

Tradename Description	Supplier
Monastral Blue Pigment (54)	Ciba Specialty Chemicals Corp.
Monastral Red B Pigment (54)	Ciba Specialty Chemicals Corp.
n-BPP® Butoxy propoxy propanol (103)	Dow
Natrosol® 250LR Hydroxyethyl cellulose (15)	Aqualon Corp.
Nuosperse® 657 Polyfunctional modified polyester polyelectrolyte (116)	Condea Servo BV
Olin-IOG® *p*-Isononylphenoxy poly(glycidol) (59)	Olin Chemicals
Paraloid® B67 Acrylic polymer (116)	Rohm & Haas
PEG DME-2000 Dimethyl poly(ethylene glycol) (MW 2000) (103)	Hoechst
Pemulen® Poly(acrylate), polymeric emulsifiers for cosmetics (14)	Lubrizol Advanced Materials, Inc.
Piccolastic® A-50 Poly(styrene) resin (54)	Eastman Chemical Resins, Inc., Hercules, Inc.
Plasdone® (Series) Vinyl pyrrolidone polymers (59, 101)	ISP
Plasdone® S-630 Vinyl pyrrolidone/vinyl acetate (60:40) copolymer (12, 59)	ISP
Pluronic® (Series) Ethylene oxide/propylene oxide block copolymer, defoamers (103)	BASF AG
Polowax® A-31 Emulsifyer wax (15)	Croda, Inc.
PolyMill® 500 500 Micron polymeric media (59)	Dow
Polyplasdone® XL Crosslinked poly(*N*-vinyl-2-pyrrolidone) (12)	ISP
Polyquat® 11 Quaternized copolymer of vinyl pyrrolidone and dimethyl aminoethylmethacrylate (103)	BASF AG

Table 9.9 (cont.)

Tradename Description	Supplier
Polyquaternium® 10 2-(2-Hydroxy-3-(trimethylammonium)propoxy)ethyl cellulose ether chloride, cationic cellulose derivative (114)	Nalco Chemical Comp.
Polyquaternium® 7 *N*,*N*-Dimethyl-*N*-2-propenyl-2-propene-1-aminium chloride, polymer with acrylamide copolymer (114)	Nalco Chemical Comp.
S-630 Vinylacetate/*N*-vinyl-2-pyrrolidone copolymer (40:60) (54)	GAF
Salcare® Poly(acrylate), ionic, rheology modifiers (14)	Ciba
Sokalan® CP (Series) Water-soluble homo- and copolymers of maleic acid and acrylic acid (103)	BASF AG
Solsperse® 13940 Hydroxystearic acid polymer with polar anchor groups; hyperdispersant-wetting agent for coatings (116)	Zeneca, Inc.
Soprophor® BSV Tristyryl phenol ethoxylate (116)	Rhone-Poulenc, Inc.
Sotex® N Long chain fatty ester (116)	Morton International, Inc.
Surfactant® 10 G *p*-Isononylphenoxy poly(glycidol) (59)	Olin Chemicals
Synfac® 8337 Potassium salt of a phosphated alkoxylated aryl phenol (116)	Milliken Chemical Co.
Synfac® TEA 97 Ethoxylated triethanolamine (116)	Milliken Chemical Co.
Teflon® Tetrafluoro polymer (116)	DuPont
Terathane® Poly(tetramethyleneoxide glycol) (PTMEG) (58)	DuPont
Tetronic® (Series) Propoxylated ethylenediamine-poly(ethylene glycol) adduct, surfactant (59, 103)	BASF-Wyandotte Corp.
TK-10 3-1-Menthoxypropane-1,2-diol, coolant in cosmetic formulations (75)	Takasago Perfumery Co., Ltd.
Trilon® FS Propylene diamine tetracetic acid (PDTA), chelate (103)	BASF AG
Triton® X-200 Alkyl aryl poly(ether sulfonate) (59)	Union Carbide Corp.

Table 9.9 (cont.)

Tradename Description	Supplier
TSMR-8800	Tokyo Ohka Kogyo Co., Ltd.
Cresol novolak resin with a naphthoquinone diazide compound, positive photoresist (115)	
Tween® (Series)	Uniqema
Ethoxylated fatty acid ester surfactants (59)	
Udel® Polysulfone	Solvay
Poly(bisphenol A sulfone) (101)	
Ultrahold® 8	BASF AG
N-tert-Butylacrylamide/ethylacrylate/acrylic acid terpolymer (114)	
Ultrason® (Series)	BASF AG
Poly(sulfone) resins (101)	
V-50®	Wako Chemicals USA, Inc.
2,2′azobis(2-amidopropane)hydrochloride (53)	
Vazo® 33	DuPont
2,2′-Azobis(2,4-dimethyl-4-methoxyvaleronitrile) (53)	
Vazo® 64	DuPont
Azobis(isobutyronitril) (53)	
Vazo® 67	DuPont
2,2′-Azobis(2-methylbutane-nitrile (11, 16)	
Witflow® 934	Witco Corp.
Modified fatty acid diethanol amide (116)	
XTJ®	Huntsman
Monocapped poly(alkoxylene glycol) (103)	
Zonyl® FSO 100	DuPont
Ethoxylated nonionic fluorosurfactant (116)	

References

1. P.-S. Kuo, S.-J. Chu, C.-C. Dai, H.-Y. Hsu, C.-T. Lin, and Y.-Y. Lin, Method for the production of *N*-vinyl-2-pyrrolidone by gas-phase dehydration reaction at atomospheric pressure, US Patent 5 569 770, assigned to Industrial Technology Research Institute (Chutung Hsinchu, TW), October 29, 1996.

2. A.P. Kahn, Process for making *N*-vinyl-2-pyrrolidone, US Patent 6 982 337, assigned to Lyondell Chemical Technology, L.P. (Greenville, DE), January 3, 2006.

3. B. Eck, J. Heilek, M. Schmidt-Radde, and H. Helfert, Method for obtaining pure *N*-vinyl pyrrolidone, US Patent 6 703 511, assigned to BASF Aktiengesellschaft (Ludwigshafen, DE), March 9, 2004.
4. J.M. Cohen and R.B. Biss, Process for the production of pure vinyl pyrrolidone, US Patent 5 329 021, assigned to ISP Investments Inc. (Wilmington, DE), July 12, 1994.
5. E.G. Malawer, V. Kabanov, K.S. Narayanan, M.A. Tallon, S.Y. Tseng, and P.F. Wolf, Crosslinked polyvinylpyrrolidone (PVPP) copolymer of vinyl pyrrolidone (VP) and monomer derived from 1-vinyl-3-(e)-ethylidene pyrrolidone (EVP), US Patent 6 124 415, assigned to ISP Investment Inc. (Wilmington, DE), September 26, 2000.
6. S. Kothrade, R. Blankenburg, I. Peters, and A. Sanner, Preparation of low molecular mass homopolymers of *N*-vinylpyrrolidone, US Patent 6 187 884, assigned to BASF Aktiengesellschaft (Ludwigshafen, DE), February 13, 2001.
7. C. Schuster, R. Sauerbier, and H. Fikentscher, Polymerization of *N*-vinyl lactams, US Patent 2 335 454, assigned to Vested in the Alien Property Custodian, Ludwigshafen-on-Rhine, November 30, 1943.
8. Y. Nakata, Vinylpyrrolidone (co)polymer, US Patent 6 642 333, assigned to Nippon Shokubai Co., Ltd. (Osaka, JP) and Dai-ichi Kogyo Seiyaku Co., Ltd. (Kyoto, JP), November 4, 2003.
9. D. Tomihisa, N. Harada, A. Naka, T. Kuriyama, Y. Shimasaki, and H. Nishibayashi, Production process for vinylpyrrolidone polymer, US Patent 6 391 994, assigned to Nippon Shokubai Co., Ltd. (Osaka, JP), May 21, 2002.
10. A. Kitada, K. Fujise, and I. Hamaguchi, Process for preparing vinylpyrrolidone polymer, US Patent 6 465 592, assigned to Dai-Ichi Kogyo Seiyaku Co., Ltd. (Kyoto, JP), October 15, 2002.
11. D. Patel, P. Taylor, and J. Mc Kittrick, Process for making vinyl pyrrolidone/vinyl acetate copolymers having a very low residual vinyl pyrrolidone monomer level within a short production cycle, US Patent 6 939 927, assigned to ISP Investments Inc. (Wilmington, DE), September 6, 2005.
12. E.G. Malawer, J.P. Cullen, C.-S. Wu, L. Senak, W.J. Drefko, D.K. Hood, and P.J. Brush, Proliferous copolymer of vinyl pyrrolidone and vinyl acetate, US Patent 6 806 334, assigned to ISP Investments Inc. (Wilmington, DE), October 19, 2004.
13. Y. Zhong and P.F. Wolf, Process for providing homogeneous copolymers of vinylpyrrolidone and 3-dimethylaminopropyl methacrylamide (DMAPMA) which form clear aqueous solutions having high cloud points, US Patent 5 684 105, assigned to ISP Investments Inc. (Wilmington, DE), November 4, 1997.

14. L.M. Brandt and J.R. Cramm, Personal care compositions containing *N,N*-diallyldialkylammonium halide/*N*-vinylpyrrolidone polymers, US Patent 7 115 254, assigned to Nalco Company (Naperville, IL), October 3, 2006.

15. K.-C. Liu, L.R. Anderson, E.E. Hardy, and Y. Reuven, Process for making substantially homogeneous copolymers of vinyl pyrrolidone and *N*-3,3-dimethylaminopropyl methacrylamide for personal care applications, US Patent 6 110 454, assigned to ISP Investments Inc. (Wilmington, DE), August 29, 2000.

16. Y. Zhong, P.F. Wolf, E.G. Malawer, and K.S. Narayanan, Homogeneous copolymers containing vinyl pyrrolidone and 2-methylene-1,3-dioxepane and process for making same, US Patent 5 912 312, assigned to ISP Investments Inc. (Wilmington, DE), June 15, 1999.

17. M.E. Benz, J.A. Alkatout, and S. Lyu, A_nB block copolymers containing poly(vinyl pyrrolidone) units, medical devices, and methods, US Patent 6 756 449, assigned to Medtronic, Inc. (Minneapolis, MN), June 29, 2004.

18. S. Benamer, M. Mahlous, A. Boukrif, B. Mansouri, and S.L. Youcef, Synthesis and characterisation of hydrogels based on poly(vinyl pyrrolidone), *Nucl. Instrum. Methods Phys. Res., Sect. B*, 248(2):284–290, August 2006.

19. A.B. Lugao, S.O. Rogero, and S.M. Malmonge, Rheological behaviour of irradiated wound dressing poly(vinyl pyrrolidone) hydrogels, *Radiat. Phys. Chem.*, 63(3-6):543–546, March 2002.

20. K.R. Park and Y.C. Nho, Preparation and characterization by radiation of poly(vinyl alcohol) and poly(*N*-vinylpyrrolidone) hydrogels containing aloe vera, *J. Appl. Polym. Sci.*, 90(6):1477–1485, November 2003.

21. H.A. Essawy and H.S. Ibrahim, Synthesis and characterization of poly(vinylpyrrolidone-*co*-methylacrylate) hydrogel for removal and recovery of heavy metal ions from wastewater, *React. Funct. Polym.*, 61(3):421–432, November 2004.

22. G.S. Chauhan, B. Singh, and S. Kumar, Synthesis and characterization of *N*-vinyl pyrrolidone and cellulosics based functional graft copolymers for use as metal ions and iodine sorbents, *J. Appl. Polym. Sci.*, 98 (1):373–382, October 2005.

23. Z. Maolin, H. Hongfei, F. Yoshii, and K. Makuuchi, Effect of κ-carrageenan on the properties of poly(*N*-vinyl pyrrolidone)/κ-carrageenan blend hydrogel synthesized by γ-radiation technology, *Radiat. Phys. Chem.*, 57(3-6):459–464, March 2000.

24. N.C. Dafader, M.E. Haque, and F. Akhtar, Synthesis of hydrogel from aqueous solution of poly(vinyl pyrrolidone) with agar by γ-rays irradiation, *Polym. Plast. Tech. Eng.*, 44(2):243–251, 2005.
25. N.C. Dafader, M.E. Haque, and F. Akhtar, Effect of κ-carrageenan on the properties of poly(vinyl pyrrolidone) hydrogel prepared by the application of radiation, *Polym. Plast. Tech. Eng.*, 44(7):1339–1346, 2005.
26. S. Jin, M. Liu, F. Zhang, S. Chen, and A. Niu, Synthesis and characterization of pH-sensitivity semi-IPN hydrogel based on hydrogen bond between poly(*N*-vinylpyrrolidone) and poly(acrylic acid), *Polymer*, 47(5):1526–1532, February 2006.
27. X. Li, W. Liu, G. Ye, B. Zhang, D. Zhu, K. Yao, Z. Liu, and X. Sheng, Thermosensitive *N*-isopropylacrylamide-*N*-propylacrylamide-vinyl pyrrolidone terpolymers: Synthesis, characterization and preliminary application as embolic agents, *Biomaterials*, 26(34):7002–7011, December 2005.
28. D. Meinhold, R. Schweiss, S. Zschoche, A. Janke, A. Baier, F. Simon, H. Dorschner, and C. Werner, Hydrogel characteristics of electron-beam-immobilized poly(vinylpyrrolidone) films on poly(ethylene terephthalate) supports, *Langmuir*, 20(2):396–401, January 2004.
29. ISP International Speciality Products, Wayne, NJ, Polyvinylpyrrolidone,
[electronic:] http://www.ispcorp.com, 2006.
30. PVP polyvinylpyrrolidone polymers, Technical Bulletin 2302-203 SM-1290, GAF(ISP), 1990.
[electronic:] www.sigmaaldrich.com/img/assets/3900/Viscosity.pdf.
31. A. Sionkowska, Interaction of collagen and poly(vinyl pyrrolidone) in blends, *Eur. Polym. J.*, 39(11):2135–2140, November 2003.
32. K. Sakurai, T. Maegawa, and T. Takahashi, Glass transition temperature of chitosan and miscibility of chitosan/poly(*N*-vinyl pyrrolidone) blends, *Polymer*, 41(19):7051–7056, September 2000.
33. E. Yilmaz, D. Ozalp, and O. Yilmaz, Miscibility study of chitosan/poly(vinyl pyrrolidone) blends in dilute solution, *Int. J. Polym. Anal. Charact.*, 10(5-6):329–339, September–December 2005.
34. T. Ohno, S. Yoshizawa, Y. Miyashita, and Y. Nishio, Interaction and scale of mixing in cellulose acetate/poly(*N*-vinyl pyrrolidone-*co*-vinyl acetate) blends, *Cellulose*, 12(3):281–291, June 2005.
35. V. Rao, P.V. Ashokan, and J.V. Amar, Studies on dielectric relaxation and ac conductivity of cellulose acetate hydrogen phthalate-poly (vinyl pyrrolidone) blend, *J. Appl. Polym. Sci.*, 86(7):1702–1708, November 2002.

36. Y. Jinghua, C. Xue, G.C. Alfonso, A. Turturro, and E. Pedemonte, Study of the miscibility and thermodynamics of cellulose diacetate-poly(vinyl pyrrolidone) blends, *Polymer*, 38(9):2127–2133, April 1997.
37. S.N. Cassu and M.I. Felisberti, Poly(vinyl alcohol) and poly(vinyl pyrrolidone) blends: Miscibility, microheterogeneity and free volume change, *Polymer*, 38(15):3907–3911, July 1997.
38. S.N. Cassu and M.I. Felisberti, Poly(vinyl alcohol) and poly(vinyl-pyrrolidone) blends: 2. Study of relaxations by dynamic mechanical analysis, *Polymer*, 40(17):4845–4851, August 1999.
39. G.B. Zhang, J.M. Zhang, X.S. Zhou, and D.Y. Shen, Miscibility and phase structure of binary blends of polylactide and poly(vinylpyrr-olidone), *J. Appl. Polym. Sci.*, 88(4):973–979, April 2003.
40. W.P. Hsu, Effect of tacticity of poly(methyl methacrylate) on the miscibility with poly(vinyl pyrrolidone), *J. Appl. Polym. Sci.*, 81(13):3190–3197, September 2001.
41. S.W. Kuo and F.C. Chang, Studies of miscibility behavior and hydrogen bonding in blends of poly(vinylphenol) and poly(vinylpyrrolidone), *Macromolecules*, 34(15):5224–5228, July 2001.
42. W.P. Hsu, Miscibility enhancement on the immiscible poly (vinyl pyrrolidone) and poly (ethyl methacrylate) with poly (vinyl phenol), *J. Appl. Polym. Sci.*, 100(2):1205–1213, April 2006.
43. A. Prinos, A. Dompros, and C. Panayiotou, Thermoanalytical and spectroscopic study of poly(vinyl pyrrolidone)/poly(styrene-*co*-vinyl phenol) blends, *Polymer*, 39(14):3011–3016, June 1998.
44. N. Shinyashiki, R.J. Sengwa, S. Tsubotani, H. Nakamura, S. Sudo, and S. Yagihara, Broadband dielectric study of dynamics of poly(vinyl pyrrolidone)-ethylene glycol oligomer blends, *J. Phys. Chem. A*, 110 (15):4953–4957, April 2006.
45. Z.K. Zhong and Y.L. Mi, Thermal characterization and solid-state C-13-NMR investigation of blends of poly(*N*-phenyl-2-hydroxytrimethylene amine) and poly(*N*-vinyl pyrrolidone), *J. Polym. Sci., Part B: Polym. Phys.*, 37(3):237–245, February 1999.
46. S.X. Zheng, Q.P. Guo, and Y.L. Mi, Examination of miscibility at molecular level of poly(hydroxyether of bisphenol A) poly(*N*-vinyl pyrrolidone) blends by cross-polarization magic angle spinning C-13 nuclear magnetic resonance spectroscopy, *J. Polym. Sci., Part B: Polym. Phys.*, 36(13):2291–2300, September 1998.
47. H. Lu, S.X. Zheng, B. Zhang, and X.Z. Tang, Miscibility and intermolecular specific interactions in blends of poly(hydroxyether sulfone) and poly (*N*-vinylpyrrolidone), *Macromol. Chem. Phys.*, 205(6):834–842, April 2004.

48. J.H. Kim, M.S. Whang, and C.K. Kim, Novel miscible blends composed of poly(ether sulfone) and poly(1-vinylpyrrolidone-*co*-styrene) copolymers and their interaction energies, *Macromolecules*, 37 (6):2287–2293, March 2004.
49. J.W. Kim, E.J. Moon, and C.K. Kim, New miscible blends composed of polysulfone and poly(1-vinylpyrrolidone-*co*-acrylonitrile) copolymers and their phase behavior, *Polymer*, 46(15):5662–5672, July 2005.
50. S. Fujihara and S. Kitta, Broadband visible emissions from sol-gel-derived silica/poly(vinyl pyrrolidone) hybrids, *Chem. Phys. Lett.*, 397 (4-6):479–483, October 2004.
51. D. Tomihisa, T. Kuriyama, T. Ueta, A. Naka, H. Nishibayashi, and Y. Nakata, Vinylpyrrolidone polymer and its stabilization and preservation processes, US Patent 6 498 231, assigned to Nippon Shokubai Co., Ltd. (Osaka, JP), December 24, 2002.
52. D. Tomihisa, T. Kuriyama, T. Ueta, A. Naka, H. Nishibayashi, and Y. Nakata, Vinylpyrrolidone polymer and its stabilization and preservation processes, US Patent 6 346 600, assigned to Nippon Shokubai Co., Ltd. (Osaka, JP), February 12, 2002.
53. J. Bock, S.J. Pace, and D.N. Schulz, Enhanced oil recovery with hydrophobically associating polymers containing *N*-vinylpyrrolidone functionality, US Patent 4 709 759, assigned to Exxon Research and Engineering Company (Florham Park, NJ), December 1, 1987.
54. P.R. West, J.E. Mitchell, G.R. Miller, P.R. Josephson, Jr., and R.W. Ryan, Jr., Radiation-sensitive composition containing a vinyl pyrrolidone polymer and use thereof in lithographic printing plates, US Patent 5 061 601, assigned to Eastman Kodak Company (Rochester, NY), October 29, 1991.
55. J.F. Kneller and D.R. Cosper, Method for preparing a ceramic oxide material, US Patent 5 266 243, November 30, 1993.
56. I. Nam, J.W. Bae, K.S. Jee, J.W. Lee, K.D. Park, and S.H. Yuk, Poly(*N*-isopropylacrylamide-*co*-*N*-vinylpyrrolidone) as a novel implant materials: Preparation and thermo-gelling behavior, *Macromol. Res.*, 10(2):115–121, April 2002.
57. F. Straub, H.-U. Schenck, S. Lang, and E. Brode, Biodegradable vinylpyrrolidone polymers, their manufacture and use, US Patent 4 254 239, assigned to BASF Aktiengesellschaft (DE), March 3, 1981.
58. M.A. Khan, M. Karakelle, M.-S. Lee, and R.A. Taller, Polyurethane-polyvinylpyrrolidone block copolymer and iodine carrier therefrom, US Patent 5 302 385, assigned to Becton, Dickinson and Company (Franklin Lakes, NJ), April 12, 1994.
59. H.W. Bosch and N.P. Ryde, Nanoparticulate compositions comprising copolymers of vinyl pyrrolidone and vinyl acetate as surface sta-

bilizers, US Patent 6 969 529, assigned to Elan Pharma International Ltd. (Dublin, IE), November 29, 2005.

60. C.T. Vogelson, Advances in drug delivery systems, *Mod. Drug Discovery*, 4(4):49–50, April 2004.
61. Y. Kaneda, Y. Tsutsumi, Y. Yoshioka, H. Kamada, Y. Yamamoto, H. Kodaira, S.-I. Tsunoda, T. Okamoto, Y. Mukai, and H. Shibata, The use of PVP as a polymeric carrier to improve the plasma half-life of drugs, *Biomaterials*, 25(16):3259–3266, July 2004.
62. Y. Yoshioka, Y. Tsutsumi, Y. Mukai, H. Shibata, T. Okamoto, Y. Kaneda, S. Tsunoda, H. Kamada, K. Koizumi, Y. Yamamoto, Y. Mu, H. Kodaira, K. Sato-Kamada, S. Nakagawa, and T. Mayumi, Effective accumulation of poly(vinylpyrrolidone-*co*-vinyl laurate) into the spleen, *J. Biomed. Mater. Res. Part A*, 70A(2):219–223, August 2004.
63. B.C. Hancock, S.L. Shamblin, and G. Zografi, Molecular mobility of amorphous pharmaceutical solids below their glass transition temperatures, *Pharm. Res.*, 12(6):799–806, 1995.
64. J. Berggren and G. Alderborn, Long-term stabilisation potential of poly(vinylpyrrolidone) for amorphous lactose in spray-dried composites, *Eur. J. Pharm. Sci.*, 21(2-3):209–215, February 2004.
65. J. Lee, Intrinsic adhesion properties of poly(vinyl pyrrolidone) to pharmaceutical materials: Humidity effect, *Macromol. Biosci.*, 5(11): 1085–1093, November 2005.
66. R. Nair, N. Nyamweya, S. Gonen, L.J. Martinez-Miranda, and S.W. Hoag, Influence of various drugs on the glass transition temperature of poly(vinylpyrrolidone): A thermodynamic and spectroscopic investigation, *Int. J. Pharm.*, 225(1-2):83–96, August 2001.
67. S. Bogdanova, I. Pajeva, P. Nikolova, I. Tsakovska, and B. Muller, Interactions of poly(vinylpyrrolidone) with ibuprofen and naproxen: Experimental and modeling studies, *Pharm. Res.*, 22(5):806–815, May 2005.
68. A.A.A. de Queiroz, E.J. Franca, G.A. Abraham, and J. San Roman, Drug complexation and physicochemical properties of vinylpyrrolidone-*N,N′*-dimethylacrylamide copolymers, *J. Appl. Polym. Sci.*, 93 (3):1337–1347, August 2004.
69. L.A. Meure, B. Warwick, F. Dehghani, H.L. Regtop, and N.R. Foster, Increasing copper indomethacin solubility by coprecipitation with poly(vinylpyrrolidone) using the aerosol solvent extraction system, *Ind. Eng. Chem. Res.*, 43(4):1103–1112, February 2004.
70. K. Khougaz and S.D. Clas, Crystallization inhibition in solid dispersions of MK-0591 and poly(vinylpyrrolidone) polymers, *J. Pharm. Sci.*, 89(10):1325–1334, October 2000.
71. H. Kamada, Y. Tsutsumi, K. Sato-Kamada, Y. Yamamoto, Y. Yoshioka, T. Okamoto, S. Nakagawa, S. Nagata, and T. Mayumi, Synthesis of

a poly(vinylpyrrolidone-*co*-dimethyl maleic anhydride) co-polymer and its application for renal drug targeting, *Nat. Biotechnol.*, 21(4): 399–404, April 2003.

72. Y. Yamamoto, Y. Tsutsumi, Y. Yoshioka, H. Kamada, K. Sato-Kamada, T. Okamoto, Y. Mukai, H. Shibata, S. Nakagawa, and T. Mayumi, Poly(vinylpyrrolidone-*co*-dimethyl maleic acid) as a novel renal targeting carrier, *J. Controlled Release*, 95(2):229–237, March 2004.
73. J.R. Gum, Jr., S.C. Crawley, J.W. Hicks, D.E. Szymkowski, and Y.S. Kim, Muc17, a novel membrane-tethered mucin, *Biochem. Biophys. Res. Commun.*, 291(3):466–475, March 2002.
74. V. Sadasivan, R. Carrier, and J. David Budil, Albert Saccco, Molecular modeling – a tool for better understanding advanced oral drug delivery, [electronic:] http://www.dac.neu.edu/cammp/drug_delivery.htm, 2006.
75. D.J. Dobrozsi, Oral liquid mucoadhesive compositions, US Patent 6 638 521, assigned to The Procter & Gamble Company (Cincinnati, OH), October 28, 2003.
76. M.K. Chun, C.S. Cho, and H.K. Choi, Characteristics of poly(vinyl pyrrolidone)/poly(acrylic acid) interpolymer complex prepared by template polymerization of acrylic acid: Effect of reaction solvent and molecular weight of template, *J. Appl. Polym. Sci.*, 94(6):2390–2394, December 2004.
77. M.-K. Chun, C.-S. Cho, and H.-K. Choi, Mucoadhesive drug carrier based on interpolymer complex of poly(vinyl pyrrolidone) and poly(acrylic acid) prepared by template polymerization, *J. Controlled Release*, 81(3):327–334, June 2002.
78. E. Barbu, I. Sarvaiya, K.L. Green, T.G. Nevell, and J. Tsibouklis, Vinylpyrrolidone-*co*-(meth) acrylic acid inserts for ocular drug delivery: Synthesis and evaluation, *J. Biomed. Mater. Res. Part A*, 74A(4):598–606, September 2005.
79. S. Gopalkrishnan and K.M. Guiney, Stable aqueous laundry detergents containing vinyl pyrrolidone copolymers, US Patent 6 743 763, assigned to BASF Corporation (Mount Olive, NJ), June 1, 2004.
80. J. Detering, C. Schade, W. Trieselt, and J. Tropsch, Use of vinylpyrrolidone and vinylimidazole copolymers as detergent additives, novel polymers of vinylpyrrolidone and of vinylimidazole, and preparation thereof, US Patent 5 846 924, assigned to BASF Aktiengesellschaft (Ludwigshafen, DE), December 8, 1998.
81. A. Roos, C. Creton, M.B. Novikov, and M.M. Feldstein, Viscoelasticity and tack of poly(vinyl pyrrolidone)-poly(ethylene glycol) blends, *J. Polym. Sci., Part B: Polym. Phys.*, 40(20):2395–2409, October 2002.

82. A.A. Chalykh, A.E. Chalykh, M.B. Novikov, and M.M. Feldstein, Pressure-sensitive adhesion in the blends of poly(*N*-vinyl pyrrolidone) and poly(ethylene glycol) of disparate chain lengths, *J. Adhes.*, 78(8):667–694, August 2002.

83. M.M. Feldstein, V.G. Kulichikhin, S.V. Kotomin, T.A. Borodulina, M.B. Novikov, A. Roos, and C. Creton, Rheology of poly(*N*-vinyl pyrrolidone)-poly(ethylene glycol) adhesive blends under shear flow, *J. Appl. Polym. Sci.*, 100(1):522–537, April 2006.

84. H. Diehl, K.-H. Schumacher, R. Fink, and M. Jung, Pressure sensitive adhesive comprising vinylpyrrolidone, US Patent 6 858 295, assigned to BASF Aktiengesellschaft (Ludwigshafen, DE), February 22, 2005.

85. G. Rehmer, A. Boettcher, and M. Portugall, Derivatives of benzophenone and their preparation, EP Patent 0 346 734, assigned to BASF AG (De), December 20, 1989.

86. J.D. Moon, Pressure-sensitive adhesive copolymers of acrylic acid ester and *N*-vinyl pyrrolidone, US Patent 4 364 972, assigned to Minnesota Mining and Manufacturing Company (St. Paul, MN), December 21, 1982.

87. C.L. Netwig and D.L. Kronmiller, Process for producing in situ polymerization of a reverse osmosis membrane and product therefrom, US Patent 5 534 146, July 9, 1996.

88. Z.-P. Zhao, J. Li, D. Wang, and C.-X. Chen, Nanofiltration membrane prepared from polyacrylonitrile ultrafiltration membrane by low-temperature plasma: 4. grafting of *N*-vinylpyrrolidone in aqueous solution, *Desalination*, 184(1-3):37–44, November 2005.

89. J.H. Kim and C.K. Kim, Ultrafiltration membranes prepared from blends of polyethersulfone and poly(1-vinylpyrrolidone-*co*-styrene) copolymers, *J. Membr. Sci.*, 262(1-2):60–68, October 2005.

90. E.J. Moon, J.W. Kim, and C.K. Kim, Fabrication of membranes for the liquid separation: Part 2: Microfiltration membranes prepared from immiscible blends containing polysulfone and poly(1-vinylpyrrolidone-*co*-acrylonitrile) copolymers, *J. Membr. Sci.*, 274(1-2):244–251, 2006.

91. J.H. Kim, M.S. Kang, and C.K. Kim, Fabrication of membranes for the liquid separation: Part 1. Ultrafiltration membranes prepared from novel miscible blends of polysulfone and poly(1-vinylpyrrolidone-*co*-acrylonitrile) copolymers, *J. Membr. Sci.*, 265(1-2):167–175, November 2005.

92. B. Smitha, S. Sridhar, and A.A. Khan, Chitosan-poly(vinyl pyrrolidone) blends as membranes for direct methanol fuel cell applications, *J. Power Sources*, 159(2):846–854, September 2006.

93. A. Sionkowska, M. Wisniewski, J. Skopinska, S. Vicini, and E. Marsano, The influence of UV irradiation on the mechanical properties of chitosan/poly(vinyl pyrrolidone) blends, *Polym. Degrad. Stabil.*, 88(2):261–267, May 2005.
94. H.L. Wu, C.C.M. Ma, H.C. Kuan, C.H. Wang, C.Y. Chen, and C.L. Chiang, Sulfonated poly(ether ether ketone)/poly(vinylpyrrolidone) acid-base polymer blends for direct methanol fuel cell application, *J. Polym. Sci., Part B: Polym. Phys.*, 44(3):565–572, February 2006.
95. J. Qiao, T. Hamaya, and T. Okada, New highly proton-conducting membrane poly(vinylpyrrolidone)(pvp) modified poly(vinyl alcohol)/2-acrylamido-2-methyl-1-propanesulfonic acid (PVA-PAMPS) for low temperature direct methanol fuel cells (DMFCs), *Polymer*, 46 (24):10809–10816, November 2005.
96. N. Chen and L. Hong, Proton-conducting membrane composed of sulfonated polystyrene microspheres, poly(vinylpyrrolidone) and poly(vinylidene fluoride), *Solid State Ionics*, 146(3-4):377–385, February 2002.
97. D. Anjali Devi, B. Smitha, S. Sridhar, and T.M. Aminabhavi, Novel crosslinked chitosan/poly(vinylpyrrolidone) blend membranes for dehydrating tetrahydrofuran by the pervaporation technique, *J. Membr. Sci.*, 280(1-2):45–53, September 2006.
98. J. Lu, Q. Nguyen, J.Q. Zhou, and Z.H. Ping, Poly(vinyl alcohol)/poly-(vinyl pyrrolidone) interpenetrating polymer network: Synthesis and pervaporation properties, *J. Appl. Polym. Sci.*, 89(10):2808–2814, September 2003.
99. Q.-T. Nguyen, R. Clement, I. Noezar, and P. Lochon, Performances of poly(vinylpyrrolidone-*co*-vinyl acetate)-cellulose acetate blend membranes in the pervaporation of ethanol-ethyl tert-butyl ether mixtures: Simplified model for flux prediction, *Sep. Purif. Tech.*, 13(3): 237–245, June 1998.
100. R.M. Wenthold, R.T. Hall, II, R.G. Andrus, P.D. Brinda, L.C. Cosentino, R.F. Reggin, and D.T. Pigott, Hollow fiber membranes and method of manufacture, US Patent 5 762 798, assigned to Minntech Corporation (Minneapolis, MN), June 9, 1998.
101. M. Fuke, T. Kuroki, and T. Tanaka, Polysulfone-base hollow-fiber hemocathartic membrane and processes for the production thereof, US Patent 6 432 309, assigned to Asahi Medical Co, Ltd (Tokyo, JP), August 13, 2002.
102. A. Higuchi, K. Shirano, M. Harashima, B.O. Yoon, M. Hara, M. Hattori, and K. Imamura, Chemically modified polysulfone hollow fibers with vinylpyrrolidone having improved blood compatibility, *Biomaterials*, 23(13):2659–2666, July 2002.

103. N.J. Gordon and M.F. Theophile, Alkaline liquid hard-surface cleaning compositions comprising *N*-vinylpyrrolidone copolymer, US Patent 6 484 735, assigned to The Procter & Gamble Company (Cincinnati, OH), November 26, 2002.
104. E.R. Bartsch, H.C. Na, and J.A. Wooton, Spraying device, US Patent 6 869 028, assigned to The Procter & Gamble Company (Cincinnati, OH), March 22, 2005.
105. M. Lappas, B. Fussnegger, G. Müller, K.H. Jung, and G. Tasser, Production and use of formulations consisting of cellulose, kalium caseinate and cross-linked vinylpyrrolidone homopolymers and/or vinylimidazol/vinylpyrrolidone copolymers, US Patent 6 232 373, assigned to BASF Aktiengesellschaft (Ludwigshafen, DE), May 15, 2001.
106. E.S. Barabas, Rust removal process, US Patent 4 424 079, assigned to GAF Corporation (New York, NY), January 3, 1984.
107. E.S. Barabas, Process for making copolymers of vinylpyrrolidone and maleic anhydride, US Patent 4 600 759, assigned to GAF Corporation (Wayne, NJ), July 15, 1986.
108. J.M. Cohen and W.D. Young, Method for inhibiting the formation of gas hydrates, US Patent 6 096 815, assigned to ISP Investments Inc. (Wilmington, DE), August 1, 2000.
109. K.N. Bakeev, J.-C. Chuang, T. Winkler, M.A. Drzewinski, and D.E. Graham, Method for preventing or retarding the formation of gas hydrates, US Patent 6 281 274, assigned to ISP Investments Inc. (Wilmington, DE), August 28, 2001.
110. Y.M. Wu, B.Q. Zhang, T. Wu, and C.G. Zhang, Properties of the forpolymer of *N*-vinylpyrrolidone with itaconic acid, acrylamide and 2-acrylamido-2-methyl-1-propane sulfonic acid as a fluid-loss reducer for drilling fluid at high temperatures, *Colloid Polym. Sci.*, 279(9):836–842, September 2001.
111. H.-S. Tong, C.-M. Hu, and Y.-C. Yu, Photoresist for cathode ray tubes that includes vinyl pyrrolidone-vinylalcohol and a di-tetraalkylammonium salt, US Patent 5 717 281, assigned to Chunghwa Picture Tubes, Ltd. (Taoyuan, TW), February 10, 1998.
112. K. Neubecker, S. Stien, and S. Kothrade, Copolymer of vinylpyrrolidone and vinylimidazole, US Patent 5 990 269, assigned to BASF Aktiengesellschaft (Ludwigshafen, DE) and Tokyo Ohka Kogyo Co., Ltd. (Kanagawa-ken, JP), November 23, 1999.
113. C. Dupuis, Cosmetic composition containing a cationic polymer and an acrylic terpolymer, and use of this composition for the treatment of keratinous material, US Patent 6 214 326, assigned to L'Oreal S.A. (Paris, FR), April 10, 2001.

114. R. Dieing, P. Hossel, and A. Sanner, Use of cationic copolymers of unsaturated acids and *N*-vinylimidazolium salts in cosmetic hair formulations, US Patent 6 355 231, assigned to BASF Aktiengesellschaft (Ludwigshafen, DE), March 12, 2002.
115. K. Wakiya, N. Kubota, S. Yokoi, and M. Kobayashi, Composition for lithographic anti-reflection coating, and resist laminate using the same, US Patent 6 416 930, assigned to Tokyo Ohka Kogyo Co., Ltd. (Kanagawa, JP), July 9, 2002.
116. C.E. Romano, Jr., K.E. Maskasky, and D. Santilli, Process for making an ink jet ink, US Patent 6 053 438, assigned to Eastman Kodak Company (Rochester, NY), April 25, 2000.
117. H.J. Klimisch, K. Deckardt, C. Gembardt, B. Hildebrand, K. Kuttler, and F.J.C. Roe, Long-term inhalation toxicity of *N*-vinylpyrrolidone-2 vapours. Studies in rats, *Food Chem. Toxicol.*, 35(10-11):1041–1060, 1997.
118. H.J. Klimisch, K. Deckardt, C. Gembardt, B. Hildebrand, K. Kuttler, and F.J.C. Roe, Subchronic inhalation and oral toxicity of *N*-vinylpyrrolidone. 2. Studies in rodents, *Food Chem. Toxicol.*, 35(10-11):1061–1074, 1997.

10

Other Cationic Polymers

As the class of cationic polymers is not unique they can be subdivided into subclasses from the chemical point of view. Of the cationic polymers where a large amount of literature has been located separate chapters have been devoted to them in this volume.

Herewith is a summary of certain cationic polymers to those less research has been devoted in the last years. The material presented here is overlapping to some extent with material presented in other chapters because many of the copolymers from monomers used are described in those chapters.

The common feature is that this class bears positive charges, either in side groups or in the backbone itself. Cationic polymers or copolymers with quaternized nitrogen or phosphorus located in the backbone of the polymer have sometimes been also addressed as ionene polymers or polyionenes (1). Polyelectrolytes have been reviewed (2–4).

10.1 Manufacture

10.1.1 Bifunctional Quaternization

A wide variety of polymeric quaternary ammonium salts has been described. They are basically produced by the reaction of a dihalide with a di-*tert*-amine. The basic reaction is illustrated be examples in Figure 10.1.

The preparation is quite simple. Equimolar amounts dihalides and tertiary diamines are heated in methanol and refluxed for 24 *h*. In the course of the reaction, the viscosity of the reaction mixture

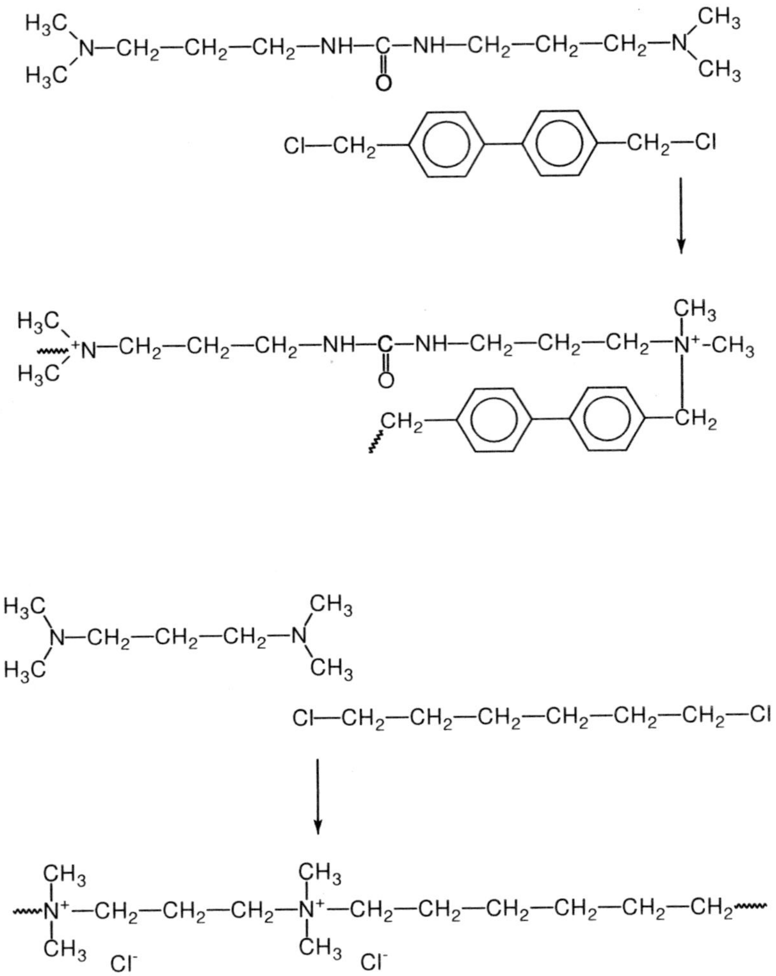

Figure 10.1 Quaternization

gradually increases. The reaction mixture is cooled and the methanol is removed in vacuo (5).

Polymeric quaternary ammonium salts are usually soluble in water and are usually obtained from monomers in quantitative yield. A wide variety of monomers that can be used is exemplified in the literature (5).

10.1.2 Addition Polymerization

Water-soluble or water dispersible cationic addition polymers are themselves well known. These polymers comprise both quaternary ammonium containing moieties and ammonium free moieties (6).

The quaternary ammonium containing units are derived from ethylenically unsaturated monomers containing either quaternary ammonium groups or tertiary amino groups which can be quaternized by conventional methods after polymerization to form the polymer.

The counter ion can be any of those commonly employed such as for example chloride, bromide, nitrate, hydrogen sulfate, methyl sulfate, sulfonate, acetate. Usually, but not necessarily, these monomers contain the acryloyl functionality, methacryloyl functionality, or vinyl functionality, although other monomers with allyl functionality or methallyl functionality may be used.

Examples of ethylenically unsaturated monomers containing quaternary ammonium groups are shown in Table 10.1. Examples

Table 10.1 Unsaturated Monomers with Quaternary Ammonium Groups (6)

Primary Amine Salt Monomers
Trimethyl-2-(methacryloyloxy)ethylammonium salt
Triethyl-2-(acryloyloxy)ethylammonium salt
N,*N*-Diethyl-*N*-methyl-2-(methacryloyloxy)ethylammonium salt
Secondary Amine Salt Monomers
Methyl-2-(methacryloyloxy)ethylammonium salt
Ethyl-2-(methacryloyloxy)ethylammonium salt
Ethyl-*p*-vinylbenzylammonium salt
Ethyl-*m*-vinylbenzylammonium salt

of ethylenically unsaturated monomers which contain at least one tertiary amino group that can be converted to a quaternary ammonium group after polymerization are shown in Table 10.2. A much

Table 10.2 Unsaturated Monomers to those Quaternary Ammonium Groups can be Attached (6)

Monmomer
N,N'-Dimethylaminoethyl methacrylate
Dimethylaminoethyl acrylate
N-(Dimethylaminoethyl) methacrylamide
N-(Dimethylaminoethyl) acrylamide
N-Ethyl-*N*-methylaminoethyl methacrylate
N,N-Dimethyl-*N*-(*p*-vinylbenzyl)amine

more extended list of monomers can be found elsewhere (6). The acronyms for the most commonly monomers for cationic polyelectrolytes are summarized in Table 10.3.

Table 10.3 Acronyms for Monomers (7,8)

Monomer	Acronym
Acryloyloxyethyl trimethyl ammonium chloride	AETAC
Acrylamidopropyl trimethyl ammonium chloride	APTAC
Diallyldiethylammonium chloride	DADEAC
Diallyldimethylammonium chloride	DADMAC
Diethylaminoethyl methacrylate	DEAEMA
N,N'-Dimethylaminoethyl methacrylate	DMAEMA
3-(Methacrylamido)propyltrimethylammonium chloride	MAPTAC
Methacryloyloxyethyl trimethyl ammonium chloride	METAC
Methacryloyloxyethyl trimethyl ammonium sulfate	METAM
Methacryloyloxyethyl trimethyl ammonium methyl sulfate	METAMS

Polymers are obtained in a conventional way by polymerizing the vinyl groups. Of course, additional comonomers can be added, such as methyl methacrylate, *n*-butyl acrylate or styrene. 2,2'-Azobis(2-methylbutyronitrile) has been used as a radical initiator (6). Polymers of this type were first reported in 1957 (9). In the course of the polymerization cyclic structures are formed, as shown in Figure 10.2. Something like a backbiting mechanism has been proposed.

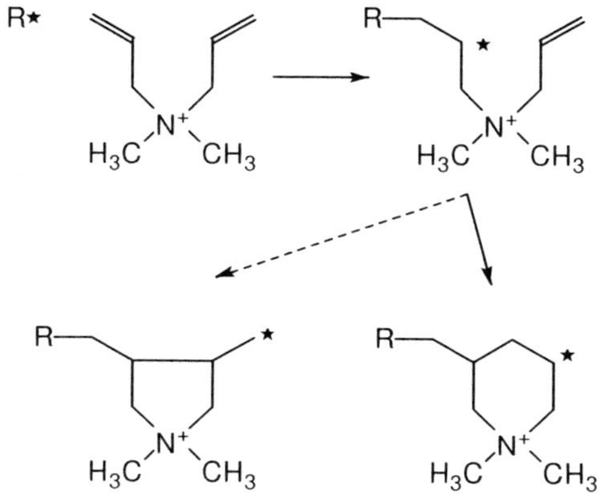

Figure 10.2 Polymerization of Diallyldimethylammonium chloride

The reaction path with the full arrow in Figure 10.2 is considered preferred since a secondary radical is formed as intermediate and a six membered ring is formed.

Copolymers from *N*-Vinyl-2-pyrrolidone and *N*-3,3-dimethylaminopropyl methacrylamide (10) or diallyldimethylammonium chloride (11) find use in hair styling compositions and are explained elsewhere. The preparation of copolymers containing diallyldimethylammonium chloride, etc. has been described in detail (12).

10.1.3 Ring Opening Polymerization

Oligomers can be formed by the reaction of, e.g., dimethylamine with epichlorohydrin. These oligomers can be quaternized. Branched structures are obtained by using multifunctional amino compounds (13).

10.1.4 Cationic Modification of Polymers

Polymers, in particular natural polymers, can be cationically modified. Cationic reagents are 3-chloro-2-hydroxypropyl trimethylammonium chloride or its epoxy version (14).

10.2 Applications

Cationic polymers are used in a wide variety of highly specialized fields of applications. Often the polymers are specifically tailored for their special application. Major applications are summarized in Table 10.4.

Table 10.4 Applications of Cationic Polymers

Monomer	References
Absorbent materials	(15,16)
Cosmetic compositions	(17–20)
Fiber and fabric treatment	(21,22)
Laundry detergents	(23)
Medical and pharmaceutical applications	(24–26)
Papermaking	(27–29)
Water treatment	(30,31)

10.2.1 Superabsorbent Hydrogels

Superabsorbent materials have been proposed that are based on poly(sodium acrylate), which is an anionic polymer (16). This polymer is surface treated with a water-soluble linear cationic polymer. Thus the charge is opposite from the base material. Examples of suitable cationic polymers for their use of surface treatment include poly(*N*-vinylamine)s, chitosan and their salts.

Poly(*N*-vinylamine)s are known to have improved adhesion to cellulose, while poly(*N*-vinylamine) coated superabsorbents have some improved permeability. Coating with linear poly(*N*-vinylamine)s shows an improved adhesion to fibers because of the high flexibility of poly(*N*-vinylamine)s. A disadvantage is that the uncrosslinked polymers are easily to extract.

On the other hand, if the poly(*N*-vinylamine) is covalently bonded to the superabsorbent, the degree of crosslinking of the superabsorbent will become higher and the absorptive capacity goes down (15).

Surface treated absorbent materials in this way exhibit enhanced liquid handling properties, in particular an enhanced gel bed permeability under load (16). Applications of these compositions include tampons, diapers, etc.

10.2.1.1 Centrifuge Retention Capacity

The centrifuge retention capacity (CRC) is a standard method for measuring the liquid uptake of a superabsorbent material (32). The CRC test measures the ability of the gel particles to retain the liquid therein after being saturated and subjected to centrifugation under controlled conditions. The resultant retention capacity is stated as grams of liquid retained per gram weight of the sample, i.e., a dimensionless number.

The sample is prepared from particles which have been sieved through a U.S. standard 30 *mesh* screen and are retained on 50 *mesh* screen. Thus, the sample comprises particles sized in the range of 300–600 μ.

The retention capacity is measured by placing 0.2 *g* sample into a water permeable bag which will contain the sample. A test solution of 0.9% sodium chloride in distilled water is allowed to saturate the sample. Empty bags serve as controls. After wetting to equilibrium, the bags are placed into the basket of a suitable centrifuge capable of subjecting the samples to a force of 0.35 *kp*. In a centrifuge of type Heraeus LaboFuge 400 this corresponds to 1,600 *rpm*. A more detailed procedure is given elsewhere (32).

10.2.2 Paper Coatings for Ink Jet Printing

Ink jet paper coatings typically comprise a silica pigment for its high absorption power and a polymeric binder, such as poly(vinyl alcohol) (PVA), for its high binding strength (33). Non-silica pigments are clays, calcium carbonate, titanium dioxide, and aluminum hydrate. Other known polymeric binders include poly(*N*-vinyl-2-pyrrolidone), styrene-butadiene rubber, poly(vinyl acetate), starch, and amine functional polymers.

Amine functional PVA is typically produced by the copolymerization of vinyl acetate (VA) with amine functional monomers, such as trimethyl-(3-methacrylamido-propyl)ammonium chloride, *N*-vinyl formamide, or acrylamide (AAm), followed by saponification to form the PVA derivative.

However, there are disadvantages to this approach. The selection of amino comonomers is very limited due to their incompatibility with the saponification conditions to produce PVA. Depending on the monomer which is copolymerized with the VA, saponification can have a deleterious effect on the comonomer. For example, when *N,N'*-dimethylaminoethyl methacrylate is the comonomer, saponification results in the hydrolysis of the ester bond, thus removing the active amine functionality from the polymer backbone. Another disadvantage which limits the number of comonomers that can be used for the preparation of amine functional PVA is the monomer reactivity ratio of the comonomer with VA. Depending on these reactivity ratios, there can be severe limitations not only on the level of amine monomer incorporation into the VA copolymer, but also on the attainable range of copolymer molecular weights.

Improved ink jet paper coatings which impart high optical density images and excellent water resistance are obtained by the incorporation of an amine functional, colloid-stabilized emulsion into the coating formulation (33).

More generally, cationic polymers are used as ingredients for ink jet printing compositions. A specific monomer feed is shown in Table 10.5. The charges in Table 10.5 are feeded one by one.

10.2.3 *Water Purification*

Polymers that have been used for water purification have been reviewed (13). Flocculation is a phase separation process which facilitates the removal of finely divided particles from a liquid by enhancing the agglomeration of the suspended particles in order to increase the particle size. Flocculation may be accomplished by the addition of a flocculating agent (34).

A polymer suitable as flocculating agent is made up from AAm, *N,N'*-methylene bisacrylamide, and acryloxyethyltrimethylammonium chloride (34). The polymer is prepared by emulsion polymerization. The aqueous phase comprises the monomers, crosslinking agent and a chain transfer agent dissolved in deionized water, and as well as other additives such as stabilizers and pH adjusters. These ingredients are ammonium sulfate, ethylenediamine tetraacetic acid, and 2-propanol as a chain transfer agent. The radical initiator is

Table 10.5 Ink Jet Printing Composition (6)

Ingredient	Charge/[*kg*]
Charge 1	
Methyl ethyl ketone	55.93
Charge 2	
Methyl ethyl ketone	28.67
2,2′-azobisisobutyronitrile	10.16
Charge3	
n-Butyl acrylate	30.44
Methyl methacrylate	87.32
2-(*tert*-Butylamino) ethyl methacrylate	40.64
Styrene	44.68
Charge 4	
Methyl ethyl ketone	2.27
Charge 5	
Methyl ethyl ketone	2.27
Charge 6	
Glacial acetic acid	9.89
Methyl ethyl ketone	2.27
Charge 7	
Deionized water	579.1
Charge 8	
Deionized water	11.1

tert-butyl hydroperoxide. The pH is adjusted with sulfuric acid.

The oil phase comprises a water-insoluble hydrocarbon solution of surfactants. The oil phase is prepared by dissolving sorbitan monooleate into paraffin oil.

Variants of the procedure of polymerization have been described in that the crosslinking agent or the chain transfer agent is not added at the beginning of the polymerization reaction but at somewhat higher conversions (35).

It is believed that polymers modified with a crosslinking agent after the start of polymerization contain a mixture of a linear high molecular weight polymer which is formed during the initial part of the reaction, and a branched polymer which is formed in the course of the latter part of the reaction. The degree of conversion at which the crosslinking agent is added is of critical importance to the flocculating ability of the resultant polymer (35).

10.2.4 Cosmetic Compositions

It is advantageous to formulate the hair compositions containing cationic polymers, which have a high viscosity and are in a thickened liquid form which spreads well, such as a styling or care cream or gel, these properties being very much appreciated by consumers since the composition does not run down the forehead, the nape of the neck, the face or into the eyes (36).

Quaternary polyammonium compounds with a structure shown in Figure 10.3 have been used in dyeing compositions (37).

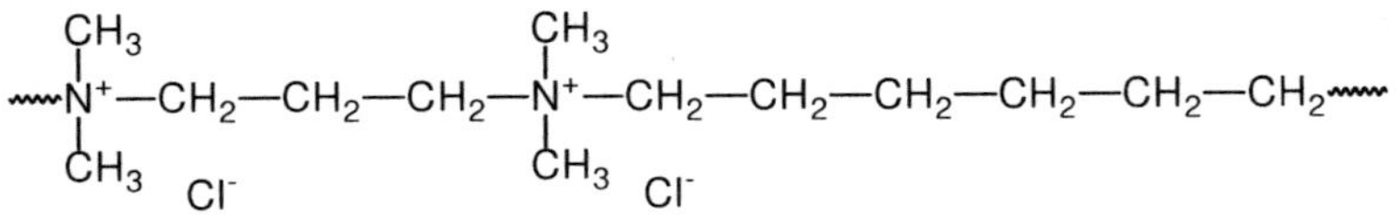

Figure 10.3 Quaternary Poly ammonium salts

A composition for a shampoo for washing and conditioning the hair is given in Table 10.6 and a composition for hair fixing is shown in Table 10.7.

This composition is used in a sequence of steps. First, a reducing composition is applied to moistened hair rolled on to curlers. This

Table 10.6 Composition for a Shampoo (19)

Component	Amount/[*g*]
Cocoglucoside	5
Cocoamidopropyl betaine	5.4
Sodium lauryl ether (5 oxyethylene) carboxylate	3
Sodium lauryl ether sulfate	4
Polyquaternium 6 (DADMAC), c.f. Table 10.3	0.4
Poly(dimethylsiloxane)	2.6
Others	
Preserving agent	
Fragrance	
Citric acid to pH 6.5	
Water up to 100 *g*	

Table 10.7 Composition for Hair Fixing (20)

Component	Amount/[*g*]
Hydrogen peroxide 50% solution	4.8
Lauryldimethylamine oxide 30% solution	2.15
Merquat 100 (polymeric DADMAC)	1.25
Citric acid	0.1
Demineralized water up to 100 g	

is followed by the application of a permanent-waving product and then the fixing composition is used (20).

10.2.5 Oil Field Applications

Monomers are grafted on to lignite which have been selected from the group consisting of 2-acrylamido-2-methyl-1-propane sulfonic acid, acrylamide, acrylic acid, vinylphosphonic acid, diallyldimethylammonium chloride and the corresponding salts. These additives act as fluid loss control in well drilling fluids and cementing compositions (38–40).

Tradenames appearing in the references are shown in Table 10.8.

Table 10.8 Tradenames in References

Tradename Description	Supplier
Abil® Wax (Series) Cosmetic water-in-silicone emulsions (18, 19)	EVONIK Goldschmidt GmbH
Accurac® 171 Acrylic Polymer (27, 29)	Cytec, Inc.
Acronal® Acrylic resins (19)	BASF
Acrylidone® LM vinylpyrrolidone/acrylic acid/lauryl methacrylate terpolymers (19)	ISP
Acrysol® RM (Series) Hydrophobically modified polyethylene oxide urethane, nonionic rheology modifier (18, 19)	Rohm & Haas
Aculyn™ (Series) hydrophobically-modified poly(acrylate) (11, 18)	Rohm and Haas
Airvol® (27, 29, 33)	Air Products and Chemicals, Inc.
Akypo® (Series) Ethoxylated acids, low foam tensides (19)	Kao
Alfonic® 1412-60 Ethoxylated linear alcohol (35)	Vista Chemical Co.
AlomarBlue® Mitochondrial redox indicator dye (1)	AccuMed International, Inc.
Alusil® 70 Aluminosilicate (30, 31)	Selecto, Inc.
Antil® 208 Oxyethylenated methyl acrylate/stearyl acrylate (18)	EVONIK Goldschmidt

Table 10.8 (cont.)

Tradename Description	Supplier
Appretan® (Series) vinyl acetate homopolymers (19)	Vinyl acetate polymes
Aristoflex® A vinyl acetate/crotonic acid/polyethylene glycol terpolymer (19)	BASF
Bermocoll® EHM 100 Modified hydroxyethylcelluloses (18)	Akzo Nobel
Berset® (Series) Starch and Protein insolubilizer (6)	Bercen Inc.
Binaquat® P 100 Copolymers of acrylamide and methacryloyloxy ethyltrimethylammonium chloride (17)	Ciba
Carbochem® CA-10 Mesoporous acidic wood-based activated carbon particles (31)	Carbochem, Inc.
Carbopol® (Sseries) Poly(acrylate) (11, 18)	Lubrizol Advanced Materials, Inc.
Cartaretine® Copolymers of adipic acid and dimethylamino-hydroxypropyl diethylenetriamine (17)	Sandoz
Cascamid® Cationic polyamide resin (27, 29)	Borden
Catiofast® (Series) Vinylformamide vinylamine Copolymers (28)	BASF
Celquat® (Series) hydroxyalkyl celluloses grafted to dimethyidiallylammonium salts (17)	National Starch
Coatex® acrylic acid/lauryl (meth)acrylate copolymers (18)	Coatex
Crepetrol® (Series) Poly(aminoamide)/epichlorohydrin, creping adhesive (27, 29)	Hercules, Inc.
Crodacel® (Series) Modified hydroxyethylcellulose (18)	Croda, Inc.
Curesan® Starch and Protein insolubilizer (6)	PPG Industries, Inc.
Cypro® Polyquaternary amines (27, 29)	Cytec Tech. Corp.
Dehyton® AB 30 Cocoylbetaine (19)	Henkel

Table 10.8 (cont.)

Tradename Description	Supplier
Delsette® 101	Hercules, Inc.
Adipic acid/epoxypropyl-diethylene-triamine copolymers (17)	
Densodrin®	BASF
Anionic silicone polymers (22)	
Disperal® Sol P3	Condea Chemie GmbH
Pseudoboehmite (6)	
Dow Corning® (Series)	Dow
Silicone Products (11)	
Elfacos® T210	Akzo
Poly(ether urethane)s (18)	
Elvanol® (Series)	DuPont
Poly(vinyl alcohol) (29)	
Estapor® LO 11	Merck Chimie Societe par Actions Simplifiee France
Poly(amide) (19)	
Eucarol® APG (Series)	Lamberti
Alkylpolyglycoside tartrates (19)	
Eudragit®	Evonik Roehm GMBH
Coating Lacquers for use on medicinal tablets (19)	
Fixate®	Noveon
Ionic acrylic copolymer (17)	
Flexan® (Series)	National Starch
Poly(styrene sulfonate) (19)	
Fluorad® (Series)	3M Comp.
Surfactant (6)	
Gaffix® VC-713	GAF
(Dimethylamino)ethyl methacrylate/vinylcaprolactam/*N*-vinyl-2-pyrrolidinone terpolymer (17)	
Gafquat® (Series)	ISP
Copolymer of vinylpyrrolidone and dimethylaminoethyl methacrylate quaternized with diethyl sulfate (17)	
Galactosol® (Series)	Aqualon
Cationic guar gums (19, 22, 23)	
Gantrez®	ISP
Methyl vinyl ether/maleic anhydride copolymer (19)	
Glydant®	Glyco Chemicals, Inc.
Antimicrobial agent (11)	

Table 10.8 (cont.)

Tradename Description	Supplier
Halad® (Series)	Halliburton Energy Services, Inc.
Fluid loss control additive (38,39)	
Hercofloc®	Hercules Inc.
Quaternized copolymers of acrylamide and dimethylaminoethyl methacrylate (17)	
Hercosett® 57	Hercules Inc.
Adipic acid/epoxypropyl-diethylene-triamine copolymers (17)	
Jaguar® (Series)	Rhodia Inc. Corp.
Cationic guar gum (17,19)	
Jeffamine® (Series)	Huntsman Petrochemical Corp.
Amine capped polyalkoxylene glycol (22,23)	
Kathon®	Rohm & Haas
Mixture or 5-chloro-2-methyl-4-isothiazoline-3-one and 2-methyl-4-isothiazoline-3-one, bactericide (11)	
Kelzan® RD	CP Kelco U.S., Inc.
Xanthan gum (22)	
Kensol®	American Refining Group, Inc.
Petroleum distillate (8)	
Kymene® 2064	Hercules Inc.
poly(amide amine epichlorohydrin), wet-strength resin (15,27,29–31)	
Kytamer® PC	Union Carbide Corp.
Chitosan pyrrolidone carboxylate (18)	
Luhydran® A 848 S	BASF AG
Butyl methacrylate copolymer (19)	
Lutensol® GD 70	BASF AG
N-decyl-*α*-*D*-glucopyranoside (19)	
Luvimer® MAEX	Methacrylic acid/ethyl acrylate copolymer
BASF AG (19)	
Luviquat FC 370	BASF AG
Vinylpyrrolidone/methylvinylimidazolium chloride copolymer (70:30) (36)	
Luviquat® FC 905	BASF
Copolymers of 1-vinyl-2-pyrrolidine and 1-vinyl-3-methyl-imidazolium salt (17)	
Luviset® CA 66	BASF AG
Vinyl acetate/crotonic acid copolymer (19)	
Merguard®	Nalco (Calgon Corp.)
Methyldibromo glutaronitrile, cosmetic ingredient (11)	

Table 10.8 (cont.)

Tradename Description	Supplier
Merquat® (Series) Copolymers of acrylic acid with dimethyl diallyl ammonium chloride (17)	Calgon Inc.
Merquat® 100 Poly(dimethyldiallyammonium chloride), antimicrobial polymer (17,19)	Calgon Corp.
Methocel® Methylcellulose (11)	Dow
Microbond™ Cement expanding additive (38,39)	Halliburton Energy Services, Inc.
Microthene® LLDPE (30,31)	Equistar Chemicals, LP
Miranol® Alkylaspartic acid, ampholytic detergent (11,17)	Rhodia Inc. Corp.
Mirapol® (Series) Polyquaternium cosmetics (17,19)	Miranol
Mirapol® 175 Copolymer of vinylpyrrolidone and methacrylamidopropyl trimethylammonium salt (17,19)	Rhodia Inc., Miranol
Mirapol® A 15 Quaternized 1,3-bis[3-(dimethylamino)propyl]urea], CAS 68555-36-2 (17)	Rhodia Inc., Miranol
N-Hance® 3196 cationic guar gums (23)	Aqualon
Nafol® 1822 C Fatty alcohol mixture (17)	Sasol Olefins & Surfactants
Nalbrite™ Powdered bentonite (35)	Nalco Chemical Comp.
Nalco® 8671 Colloidal silica in water (35)	Nalco Chemical Comp.
Nalco® 8677 PLUS copolymers of acrylic acid and acrylamide useful as microparticles (35)	Nalco Chemical Comp.
Nalco® 8678 Naphthalene sulfonate/formaldehyde polymers (35)	Nalco Chemical Comp.
Nalco® 8692 Borosilicate dispersed in water (35)	Nalco Chemical Comp.

Table 10.8 (cont.)

Tradename Description	Supplier
Natrosol® Plus Grade 330 CS	Aqualon
Modified hydroxyethylcellulose (18)	
Nipol® LX 531 B	Nipon Zeon Co.
Latex from butadiene and alkyl (meth)acrylates (19)	
Norpar™	Exxon Mobil
Dearomatized hydrocarbon fluids (8)	
Nuchar® RGC	MeadWestvaco Corp.
Mesoporous activated carbon (30,31)	
Oramix® CG 110	Seppic
Capryl glucoside (19)	
Oramix® NS 10	Seppic
Decyl Glucoside (19)	
Parez® (Series)	Kemira Oyj Comp.
Cationic poly(acrylamide) copolymers (27,29)	
Pegosperse® (Series)	Lonza, Inc.
PEG fatty esters (27)	
Pemulen®	Lubrizol Advanced Materials, Inc.
Poly(acrylate), polymeric emulsifiers for cosmetics (11,18)	
Peox® (Series)	Dow
poly(oxazoline)s (19)	
Plantacare®	Cognis GmbH
Alkyl poly(glucoside)s (19)	
Plantapon® LGC	Cognis GmbH
Alkyl poly(glycoside) tartrates (19)	
Plantaren®	Cognis GmbH
Alkyl poly(glycoside) (19)	
Polyquart® H	Cognis GmbH
Poly(amine)s (17)	
Poval®	Kuraray Co. Ltd.
Cationic poly(vinyl alcohol) (33)	
Primal®	Rohm & Haas Comp.
Acrylic polymer (19)	
Quatrisoft® LM 200	Union Carbide Corp.
Polyquaternium 24 (18)	
Remsol®	REMET UK Ltd.
Colloidal silica (8)	

Table 10.8 (cont.)

Tradename Description	Supplier
Reten® (Series) Flocculants (27,29)	Hercules, Inc.
Rheolate® (Series) Acrylic Rheological Additives (18)	Elementis Specialties, Inc.
Rhodopas® A 012 Vinyl acetate homopolymer (19)	Rhone-Poulenc
Rhodopas® AD 310 EVA (19)	Rhone-Poulenc
Rohadon® (Series) Acrylic polymer (6)	Rohm & Haas
Rohamere® (Series) Acrylic Resins (6)	Evonik Rohm GmbH
Salcare® Poly(acrylate), ionic, rheology modifiers (11,17,18,20,21)	Ciba
Sandopan® DTC Acid Trideceth-7-carboxylic acid, anionic surfactant (19)	Clariant
Sandopan® LS 24 N Sodium laureth-13 carboxylate (19)	Clariant
Santicizer® (Series) Alkyl benzyl phthalates (6)	Solutia, Inc.
Scripset® 700 Copolymer of styrene maleic anhydride disodium salt (27,29)	Hercules Inc.
Sequarez® (Series) Insolubilizer for paper coatings (6)	Omnova Solutions, Inc. Corp.
Silbione® 70045 V 2 Octamethylcyclotetrasiloxane (19)	Rhodia
Silicalite® High surface area amorphous silica (38,39)	Halliburton Energy Services, Inc.
Silwet® Organo silicone surfactants (18,19,22)	O Si Specialities, Inc.
Styleze® CC 10 Vinylpyrrolidone/methacrylamidopropyidimethylamine copolymers (17)	ISP
Taxotere® Anticancer drug (25)	Rhone-Poulenc Rorer S.A.
Teflon® Tetrafluoro polymer (30,31)	DuPont
Tegobetaine® F50 Cocamidopropylbetaine (19)	Goldschmidt GmbH

Table 10.8 (cont.)

Tradename Description	Supplier
Texapon® N 702	Cognis IP Management GmbH
Poly(oxy-1,2-ethanediyl), α-sulfo-ω-(dodecyloxy)-, sodium salt (19)	
Thixcin®	Elementis Specialties, Inc.
Castor oil, hydrogenated (22, 23)	
Triton® CG 110	Union Carbide Corp.
C8-10-alkyl ethers of oligomeric polyglucose (19)	
Triton® X (Series)	Union Carbide Corp. (Rohm & Haas)
Poly(alkylene oxide), nonionic surfactants (6)	
Ultrahold® strong	BASF
Acrylic acid/ethyl acrylate/N-tert-butylacrylamide terpolymer (19)	
Ultrasil® CA 1	Noveon/BF Goodrich
Polysiloxane (22)	
Unipure® Red	Daito Kasei Kogyo Co., Ltd.
Octylsilylated red (19)	
Varisoft® (Series)	Goldschmidt Chemical Corp.
Fatty amide amides (creping agents) (27, 29)	
Vazo® (Series)	DuPont
Azonitriles, radical initiators (6)	
Vazo® 67	DuPont
2,2′-Azobis(2-methylbutane-nitrile (6)	
Vidogum® GH 175	Unipectine
Guar Gum (19)	
Vinylon®	Various Companies
Vinal (PVAl) fibers (27, 29)	
Wacker Belsil® ADM 1100	Wacker
Amino silicone (22, 23)	
Wacker Finish® WR 1100	Wacker
Amino silicones (22, 23)	
Xanflood®	CP Kelco U.S. Inc.
Xanthan gum (28)	
Zonyl® FS-300	DuPont
Nonionic fluorosurfactant (6)	

References

1. R.J. Fitzpatrick, K.K. Shackett, and J.D. Klinger, Ionene polymers and their use as antimicrobial agents, US Patent 6 955 806, assigned to Genzyme Corporation (Cambridge, MA), October 18, 2005.
2. M. Schmidt, ed., *Polyelectrolytes with Defined Molecular Architecture I*, Vol. 165 of *Advances in Polymer Science*, Springer, Heidelberg, 2004.
3. M. Schmidt, ed., *Polyelectrolytes with Defined Molecular Architecture II*, Vol. 166 of *Advances in Polymer Science*, Springer, Heidelberg, 2004.
4. A.B. Lowe and C.L. McCormick, eds., *Polyelectrolytes and Polyzwitterions: Synthesis, Properties, and Applications*, Vol. 937 of *ACS Symposium Series*, Division of Polymer Chemistry, American Chemical Society, Washington, DC, 2006.
5. J. Haase, U. Horn, and H.-U. Berendt, Polymeric quaternary ammonium salts containing specific cationic recurring units, US Patent 4 247 476, assigned to Ciba-Geigy Corporation (Ardsley, NY), January 27, 1981.
6. R.G. Swisher and H. Li, Inkjet printing media containing substantially water-insoluble plasticizer, US Patent 6 265 049, assigned to Hewlett-Packard Company (Palo Alto, CA), July 24, 2001.
7. C.L. McCormick, J. Bock, and D. Schulz, "Water-soluble polymers," in H.F. Mark, N. Bikales, C.G. Overberger, and G. Menges, eds., *Encyclopedia of Polymer Science and Engineering*, Vol. 17, pp. 730–784. Wiley Interscience, New York, 2nd edition, 1988.
8. R.R. Reese and P. Rey, Method of fracturing subterranean formations utilizing emulsions comprising acrylamide copolymers, US Patent 7 482 310, assigned to Kroff Chemical Company, Inc. (Pittsburgh, PA) Superior Well Services, Inc. (Indiana, PA), January 27, 2009.
9. G.B. Butler and R.J. Angelo, Preparation and polymerization of unsaturated quaternary ammonium compounds. VIII. A proposed alternating intramolecular-intermolecular chain propagation, *J. Am. Chem. Soc.*, 79(12):3128–3131, June 1957.
10. Y. Zhong and P.F. Wolf, Process for providing homogeneous copolymers of vinylpyrrolidone and 3-dimethylaminopropyl methacrylamide (DMAPMA) which form clear aqueous solutions having high cloud points, US Patent 5 684 105, assigned to ISP Investments Inc. (Wilmington, DE), November 4, 1997.
11. L.M. Brandt and J.R. Cramm, Personal care compositions containing *N,N*-diallyldialkylammonium halide/*N*-vinylpyrrolidone polymers, US Patent 7 115 254, assigned to Nalco Company (Naperville, IL), October 3, 2006.

12. J.B. Wong Shing, A. Gerli, X.S. Cardoso, A.P. Zagala, P. Pruszynski, and C.C. Doucette, Method of preparing modified diallyl-n,n-disubstituted ammonium halide polymers, US Patent 7 473 334, assigned to Nalco Company (Naperville, IL), January 6, 2009.
13. B.A. Bolto, Soluble polymers in water purification, *Prog. Polym. Sci.*, 20(6):987–1041, 1995.
14. L.M. Popplewell, K.D. Lee, J.G.L. Pluyter, J. Brain, and Y. Zhen, Encapsulated fragrance chemicals, US Patent 7 585 824, assigned to International Flavors & Fragrances Inc. (New York, NY), September 8, 2009.
15. M.M. Azad, N. Herfert, M. Mitchell, and J. Robinson, Crosslinked polyamine coating on superabsorbent hydrogels, US Patent 7 396 584, assigned to BASF Aktiengesellschaft (Ludwigshafen, DE), July 8, 2008.
16. J. Qin, X. Zhang, and D.A. Miller, Absorbent materials and absorbent articles incorporating such absorbent materials, US Patent 7 696 401, assigned to Evonik Stockhausen, Inc. (Greensboro, NC), April 13, 2010.
17. M. Maubru, Cosmetic composition comprising at least one anionic surfactant, at least one cationic polymer and at least one amphiphilic, branched block acrylic copolymer and method for treating hair using such a composition, US Patent 7 498 022, assigned to L'Oreal S.A. (Paris, FR), March 3, 2009.
18. M. De Boni and B. Laguitton, Process for dyeing the hair using an anionic coloured polymer, US Patent 7 744 655, assigned to L'Oreal S.A. (Paris, FR), June 29, 2010.
19. L. Paul, F. Giroud, and H. Samain, Detergent cosmetic compositions comprising four surfactants, a cationic polymer, and a beneficial agent and uses thereof, US Patent 7 829 514, assigned to L'Oreal S.A. (Paris, FR), November 9, 2010.
20. T. Fondin and A. Sabbagh, Reducing composition for permanently reshaping keratin fibers and permanent-reshaping process, US Patent 7 754 193, assigned to L'Oreal S.A. (Paris, FR), July 13, 2010.
21. F. Simonet, L. Nicolas-Morgantini, F. Cottard, and C. Rondeau, Oxidation dye composition for keratin fibers, comprising at least one oxidation dye, at least one associative polymer, at least one nonionic cellulose-based compound not comprising a C8-C30 fatty chain, and at least one cationic polymer with a charge density of greater than 1 meq/g and not comprising a c8-c30 fatty chain, US Patent 7 771 491, assigned to L'Oreal S.A. (Paris, FR), August 10, 2010.
22. P.F.A. Delplancke, J.-P. Boutique, and R. Wagner, Fabric treatment compositions comprising oppositely charged polymers, US Patent 7 737 105, assigned to The Procter & Gamble Company (Cincinnati, OH), June 15, 2010.

23. J.-P. Boutique, P.F.A. Delplancke, R. Wagner, M.D. Butts, S.E. Genovese, and S. Scialla, Liquid laundry detergent comprising a cationic silicone polymer and a coacervate phase forming cationic polymer, US Patent 7 439 217, assigned to The Procter & Gamble Company (Cincinnati, OH), October 21, 2008.
24. K. Esuvaranathan, R. Mahendran, and C. Lawrencia, Methods and compositions for delivery of pharmaceutical agents, US Patent 7 709 457, assigned to Genecure Pte Ltd. (Singapore, SG), May 4, 2010.
25. E.H. Chang and K.F. Pirollo, Simplified and improved method for preparing an antibody or an antibody fragment targeted immunoliposome for systemic administration of a therapeutic or diagnostic agent, US Patent 7 780 882, assigned to Georgetown University (Washington, DC), August 24, 2010.
26. H. Yamada, J. Futami, and H. Nakanishi, Method of transducing a protein into cells, US Patent 7 824 910, assigned to Nippon Shokubai Co., Ltd. (Osaka-Shi, Osaka, JP), November 2, 2010.
27. C.W. Neal, E. Aprahamian, Jr., and J.A. Cain, Creping aid composition and methods for producing paper products using that system, US Patent 7 683 126, assigned to The Procter & Gamble Company (Cincinnati, OH), March 23, 2010.
28. W.L. Griffith, A.L. Compere, and C.F. Leitten, Jr., Method for improving separation of carbohydrates from wood pulping and wood or biomass hydrolysis liquors, US Patent 7 699 958, assigned to UT-Battelle, LLC (Oak Ridge, TN), April 20, 2010.
29. C.W. Neal, E. Aprahamian, Jr., and J.A. Cain, Creping aid composition and methods for producing paper products using that system, US Patent 7 700 027, assigned to The Procter & Gamble Company (Cincinnati, OH), April 20, 2010.
30. J.R. Bahm, A.T. Pearks, G.M. Vidal, D.I. Collias, M.D. Mitchell, R.E. Astle, K.L.K. Faye, R.A. Governal, T.J. Hamlin, R.A. Lucht, and H. Patel, Water filter materials and water filters containing a mixture of microporous and mesoporous carbon particles, US Patent 7 712 613, assigned to PUR Water Purification Products, Inc. (Cincinnati, OH), May 11, 2010.
31. M.D. Mitchell, D.I. Collias, D.W. Bjorkquist, P.N. Zaveri, and M.M. Woolley, Methods for treating water, US Patent 7 740 766, assigned to The Procter & Gamble Company (Cincinnati, OH), June 22, 2010.
32. J. Qin, K.R. Schueler, Jr., H.L. Wilhelm, and D.A. Soerens, Damage-resistant superabsorbent materials, US Patent 7 179 851, assigned to Kimberly-Clark Worldwide, Inc. (Neenah, WI), February 20, 2007.

33. J.J. Rabasco, Ink jet media comprising a coating containing amine functional emulsion polymers, US Patent 6 455 134, assigned to Air Products Polymers, L.P. (Allentown, PA), September 24, 2002.
34. R.E. Neff, J.J. Pellon, and R.G. Ryles, High performance cationic polymer flocculating agents, US Patent 5 945 494, assigned to Cytec Technology Corp. (Wilmington, DE), August 31, 1999.
35. W.L. Whipple, C. Maltesh, C.C. Johnson, A. Sivakumar, T.M. Guddendorf, and A.P. Zagala, Structurally-modified polymer flocculants, US Patent 6 753 388, assigned to Nalco Company (Naperville, IL), June 22, 2004.
36. C. Dupuis, Cosmetic composition containing a cationic polymer and an acrylic terpolymer, and use of this composition for the treatment of keratinous material, US Patent 6 214 326, assigned to L'Oreal S.A. (Paris, FR), April 10, 2001.
37. F. Cottard and C. Rondeau, Compositions for oxidation dyeing keratin fibers comprising at least one thickening polymer comprising at least one fatty chain and at least one fatty alcohol chosen from monoglycerolated fatty alcohols and polyglycerolated fatty alcohols, US Patent 7 771 492, assigned to L'Oreal S.A. (Paris, FR), August 10, 2010.
38. S. Lewis, J. Chatterji, B. King, and C. Brennels, Cement compositions comprising lignite grafted fluid loss control additives, US Patent 7 388 045, assigned to Halliburton Energy Services, Inc. (Duncan, OK), June 17, 2008.
39. S. Lewis, J. Chatterji, B. King, and D.C. Brenneis, Cement compositions comprising humic acid grafted fluid loss control additives, US Patent 7 576 040, assigned to Halliburton Energy Services, Inc. (Duncan, OK), August 18, 2009.
40. K.W. Smith, Well drilling fluids, US Patent 7 576 038, assigned to Clearwater International, L.L.C. (Houston, TX), August 18, 2009.

11

Other Anionic Polymers

11.1 2-Acrylamido-2-methyl-1-propane sulfonic acid

2-Acrylamido-2-methyl-1-propane sulfonic acid (AMPS) is prepared by the reaction of acrylonitrile and *i*-butene in presence of concentrated sulfuric acid. The reaction is sketched out in Figure 11.1.

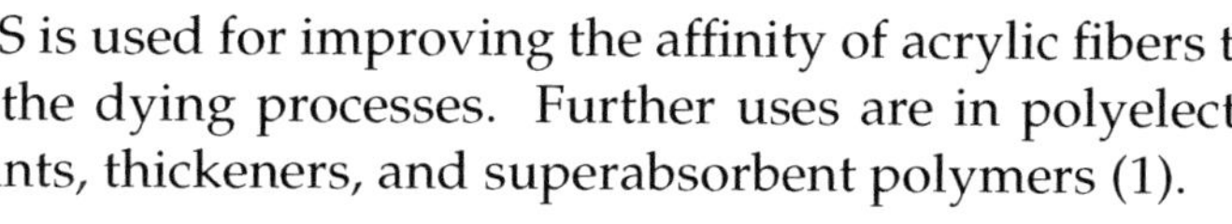

Figure 11.1 Preparation of 2-Acrylamido-2-methyl-1-propane sulfonic acid

AMPS is used for improving the affinity of acrylic fibers towards dyes in the dying processes. Further uses are in polyelectrolytes, dispersants, thickeners, and superabsorbent polymers (1).

11.1.1 Copolymers

Copolymers containing the vinyl alcohol moiety and a minor amount of AMPS is used for the preparation of aqueous dispersions suitable for drilling fluids, hydraulic cement compositions, mineral pigment containing coatings, and papermaking furnishes.

It is difficult to produce a copolymer with a sufficiently high content of AMPS moieties while avoiding a compositional drift, i.e., unacceptable variations in the content of AMPS in the copolymer from one batch to the next (2).

These drawbacks can be overcome by a continuous polymerization process. The process is conducted in two consecutive reactions zones. To the first reaction zone, vinyl acetate (VA) and AMPS, di(ethylhexyl) peroxydicarbonate as a free radical polymerization initiator are added together with methanol a solvent for the monomers. Acetaldehyde can be used as chain transfer agent. In the second reaction zone, an additional supply of AMPS is provided. The average residence time in the first reaction zone is 45–70 *min* with a temperature of 60–80°C. The conditions in the second reaction zone are quite similar. Behind the second reaction zone, the copolymer is separated and saponified by hydrolysis or alcoholysis. Eventually, the copolymer is dispersed in water, or else a melt extrudate is formed (2).

The copolymers have been characterized with respect to a variety of properties. A stable solubility in brines and sweater is important for in oil field applications. It has been demonstrated that the copolymers are suitable as components in drilling fluids and in hydraulic cement compositions. In particular, the resins are useful in a broad range of hydraulic cement compositions to control the fluid loss (2).

Further, the copolymers can be added to papermaking furnish at the wet end of a paper machine as a strength agent, or dispersant, either alone or in combination with other conventional agents. This has the advantage of being repulpable, which is a highly desirable feature with respect to recycling (2).

Melt extruded articles have been prepared utilizing poly(ethylene glycol) as a plasticizer. Another expedient for the smooth melt extrusion of the copolymer is a water aided extrusion which lowers the melting point somewhat and enables the extrusion of a more uniform melt at a relatively low temperature. This technique involves the formation of a high solids solution of the copolymer which is then extruded into a film from which the moisture is removed through evaporation to form a solid water free sheet (2).

11.1.2 Oil Field Applications

11.1.2.1 Filtration Control

Filtration control during drilling operations can be provided by the addition of a copolymer made from acrylamide (AAm) and AMPS

(3). In addition, terpolymers from *N*-vinyl-2-pyrrolidone, AMPS, and AAm have been described. The polymers are crosslinked with poly(ethyleneimine) (4).

11.1.2.2 Emulsion Stabilizers

Terpolymers based on AMPS and acrylic acid (AA) are prepared in two steps. A binary AA/AMPS copolymer is first prepared by a radical polymerization and a hydrophobic modification. Ammonium peroxodisulfate is used as a radical initiator. In the second step the carboxyl groups are amidified with didodecylamine (5).

These polymers can be used for stabilizing direct or invert emulsions likely to be destabilized or inverted by a modification in the degree of salinity of the aqueous phase or a pH modification. This method is particularly useful for stabilizing oil drilling fluids, fracturing fluids, or completion fluids (5).

11.1.2.3 Fluid Recovery

The flocculation of very finely divided solids from completion and workover fluids targets to the recovery of the fluid so that it can be reused. This is accomplished by admixing AMPS to the fluid. Diatomaceous earth is added as a filter aid for filtering the completion and workover fluid after flocculation is achieved.

AMPS flocculates fines from 0.1 μm from 0.25 *in*. Using AMPS as a flocculant, no gravimetrically measurable solids remain in the solution after filtration (6).

11.1.2.4 Cementing

Well cement compositions are commonly utilized in subterranean operations, particularly subterranean well construction and remedial operations (7). For example, in subterranean well construction, a pipe string may be run into a well bore and cemented in place. The process of cementing the pipe string in place is commonly referred to as *primary cementing*.

In a typical primary cementing operation, a cement composition is pumped and disposed into an annulus between the walls of the well bore and the exterior surface of the pipe string. The cement composition sets in the annular space, thereby forming an annular sheath of hardened, substantially impermeable cement that supports and positions the pipe string in the well bore and bonds the exterior surface of the pipe string to the subterranean formation.

The annular sheath of set cement surrounding the pipe string functions to prevent the migration of fluids in the annulus, as well as protecting the pipe string from corrosion. In addition, cement compositions may be used in plugging and abandonment operations as well as remedial cementing operations such as squeeze cementing and the placement of cement plugs.

For subterranean cementing operations to be successful, the cement compositions typically include a fluid loss control additive to reduce the loss of fluid from the cement compositions, e.g., when they contact permeable subterranean formations and zones. Excessive fluid loss may cause a cement composition to become prematurely dehydrated which limits the ability of pumping. Due to this premature dehydration, an excessive pump pressure may be required to place the cement composition, potentially resulting in breakdown of the formation or destabilization of the well bore (7).

In cement compositions fluid loss can be achieved using humic acid on with grafted AMPS (8). These graft copolymers are effective at reducing fluid loss from cement compositions in a variety of cementing applications. It is believed that they are particularly useful in high temperature environments, for example, in wells having a bottom hole circulating temperature of 260°C.

In the preparation, sodium humate, water, a defoamer, and ethylenediamine tetraacetic acid are added to a reactor. To this vessel, 2-acrylamido-2-methyl-1-propane sulfonic acid sodium salt, AAm and DADMAC are added in solution. The mixture is then heated to 70°C for 1 *h* while purging with nitrogen. After 1 *h*, ammonium persulfate is added to initiate the reaction.

Similarly to humic acids, AMPS can be grafted on to lignite (7). Lignite refers to a variety of low rank coals, including oxidized lignite, e.g., leonardite, mined lignin, brown coal or slack. Lignite may be treated with potassium hydroxide to solubilize the lignite in water. In general, only a portion of the lignite goes into solution. The soluble part is used for grafting. The preparation follows essentially the same prescription as given above.

11.1.3 Electroluminescent Devices

Compositions from poly(aniline) (PANI), poly(styrene sulfonic acid), and poly(2-acrylamido-2-methyl-1-propane sulfonic acid) (PAMPS) are useful as high resistance buffer layers for electroluminescent devices such as organic light-emitting diodes (LED)s (9).

Conductive polymers have been used in the development of electroluminescent devices for use in light emissive displays. Organic LEDs with conductive polymers generally have the following configuration: anode/buffer layer/electroluminescent polymer/cathode.

The anode is typically any material that has the ability to inject holes into the otherwise filled π-band of the semiconducting electroluminescent polymer mostly indium tin oxide. The anode is supported on a glass or plastic substrate. The electroluminescent polymer is typically a conjugated semiconducting polymer such as poly(*p*-phenylenevinylene) or poly(fluorene).

Last but not least, the cathode is made from Ca or Ba. These materials have the ability to inject electrons into the otherwise empty π-band of the semiconducting electroluminescent polymer (9).

The buffer layer is typically a conductive polymer and facilitates the injection of holes from the anode into the electroluminescent polymer layer. The buffer layer can be also addressed as the hole injection layer or as hole transport layer. Typical conductive polymers employed as buffer layers are the emeraldine salt form of PANI or a poly(dioxythiophene) doped with a sulfonic acid.

Although PANI has been used successfully as the buffer layer in certain types of organic LED, the low electrical resistivity typical of PANI inhibits its use in pixellated displays. For pixellated displays, a buffer layer having a higher resistance, i.e., lower conductivity is desired to eliminate or minimize crosstalk between neighboring pixels. Inter-pixel current leakage significantly reduces power efficiency and limits both the resolution and clarity of the display.

It has been found that the addition of poly(styrene sulfonic acid) to compositions of PANI and PAMPS results in a marked decrease in conductivity (9).

11.1.4 Chemoembolotherapy

A composition for chemoembolotherapy of solid tumors comprises particles of a water-insoluble water-swellable synthetic anionic polymer on that an anthracycline compound is absorbed (10). The anionic copolymer has VA and AMPS moieties. Microspheres are prepared from this copolymer in order to absorb the drug.

11.2 Poly(sulfonic acid)s

11.2.1 Poly(vinylsulfonic acid)

11.2.1.1 Corrosion Protection

Certain polymers of vinylsulfonic acid and methacrylic acid (MA) promote the formation of passive, protective oxide films on steel surfaces under alkaline boiler conditions and at high pressures i.e., at pH 11 and 7–10 *MPa* (11).

The polymers are added to the boiler water with a preferred dosage of about 5–70 *ppm*. At these dosages, a good black metal oxide film is formed on the boiler. The protective layers give a good control of the mineral deposition in the boiler.

The passivating effect of the vinylsulfonic acid comonomer is only observed in its copolymers with MA. The corresponding vinylsulfonic acid/AA copolymer shows no passivation enhancement compared with poly(acrylic acid).

11.2.1.2 Electrolyte Membranes

A method for the preparation of electrolyte membranes consists of filling a porous base material with suitable monomers that are polymerized *in situ* (12).

The base material is a mixture of a norbornene polymer and a thermoplastic elastomer with a paraffin. The mixture is kneaded and then sandwiched with rolls. After the sheets are formed the paraffin is removed by means of heptane. In this way, after some additional steps, a microporous film is obtained.

This microporous film is impregnated with vinylsulfonic acid and *N,N'*-methylene bisacrylamide as a crosslinking agent, together with a water-soluble radical azo initiator. After polymerization, the excess of polymers on the surface of the membrane is removed, the membrane is then washed with distilled water and dried in an oven at 50°C.

The electrolyte membrane is used in a solid polymer fuel cell, i.e., in a methanol fuel cell including a direct type methanol solid polymer fuel cell, or a modified type methanol solid polymer fuel cell and a pure hydrogen gas type fuel cell using hydrogen gas (12).

11.2.1.3 *Polishing Slurries*

In the chemical or mechanical polishing of microelectronic devices, polishing slurries are used. The abrasive particles are metal oxides selected from ceria, silica, alumina, titania, zirconia, and germania. Two passivation agents are used. The first passivation agent may be an anionic surfactant, i.e., poly(vinylsulfonic acid) and the second passivation agent may be phthalic acid and its salts (13).

11.2.1.4 *Drive Fluids*

Vinylsulfonic acid can be alkoxylated with glycidol under alkaline conditions (14). The alkoxylated vinylsulfonic acid has been copolymerized with AAm using ethanol as solvent and 2,2′-azobisisobutyronitrile as radical initiator. These copolymers are used as additives for drive fluids in enhanced oil recovery operations (14).

11.2.1.5 *Dye-fixative Compositions*

Dye-fixative compositions typically used in industry contain residual phenols and formaldehyde. The environmental hazards associated with such toxic substances are well-known. However, these substances also cause the discoloration or, more particularly, shade variation of the dye with which they come into contact (15).

For example, rhodamine based dyestuffs, treated with a x containing one or both of such compounds, have a tendency to experience a variation in shade which ultimately results in the substrate either being damaged or necessitating further dyeing to replace the lost dyes. This phenomenon is caused by a chemical reaction between the dye and the phenols present in the dye-fixative (15).

Improvements have been reported with dye-fixative composition comprising poly(methacrylic acid), and copolymers of MA and AMPS, sodium vinyl sulfonate, sodium styrene sulfonate, etc. (15).

The presence of vinylsulfonic acid in the copolymer serve for application of the copolymer at a lower pH in comparison with a copolymer bearing only exclusively carboxylic acid groups. This arises, because the presence of the vinylsulfonic acid enables the copolymer to become soluble at a lower pH. The preparation of dye-fixative compositions is given elsewhere (15).

11.2.2 *Poly(4-vinylbenzoic acid)*

11.2.2.1 *Fabrication of Photoresists*

In a multilayer photoresist film a layer of a photosensitive resist material is disposed on a layer of poly(vinylbenzoic acid) (16). Preferred photosensitive materials form a positive tone image and a preferred material is poly(4-trimethylsilylphthalaldehyde).

For etching, the film is exposed to radiation in the deep UV region. The poly(vinylbenzoic acid) has a high absorbance in this region and effectively prevents radiation backscattering during exposure. The radiation causes a chemical change in the exposed area of the photosensitive material rendering this area more or less soluble in the development step.

11.2.2.2 *Anticaries Compositions*

It has been found that the rate of development of dental caries, as characterized by proximal, smooth surface, pit and fissure caries, can be prevented or substantially retarded. This is achieved by the daily application to the teeth of a composition comprising a fluoride salt and a metal salt of poly(vinylbenzoic acid) (17).

The poly(vinylbenzoic acid) required as intermediates for the preparation of their salts are readily prepared by the free radical polymerization of 4-vinylbenzoic acid.

In such compositions, the fluoride ion concentration should be 0.01–1%. The concentration of the poly(vinylbenzoic acid) should be 0.05–3%. In concentrate formulations, the poly(vinylbenzoic acid) may have a concentration as high as 80%. A typical gel dentifrice has the composition is described in Table 11.1.

Table 11.1 Dentifrice Composition (17)

Compound	Amount/[%]
Sodium fluoride	0.22
Poly(sodium 4-vinylbenzoate)	1.00
Silicon dioxide	1.00
Pluronic F-127 (Ethoxylated PPG)	18.00
Sweetener	0.80
Sodium benzoate	0.30
Glycerol	10.00
Sorbitol solution, 70%	2.00
Flavoring	0.80
Dye (0.5% in water)	0.70
Deionized water q.s. to 100%	

In addition, an abrasive paste dentifrice and a mouth rinse solution formulated with poly(vinylbenzoic acid) have been described. A mouth rinse composition is described in Table 11.2.

Table 11.2 Mouth Rinse Composition (17)

Compound	Amount/[%]
Sodium fluoride	0.05
Copper poly(4-vinylbenzoate)	0.02
Ethanol	7
Pluronic F-108 (Ethoxylated PPG)	2.00
Sweetener	0.20
Sorbitol solution, 70%	10.00
Flavoring	0.20
Deionized water q.s. to 100%	

The enhancing properties of poly(vinylbenzoic acid) on the fluoride ion activity have been demonstrated. An in vitro assay technique has been used. This technique is based on the measurement of organic acids produced from sucrose by the cariogenic bacterium S. mutans, simply by acid base titration (17).

Moreover, the alkali metal salts of the poly(vinylbenzoic acid) are highly effective in reducing the deposition of plaque during in vitro testing. For example, poly(sodium 4-vinylbenzoate) exhibited a 91% reduction in the deposition of plaque (18).

11.2.2.3 *Analysis of Nucleic Acids*

Vinylbenzoic acid is used in the analysis of nucleic acids by electrochemical methods (19). It serves as the functionalized moiety in electrode films containing a ruthenium catalyst to which the nucleic acid is fixed by a carbodiimide reaction followed by amidation. The catalyst is used to catalyze the DNA oxidation. If specific moieties are present in the DNA that are easily oxidized, a dramatic enhancement in the oxidative current can be observed.

11.2.3 *Poly(styrene sulfonic acid)*

Sulfonated poly(styrene) (PS) can prepared through the post sulfonation of PS. On the other hand, another path to poly(styrene sulfonic acid) has been described (20). The polymer is obtained by the polymerization of *p*-styrene sulfonyl chloride and subsequent hydrolysis. *p*-Styrene sulfonyl chloride itself is prepared from *p*-styrene sulfonic acid sodium salt and thionyl chloride.

Using a monomer with sulfonyl chloride causes the monomer composition to be changed easily, and further various random or alternating polymers can be produced (20).

11.2.3.1 *Separation Membranes*

The selectivity of poly(benzoxazole) membranes with respect to gas, vapor, and liquid separations can be improved by the introduction of a poly(styrene sulfonic acid) (21).

The poly(benzoxazole) membranes are prepared by thermal cyclization of aromatic poly(imide) membranes containing poly(styrene sulfonic acid) in a temperature range of 350–450°C under inert atmosphere. The aromatic polyimide membranes are prepared from a mixture of poly(styrene sulfonic acid) polymer and aromatic polyimide polymers comprising pendent hydroxyl groups in ortho position to the imide nitrogen in the polymer backbone. Details and a series of examples of the method of preparation are given in the literature (21).

11.2.3.2 Dispersants for Ink Jet Formulations

The composition of the ink is traditionally comprised of deionized water, a water-soluble organic solvent, and a colorant. The colorant may be a soluble dye or an insoluble pigment. Several problems, however, are associated with soluble dyes in contrast to insoluble pigments. These problems include poor water fastness, poor light fastness, poor thermal stability, facile oxidation, dye crystallization, and ink bleeding and feathering on the print medium.

To circumvent these problems, the use of a pigment as the colorant is preferred. Pigments generally have better light fast and water fast properties, are more resistant to oxidation, and have higher thermal stability. The use of a pigment instead of an aqueous dye presents solubility problems since the pigments are insoluble in aqueous media. As a result, the insoluble pigment is generally stabilized in a dispersion by a polymeric dispersant (22).

Dispersants for ink jet compositions comprise two structurally distinct segments: a hydrophilic segment and a hydrophobic segment. The preferred hydrophilic segment is a methacryloyl terminated poly(dimethylsiloxane), or a copolymer with *p*-styrene sulfonic acid. The monomeric hydrophobic segment is (ethylene glycol) 2,4,6-tris-(1-phenylethyl)phenyl ether on to that a methacryloyl group is attached (22).

11.3 Sulfonated Asphalt

Sulfonated asphalt is produced by heating an asphaltic material with a softening point of 160–180°C by mixing the asphalt with a solvent, such as hexane. Then the asphalt is treated with a liquid sulfonating agent, such as liquid sulfur trioxide. By neutralization with alkali hydroxides, such as NaOH or ammonia, the corresponding sulfonate salts result. The evaporated solvent can be recovered for reuse (23). Only a limited portion of the sulfonated product can be extracted by hot water extraction.

However, the fraction thus obtained, which is more or less water soluble or water dispersible, is crucial for the quality and, in terms of utilizing technical application methods, is directly correlated with the desired properties of the sulfonated asphalt as a drilling fluid additive (24).

11.3.1 Drilling Fluids

Drilling fluids are complex, often circulated liquid systems, used during the drilling of a wellbore into or through subterranean formations, thereby supporting the drilling process. A drilling fluid has a number of tasks which include transporting the cuttings out of the borehole, simultaneously cooling and lubricating the drill bit, stabilizing the borehole wall or also compensating the hydrostatic pressure exerted by the formation.

Due to the large number of tasks, specific additives have been developed for drilling fluids in order to meet the specific requirements of a particular well. Based on the so-called continuous phase, drilling fluids can be divided into water-based or oil-based drilling fluid systems (24).

Sulfonated asphalt is predominantly utilized for water-based drilling fluids. Apart from reduced filtrate loss and improved filter cake properties, good lubrication of the drill bit and decreased formation damage are important features assigned to sulfonated asphalt as drilling fluid additive.

The mechanism of action of sulfonated asphalt as a drilling fluid additive, and as a clay inhibitor, is explained as the electronegative sulfonated macromolecules and aggregates of the substantially water-soluble or water dispersible fraction attach to the electropositive ends of the clay platelets. Thereby, a neutralization barrier is created, which suppresses the absorption of water into the clay. Because the sulfonated asphalt is lipophilic, and therefore water repellent, the water influx into the clay is restricted by purely physical principles.

11.4 Lignosulfonate

11.4.1 Biopenetrants

Ferrous sulfide deposits are a major source of economic loss in the oil industry. The deposits are mainly the result of a reaction between hydrogen sulfide, formed by sulfate-reducing bacteria, and ferrous metal oil field equipment and iron compound in the formation.

However, problems due to iron sulfide deposits are not confined to the oil industry but are encountered in a wide range of industrial water systems. For instance, ferrous sulfide deposits are a serious problem in the paper industry, causing the scaling of Fourdriniers and other papermaking equipment (25).

Tris(hydroxyorgano)phosphines and aminocarboxylic acids or amino phosphonic acids act synergistically to dissolve iron sulfide deposits. The formulations may contain further biocides, water dispersants, demulsifiers, antifoaming agents, solvents, scale inhibitors, corrosion inhibitors, oxygen scavengers, flocculants and non-surfactant biopenetrants.

Biopenetrants enhance the biocidal efficacy of commodity biocides. Biopenetrants are water-soluble polymers. A particularly useful biopenetrant is poly [oxyethylene(dimethyliminio) ethylene(dimethyliminio)ethylene dichloride]. This is a copolymer of *N,N,N′N′*-tetramethyl-1,2-diamino ethane and bis(2-chloroethyl) ether (25).

Tradenames appearing in the references are shown in Table 11.3.

Table 11.3 Tradenames in References

Tradename Description	Supplier
ALL-TEMP®	Baker Hughes Drilling Fluids
Acrylate tetrapolymer (3)	
Amres®	Georgia-Pacific Resins, Inc.
Polyamide-epichlorohydrin wet strength resin (2)	
Baytron® P	Bayer AG
Complex of poly(3,4-ethylenedioxythiophene) and poly(styrenesulfonic acid) (9)	
Blankophor®	Bayer
Optical brightener (2)	
BORE-DRILL™	Borden Chemicals
Anionic polymer (3)	
Briquest® 543	Rhodia Consumer Specialties Ltd.
Sodium diethylene triamine pentakis(methylene phosphonate) (25)	
Celvol® (Series)	Celanese
Poly(vinyl alcohol) (2)	

Table 11.3 (cont.)

Tradename Description	Supplier
Chek-Loss® PLUS Ultra-fine lignin (3)	Baker Hughes
Chemtrol® X Blend of ground lignitic earth and synthetic maleic anhydride copolymers (3)	Baker Hughes
CO-BOND® (Series) Modified starches (2)	National Starch and Chemical Comp.
DFE-129™ Acrylamide/AMPS copolymer (3)	Baker Hughes Drilling Fluids
Driscal® D Water soluble polymer (3)	Drilling Specialties Comp.
Halad® (Series) Fluid loss control additive (7,8)	Halliburton Energy Services, Inc.
Hostalux® Optical Brightener (2)	Hoechst
KEM-SEAL® PLUS NaAMPS/*N*,*N*-dimethylacrylamide copolymer (3)	Baker Hughes Drilling Fluids
Kemseal® Fluid loss additive (3)	Baker Hughes Norge
Leucophor® Optical brightener (2)	Clariant
Ligco® Lignite (3)	Baker Hughes
Ligcon® Causticized lignite (3)	Milchem Inc.
Lomar® D Sodium salt of the formaldehyde condensation product of naphthalene sulfonic acid (2)	Geo Specialty Chemicals, Inc. (Henkel)
MAX-TROL® Sulfonated resin (3)	Baker Hughes Drilling Fluids
Microbond™ Cement expanding additive (7,8)	Halliburton Energy Services, Inc.
Mil-Bar® Barite weighting agent (3)	Baker Hughes
Mil-Carb® Ground marble (3)	Baker Hughes Drilling Fluids

Table 11.3 (cont.)

Tradename Description	Supplier
Mil-Gel-NT® Bentonite quartz mixture (3)	Baker Hughes
Mil-Gel™ Ground montmorillonite (3)	Baker Hughes
Mil-Temp® Maleic anhydride copolymer (3)	Baker Hughes
Polydrill® Anionic polymer (3)	Degussa AG
Protecto-Magic™ Ground asphalt (3)	Baker Hughes
PYRO-TROL® Acrylamide/AMPS copolymer (3)	Baker Hughes
Rev Dust Artificial drill solids (3)	Milwhite, Inc.
Silicalite® High surface area amorphous silica (7,8)	Halliburton Energy Services, Inc.
Silwet® Organo silicone surfactants (22)	O Si Specialities, Inc.
Soltex® Sulfonated asphalt (3)	Chevron Phillips Chemical Comp.
SULFA-TROL® Sulfonated asphalt (3)	Baker Hughes Drilling Fluids
Teflon® Tetrafluoro polymer (4)	DuPont
Tinopal® Optical brightener (2)	Ciba-Geigy

References

1. P.P. Barve, S.S. Joshi, R.W. Shinde, M.Y. Gupte, C.N. Joshi, S.M. Ghike, R.V. Naik, R.A. Kulkarni, and A.N. Bote, Process for the preparation of 2-acrylamido-2-methyl-1-propanesulfonic acid, US Patent 6 504 050, assigned to Council of Scientific and Industrial Research (New Delhi, IN), January 7, 2003.
2. R. Vicari, Vinyl alcohol copolymers for use in aqueous dispersions and melt extruded articles, US Patent 7 790 815, assigned to Serisui

Specialty Chemicals America, LLC (Dallas, TX), September 7, 2010.

3. M. Jarrett and D. Clapper, High temperature filtration control using water based drilling fluid systems comprising water soluble polymers, US Patent 7 651 980, assigned to Baker Hughes Incorporated (Houston, TX), January 26, 2010.
4. B.R. Reddy, L.S. Eoff, J. Chatterji, S.T. Tran, and E.D. Dalrymple, Preventing flow through subterranean zones, US Patent 6 176 315, assigned to Halliburton Energy Services, Inc. (Duncan, OK), January 23, 2001.
5. N. Monfreux-Gaillard, P. Perrin, F. LaFuma, and C. Sawdon, Reversible emulsions stabilized by amphiphilic polymers and application to drilling fluid, US Patent 7 262 152, assigned to M-I L.L.C. (Houston, TX), August 28, 2007.
6. S.R. Luxemburg, Process for cleaning fluids and particulate solids, US Patent 6 267 893, July 31, 2001.
7. S. Lewis, J. Chatterji, B. King, and C. Brennels, Cement compositions comprising lignite grafted fluid loss control additives, US Patent 7 388 045, assigned to Halliburton Energy Services, Inc. (Duncan, OK), June 17, 2008.
8. S. Lewis, J. Chatterji, B. King, and D.C. Brenneis, Cement compositions comprising humic acid grafted fluid loss control additives, US Patent 7 576 040, assigned to Halliburton Energy Services, Inc. (Duncan, OK), August 18, 2009.
9. C. Zhang, High resistance polyaniline blend for use in high efficiency pixellated polymer electroluminescent devices, US Patent 7 033 646, assigned to E. I. du Pont de Nemours and Company (Wilmington, DE), April 25, 2006.
10. A.L. Lewis, P.W. Stratford, S. Leppard, P. Garcia, B. Hall, and M.V. Fajardo Gonzalez, Chemoembolisation, US Patent 7 442 385, assigned to Biocompatibles UK Limited (Surrey, GB), October 28, 2008.
11. R.S. Robinson, Vinyl sulfonic acid - methacrylic acid copolymer passivators for high pressure boilers, US Patent 4 719 082, assigned to Nalco Chemical Company (Naperville, IL), January 12, 1988.
12. K. Yamamoto, H. Emori, M. Abe, and K. Sho, Production method of electrolyte membrane, electrolyte membrane and solid polymer fuel cell using same, US Patent 7 785 751, assigned to Nitto Denko Corporation (Osaka, JP), August 31, 2010.
13. J.-W. Lee, J.-D. Lee, B.-U. Yoon, and S.-R. Hah, Chemical/mechanical polishing slurry, and chemical mechanical polishing process and shallow trench isolation process employing the same, US Patent 6 914 001, assigned to Samsung Electronics Co., Ltd. (Suwon-Si, KR), July 5, 2005.

14. W.D. Hunter, Bipolymer of vinyl sulfonic acid-glycidol adduct and acrylamide, US Patent 4 297 469, assigned to Texaco Development Corp. (White Plains, NY), October 27, 1981.
15. A.H. Cole, S.C. Glenn, and G.S. Johnson, Process for fixing dyes in textile materials, US Patent 5 525 125, assigned to Henkel Corporation (Plymouth Meeting, PA), June 11, 1996.
16. H. Ito, Multilayer photoresist comprising poly-(vinylbenzoic acid) as a planarizing layer, US Patent 5 260 172, assigned to International Business Machines Corporation (Armonk, NY), November 9, 1993.
17. T. Sipos, Anticaries compositions, US Patent 4 459 282, assigned to Johnson & Johnson Products Inc. (New Brunswick, NJ), July 10, 1984.
18. R.J. Gander, C.J. Buck, and T. Sipos, Alkali metal salts of poly(vinylbenzoic acid) as dental plaque barrier agents, US Patent 4 364 924, assigned to Johnson & Johnson Products, Inc. (New Brunswick, NJ), December 21, 1982.
19. H.H. Thorp and A.C. Ontko, Electropolymerizable film, and method of making and use thereof, US Patent 6 180 346, assigned to The Universtiy of North Carolina at Chapel Hill (Chapel Hill, NC), January 30, 2001.
20. H.-J. Kim, H.-J. Kweon, Y.-C. Eun, and S.-Y. Cho, Method for preparing sulfonated polystyrene for polymer electrolyte of fuel cell, US Patent 7 244 791, assigned to Samsung SDI Co., Ltd. (Suwon-Si, Gyeonggi-Do, KR), July 17, 2007.
21. C. Liu, R. Minkov, M.-W. Tang, L. Zhou, and J.C. Bricker, Method to improve the selectivity of polybenzoxazole membranes, US Patent 7 810 652, assigned to UOP LLC (Des Plaines, IL), October 12, 2010.
22. C.E. Akers, Jr., T.E. Franey, J.X. Sun, and C.M. Butler, Polymeric dispersants used for aqueous pigmented inks for ink-jet printing, US Patent 6 652 634, assigned to Lexmark International, Inc. (Lexington, KY), November 25, 2003.
23. P. Rooney, J.A. Russell, and T.D. Brown, Production of sulfonated asphalt, US Patent 4 741 868, assigned to Phillips Petroleum Company (Bartlesville, OK), May 3, 1988.
24. J. Huber, J. Plank, J. Heidlas, G. Keilhofer, and P. Lange, Additive for drilling fluids, US Patent 7 576 039, assigned to BASF Construction Polymers GmbH (Trostberg, DE), August 18, 2009.
25. S.D. Fidoe, R.E. Talbot, C.R. Jones, and R. Gabriel, Treatment of iron sulphide deposits, US Patent 6 926 836, assigned to Rhodia Consumer Specialties Limited (Watford, GB), August 9, 2005.

Index

Tradenames

Acronyms

Chemicals

General Index

Also of Interest

Check out these published and forthcoming related titles from Scrivener Publishing

Handbook of Engineering and Specialty Thermoplastics

Part 1: Polyolefins and Styrenics by Johannes Karl Fink
Published 2010. ISBN 978-0-470-62483-5

Part 2: Water Soluble Polymers by Johannes Karl Fink
Published 2011. ISBN 978-1-118-06275-3

Part 3: Polyethers and Polyesters edited by Sabu Thomas and Visakh P.M.
Forthcoming June 2011. ISBN 978-0-470-63926-9

Part 4: Nylons edited by Sabu Thomas and Visakh P.M.
Forthcoming June 2011. ISBN 978-0-470-63925-2

Introduction to Industrial Polyethylene: Properties, Catalysts, Processes by Dennis P. Malpass.
Published 2010. ISBN 978-0-470-62598-9

A Concise Introduction to Additives for Thermoplastic Polymers by Johannes Karl Fink.
Published 2010. ISBN 978-0-470-60955-2

Polymer Nanotube Nanocomposites: Synthesis, Properties, and Applications edited by Vikas Mittal.
Published 2010. ISBN 978-0-470-62592-7

Carbon Dioxide Thermodynamics Properties Handbook by Sara Anwar and John Carroll
Published 2010. ISBN 978-1-118-01298-7

Miniemulsion Polymerization Technology edited by Vikas Mittal
Published 2010. ISBN 978-0-470-62596-5

A Guide to Safe Material and Chemical Handling by Nicholas P. Cheremisinoff and Anton Davletshin.
Published 2010. ISBN 978-0-470-62582-8

Reverse Osmosis: Design, Processes, and Applications for Engineers by Jane Kucera
Published 2010. ISBN 978-0-470-61843-1